6th International Conference on Materials and Applications for Sensors and Transducers (IC-MAST 2016)

Journal of Physics: Conference Series Volume 939

Athens, Greece
27 - 30 September 2016

Editors:

Evangelos Hristoforou
Dimitrios S. Vlachos

ISBN: 978-1-5108-5519-9
ISSN: 1742-6588

Printed from e-media with permission by:

Curran Associates, Inc.
57 Morehouse Lane
Red Hook, NY 12571

Some format issues inherent in the e-media version may also appear in this print version.

Printed by Curran Associates, Inc. (2018)

Institute of Physics
Dirac House, Temple Back
Bristol BS1 6BE UK

Phone: 44 1 17 929 7481
Fax: 44 1 17 920 0979

techtracking@iop.org

Additional copies of this publication are available from:

Curran Associates, Inc.
57 Morehouse Lane
Red Hook, NY 12571 USA
Phone: 845-758-0400
Fax: 845-758-2633
Email: curran@proceedings.com
Web: www.proceedings.com

Table of contents

Volume 939

6th International Conference on Materials and Applications for Sensors and Transducers - IC-MAST

27–30 September 2016, Athens, Greece

Accepted papers received: 13 December 2017
Published online: 30 December 2017

Preface

Papers

IOP Conf. Series: Journal of Physics: Conf. Series **939** (2017) 011001 doi:10.1088/1742-6596/939/1/011001

Preface

The 6th International Conference on Materials and Applications for Sensors and Transducers, was held in Athens, Greece, from September 27 to September 30 2016. The Conference, hosted about 70 oral and poster papers and more than 90 participants. IC-MAST as an international annual conference which tries to meet the needs for various types of sensors, particularly those ones which may be manufactured by low cost methods (i.e. hybrid sensors, smart specialization devices, particular applications not necessarily requiring integrated micro-nano technologies), covering all types of materials and physical effects, appears to be a necessity. IC-MAST has been established as a high quality international conference by:

I. Gathering together multinational researchers from all over the world, working in different materials for sensors and transducers and technical applications of sensors, but also in some cases in the management of the data coming from sensors and transducers; the careful selection of the conference place (like Aegean Sea, Budapest, Prague, Bilbao, Mykonos, Athens etc) allows for enjoying the local hospitality and sightseeing

II. Emphasizing in hybrid sensors and smart specialization devices produced by inexpensive methods, without excluding of course micro-nano technology, from all kinds of solid state, liquid and gaseous materials, as well as in particular transducer applications (design and development, as well as use of sensing data)

III. Innovatively implementing the Virtual Paper Concept, allowing for large impact of research works presented in the conference by authors who either have no time or no funding support for visiting a conference; this year more than 10 virtual papers are presented in the 6th IC MAST, following a standardized procedure via the our robust and reliable Conference Site (www.icmast.net)

IV. Allowing for lengthy technical and managerial discussions in terms of sensor, material and instrumentation development; furthermore, the different research groups gathered together are offered the particular advantage of arranging and concluding research proposals and projects, otherwise not having a visible possibility of such realization

The IC-MAST 2016 organizers believe that the target of the Conference has been successfully met by enhancing knowledge in sensors by all participants, accelerating the achievement of results and optimizing the under design products, in a quite friendly way!

The Conference Chairmen

Evangelos Hristoforou
National Technical University of Athens, Greece

Dimitrios Vlachos
University of Peloponnese

IOP Conf. Series: Journal of Physics: Conf. Series **939** (2017) 011001 doi:10.1088/1742-6596/939/1/011001

International Advisory Committee

- Shehab Ahmed, Texas A&M University at Qatar, Doha, Qatar
- Faramarz Hossein-Babaei, K.N. Toosi University of Technology, Iran
- José Manuel Barandiarán, BCMAN, Basque Center for Materials, Applications and Nanostructures

International Scientific Committee

- Mihai Popescu, Editor-in-Chief, Journal of Optoelectronics and Advanced Materials
- Spyridon Diplas, Sintef Oslo, Norway
- Jon Gutiérrez, Universidad del País Vasco UPV/EHU, Spain
- C.N.R. Rao, Jawaharlal Nehru Centre for Advanced Scientific Research, India
- Gascon, J., Catalysis Engineering / Delft University of Technology
- Athouël, Laurence, Dr., IMN-CNRS, University of Nantes, France
- Peter Svec, Slovak Academy of Sciences
- Horia Chiriac, National Institute for R&D in Technical Physics, Iasi, Romania
- Dimitris Kouzoudis, University of Patras, Greece
- Saci Messaadi, Université de Batna, Algeria
- Pavel Ripka, Czech Technical University, Czech Republic
- Christophe Dolabdjian, University of CAEN, France
- Raffaele Battaglia, Novaetech Srl
- Ioannis Raptis, NCSR "Demokritos", Athens, Greece
- Roberto Germano, Promete, Naples, Italy
- Papakonstantinou, P., University of Ulster, School of Engineering, Engineering Research Institute
- Morent, Rino, Applied Physics - Ghent University
- Anthopoulos, T., Imperial College London
- Ajay Sood, Indian Institute of Science, India
- Valiyaveettil, S., National University of Singapore
- Carlos, L.D., Department of Physics and CICECO, University of Aveiro, Portugal
- Qifeng Zhang, University of Washington
- Jiujun Zhang, Dr., Prof., National Reserach Council of Canada, Vancouver, BC, V6T 1W5 Canada
- Yong Zhang, Department of Bioengineering, National University of Singapore
- Francesco Stellacci, Ecole Polytechnique Fédérale de Lausanne, Switzerland
- Wang, E., Changchun Institute of Applied Chemistry,Chinese Academy of Sciences

- Dr. Zhigadlo, N.D., Laboratory for Solid State Physics, ETH Zurich, Switzerland
- Dediu, Valentin Alek, ISMN-CNR, via Gobetti101, 40129 Bologna, Italy
- Velonia, K., Department of Materials Science and Technology, University of Crete
- Dr. Vladimir Sivakov, Institute of Photonic Technology, Albert Einstein Str. 9, 07745 Jena, Germany
- Woo Lee, Korea Research Institute of Standards and Science
- Mei, Yongfeng, Fudan University, Shanghai China
- Chunyi Zhi, City University of Hong Kong, Hong Kong
- Horowitz, Gilles, CNRS - Ecole Polytechnique, France

Editors:

Evangelos Hristoforou, National Technical University of Athens, Greece.
Dimitrios S. Vlachos, University of Peloponnese, Greece.

IOP Conf. Series: Journal of Physics: Conf. Series **939** (2017) 011002 doi:10.1088/1742-6596/939/1/011002

Peer review statement

All papers published in this volume of *Journal of Physics: Conference Series* have been peer reviewed through processes administered by the proceedings Editors. Reviews were conducted by expert referees to the professional and scientific standards expected of a proceedings journal published by IOP Publishing.

IOP Conf. Series: Journal of Physics: Conf. Series **939** (2017) 012001 doi:10.1088/1742-6596/939/1/012001

Monitoring and Imaging of Magnetic Nanoparticles:state-of-the-art and prospective

P. Tselou

School of Electrical and Computer Engineering, Laboratory of Electronic Sensors, National Technical University of Athens, Zografou Campus, Athens, 15780, Greece

E-mail: ptselou@central.ntua.gr

Abstract. This paper describes the most common techniques to monitor magnetic particles and perform magnetic particle imaging (MPI) for biomedical applications. Then a new and innovative device is present which is able to enhance the image quality of magnetic particle. It will be also discussed how to ensure tracer stability and biocompatibility acting on the tracers' properties and by understanding in depth the physics of tracer response. Furthermore, it is highlighted that to achieve its full potential the following parameters: MPI sensitivity, signal-to-noise ratio (SNR) and spatial resolution must be optimized. Finally, it will be shown how the use of high-quality iron-based tracers with tailored magnetic and surface properties is able to enhance the performance for both detection and imaging.

1. Introduction

Magnetic Particle Imaging (MPI) is a technique introduced in 2005 by Gleich and Weizenecker that directly images the intense magnetization of superparamagnetic iron nanoparticles (SPIOs) rather than indirectly detecting SPIOs via magnetic resonance imaging (MRI) signal dropouts [1]. This technique produces quantitative images of the tracer materials distribution (SPIOs) without interference from the anatomical background of the imaging objects, such as tissues, bones and organs [2]. Indeed, since no biological tissue contains a magnetic signature similar to magnetic tracers, the background signal in MPI is virtually absent, unlike other nuclear medicine techniques and fluorine MRI.

Specifically, MPI has the advantage of three-dimensioal visualization with very high sensitivity, temporal and spatial resolution [3] and the absence of hazardous ionizing radiation. It exploits the unique characteristics of superparamagnetic nanoparticles and is the first medical imaging method in which the nanoparticles are not just supportive contrast agents but the only source for signal and therefore the only visualized element. This is why they are referred to as "tracers"; MPI visualizes anatomical structures only if they are labelled by the tracer. As a result, MPI is an emerging imaging technique with promising applications in diagnostic and guided therapy [4] e.g. vascular imaging and ultra-sensitive cancer therapeutics, using advanced manufacturing methods, that can be scaled up to mass production.

2. Operating principles

The main purpose of the use of MPI scanner with high resolution, sensitivity and signal-to-noise ratio (SNR) is the unique feature of magnetic materials called magnetic hysteresis, which make magnetic nanoparticles ideal nanotherapeutics [5] [6]. Magnetic hysteresis [7] represents a sort of energy loss that inevitably is dissipated in the environment as thermal energy when magnetic materials are subjected to alternating magnetic field. Though for many industrial applications this energy loss represents a disadvantage, in biomedical field it is indeed a great opportunity. In fact, this phenomenon is the base of classic hyperthermia against, e.g., cancer diseases [8]. This elevation in the temperature (up to 45-47 °C) may kill cancer cells directly or render them more susceptible to other kind of treatments [9]. However, classical hyperthermia based on the use of magnetic nanoparticles still presents some drawbacks. Among them, two are the most critical: I) lack of specificity to discriminate tumor tissue surround by health tissue; II) moderate effect on cell viability when used a solely therapeutic regime. As it has been shown in 2016 by Hristoforou et al. over-hyperthermia is a tool for precise cellular ablation that does not suffered of the above mentioned drawbacks. As it has been demonstrated, this method generated high temperature elevation that caused an efficient cell death by provoking holes in the cell protoplasm, figure 1 [10].

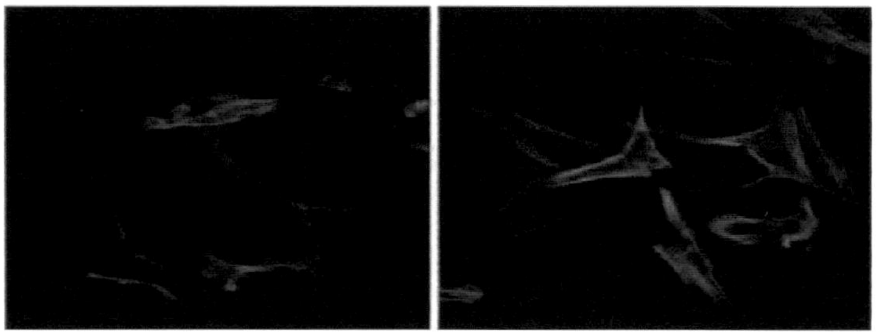

Figure 1: Over-hyperthermia resulted in detrimental effects to the cell cytoskeleton (Picture from Hristoforou et al. 2016, [10]).

Except for magnetic hysteresis, MPI scanners will track the electron paramagnetic resonance (EPR) of magnetic tracers instead of their magnetic moments (Gleich, B. & Weizenecker idea, [1]) that requires ad hoc magnetic scanners. It is known that magnetic moments of magnetic nanoparticles are much bigger than those of protons. In magnetic fields of 1.5 to 3 Tesla (T) we could modify an MRI machine to detect EPR in order to track magnetic tracers, especially those made of pure iron (Fe), because iron oxide nanoparticles have lower magnetic moments than pure Fe.

Electron paramagnetic resonance is a method for studying materials with unpaired electrons. The basic concepts of EPR are electron spins that are excited instead of the proton spins of atomic nuclei. EPR measurements are made with microwaves in the 9000–10000 MHz (9–10 GHz) region, with fields corresponding to about 3500 Gauss (G) (0.35 T). The ferromagnetic resonance frequency of Fe is 9-23 GHz, so there is a net absorption of energy when Fe-based tracers are used. Another advantage is that microwaves are able to reach deeper in patient's body giving high resolution and good quality images of tissues and cells [11].

As mentioned above, like protons also electrons have a spin, which gives to them a magnetic property known as magnetic moment. When an external magnetic field is supplied, the paramagnetic electrons can either orient in a direction parallel or antiparallel to the direction of the magnetic field. This creates two distinct energy levels for the unpaired electrons and measurements are taken as they are driven between the two levels [12].

EPR spectroscopy is the measurement and interpretation of the energy differences between the atomic or molecular states. These measurements are obtained thanks to the relationship between the energy differences and the absorption of electromagnetic radiation. To acquire a

spectrum, the frequency of the electromagnetic radiation is changed and the amount of radiation which passes through the sample is measured with a detector, which detects the spectroscopic absorptions.
In addition, to further improve the sensitivity of magnetic tracers for certain tissues (e.g. cancer), magnetic nanoparticles can be decorated with specific ligands, e.g. antibodies and non-antibody scaffold [13], [14] able to find specific cell-type.

3. Development of the method

The development of MPI scanner as a reliable and inexpensive device for monitoring the spatial distribution of Super-Paramagnetic Nanoparticles (SPANs) in the body [15] of patient affected by disease such a s cancer is based on the attachment of SPANs to the cancer cell surface through the binding of specific protein molecules involved in cancer development (with overexpressed receptors on the cancer cell membrane). The sensitivity of the method could touch the limits of single-cell measurement under given conditions, which is very important for the diagnosis of micro-metastases. Thus, measuring the density and the distribution of density of SPANs into the body, the recognition of early stage malignancies may be possible [16]. The better the magnetic sensitivity of the device is, the earlier the malignancy recognition in the body. The main target is to obtain the highest possible sensitivity in the measurement of SPANs with the minimum possible cost.
Furthermore, apart from an early stage tumour diagnosis tool, the device will offer a staging method for malignancies, with simultaneous killing of the cancer cells applying inductive heating techniques. Possible applications could be breast and prostate malignancies. This process can be feasible by controlling the rate of temperature increase; this can be achieved by controlling the structure and composition of the SPANs [17].
The motivation of the MPI scanner is the development of a novel diagnostic/therapeutic technique for the reliable diagnosis and treatment of malignancies. The novelty of the method is based on the suggested magnetic imaging technique, accompanied by the selective coating/decorating of SPANs with the appropriate protein which targets to specific cell categories and the consequent inductive heating, which may be effective due to the temperature gradient caused by the SPAN [18].
The magnetic imaging device and method are mapping SPANs distribution in space. This identification of magnetic dots reveal the existence of cellular areas with high number of bound magnetic nanoparticles and as a result the presence of a malignancy, offering possibly the ability of detecting one single malignant cell. Consequently, high inductive heating is destroying the malignant cells ($\Delta T > 50^{\circ}C$), without being able to destroy the healthy ones due to the very small increase of temperature to them ($\Delta T < 2^{\circ}C$) [19].

4. Conclusions

Magnetic resonance imaging (MRI) is an imaging technique used primarily in medical settings to produce high quality images of the soft tissues of the human body. MRI employs powerful magnets which produce a strong magnetic field (in clinical diagnosis, magnetic fields of 1.5 or 3 T are usually used) that forces protons in the body to align with that field. When a radiofrequency (5–100 MHz) current is then pulsed through the patient, the protons are stimulated, and spin out of equilibrium, straining against the pull of the magnetic field.
Major drawbacks are associate with voluntary or involuntary (heart beating, bowel movement, e.g.) patient's movements, use of contrast agent for high quality images of soft tissues and MRI may not always distinguish between cancer tissue and fluid, as well as small tumours surrounded by healthy tissues. Even though, some of these drawbacks (movements) are efficiently controlled, and in same case eliminated, by professional MR technologists, it is unquestionable that others, such as the ability to discriminate tumour tissues surrounded by

the healthy one or few cancer cells, need to solved. In fact, it takes millions of cells to make a tumour big enough to show up on MRI test.
The development of the innovative magnetic measuring (and heating) device capable of measuring the distribution of SPANs with sensitivity quite better than 1 μm³ (sub-micron tumour cell diameter) could be considered as a universal breakthrough in tumour diagnosis. In addition, the development of the structurally well controlled biocompatible SPANs decorated with the appropriate ligands would permit for targeted SPAN delivery and subsequent immediate killing of the cancer cells by applying controlled inductive heating techniques.

5. References

[1] B. Gleich and J. Weizenecker, "Tomographic imaging using the nonlinear response of magnetic particles," *Nature*, vol. 435, no. 7046, pp. 1214–1217, 2005.
[2] L. M. Bauer, S. F. Situ, M. A. Griswold, and A. C. S. Samia, "Magnetic Particle Imaging Tracers: State-of-the-Art and Future Directions," *J. Phys. Chem. Lett.*, vol. 6, no. 13, pp. 2509–2517, Jul. 2015.
[3] B. Zheng *et al.*, "Magnetic Particle Imaging tracks the long-term fate of in vivo neural cell implants with high image contrast.," *Sci. Rep.*, vol. 5, no. March, p. 14055, 2015.
[4] J. H. Lee, J. W. Kim, and J. Cheon, "Magnetic nanoparticles for multi-imaging and drug delivery," *Mol. Cells*, vol. 35, no. 4, pp. 274–284, 2013.
[5] R. Singh and H. S. Nalwa, "Medical applications of nanoparticles in biological imaging, cell labeling, antimicrobial agents, and anticancer nanodrugs," *J. Biomed. Nanotechnol.*, vol. 7, no. 4, pp. 489–503, 2011.
[6] P. W. Goodwill and S. M. Conolly, "The X-space formulation of the magnetic particle imaging process: 1-D signal, resolution, bandwidth, SNR, SAR, and magnetostimulation," *IEEE Trans. Med. Imaging*, vol. 29, no. 11, pp. 1851–1859, 2010.
[7] H. Chiriac, M. Pletea, and E. Hristoforou, "Magnetoelastic characterization of thin films dedicated to magnetomechanical microsensor applications," *Sensors Actuators, A Phys.*, vol. 68, no. 1–3 pt 2, pp. 414–418, 1998.
[8] G. Andocs, O. Szasz, and A. Szasz, "Oncothermia treatment of cancer: From the laboratory to clinic," *Electromagn. Biol. Med.*, vol. 28, no. 2, pp. 148–165, 2009.
[9] J. Beik *et al.*, "Nanotechnology in hyperthermia cancer therapy: From fundamental principles to advanced applications," *J. Control. Release*, vol. 235, pp. 205–221, 2016.
[10] A. Mamalis, *Over-hyperthermia using magnetic Fe nanoparticles*, vol. 6. 2016.
[11] T. Prisner, M. Rohrer, and F. MacMillan, "PULSED EPR SPECTROSCOPY: Biological Applications," *Annu. Rev. Phys. Chem.*, vol. 52, no. 1, pp. 279–313, 2001.
[12] D. J. Lurie and K. Mäder, "Monitoring drug delivery processes by EPR and related techniques - Principles and applications," *Adv. Drug Deliv. Rev.*, vol. 57, no. 8, pp. 1171–1190, 2005.
[13] C. Tiede *et al.*, "Adhiron: A stable and versatile peptide display scaffold for molecular recognition applications," *Protein Eng. Des. Sel.*, vol. 27, no. 5, pp. 145–155, 2014.
[14] N. A. Usov and J. M. Barandiarán, "Magnetic nanoparticles with combined anisotropy," *J. Appl. Phys.*, vol. 112, no. 5, 2012.
[15] G. S. E. Antipas, E. Statharas, P. Tserotas, N. Papadopoulos, and E. Hristoforou, "Experimental and first-principles characterization of functionalized magnetic nanoparticles," *ChemPhysChem*, vol. 14, no. 9, pp. 1934–1942, 2013.
[16] T. Shen, R. Weissleder, M. Papisov, A. Bogdanov, and T. J. Brady, "Monocrystalline iron oxide nanocompounds (MION): Physicochemical properties," *Magn. Reson. Med.*, vol. 29, no. 5, pp. 599–604, 1993.
[17] A. Ferraro, A. Mamalis, and E. Hristoforou, "Monitoring Magnetic Nanoparticles in the Body," vol. 856, pp. 85–91, 2016.

IOP Conf. Series: Journal of Physics: Conf. Series **939** (2017) 012001 doi:10.1088/1742-6596/939/1/012001

[18] C. Corot, P. Robert, J. M. Idée, and M. Port, "Recent advances in iron oxide nanocrystal technology for medical imaging," *Adv. Drug Deliv. Rev.*, vol. 58, no. 14, pp. 1471–1504, 2006.

[19] Hristoforou. E., «Device and method of volume distribution measurement and selective heating of super-paramagnetic nanoparticles». European Patent 20110100260, 2011.

Development of metal oxide gas sensors for very low concentration (ppb) of BTEX vapors

A FAVARD[1, 2], K AGUIR[1], T CONTARET[1], L CARIS[2] and M BENDAHAN[1]

[1] Aix Marseille Université, Université de Toulon, CNRS, IM2NP, Marseille, France
[2] EcoLogicSense, Rousset, France

Khalifa.aguir@im2np.fr

Abstract. The control and analysis of air quality have become a major preoccupation of the last twenty years. In 2008, the European Union has introduced a Directive (2008/50/EC) to impose measurement obligations and thresholds to not exceed for some pollutants, including BTEX gases, in view of their adverse effects on the health. In this paper, we show the ability to detect very low concentrations of BTEX using a gas microsensor based on metal oxide thin-film. A test bench able to generate very low vapors concentrations has been achieved and fully automated. Thin metal oxides layers have been realized by reactive magnetron sputtering. The sensitive layers are functionalized with gold nanoparticles by thermal evaporation technique. Our sensors have been tested on a wide range of concentrations of BTEX (5 - 500 ppb) and have been able to detect concentrations of a few ppb for operating temperatures below 593 K. These results are very promising for detection of very low BTEX concentration for indoor as well as outdoor application. We showed that the addition of gold nanoparticles on the sensitive layers decreases the sensors operating temperature and increases the response to BTEX gas. The best results are obtained with a sensitive layer based on ZnO.

1. Introduction

Monitoring and analysis of the outdoor air quality has become a major concern of the past 20 years. Several toxicological and epidemiological studies have shown that air pollution causes respiratory problems, asthma, cardiovascular disease and cancer. Air pollution is estimated to be responsible for more than 300,000 premature deaths in the Europe per year [1] [2]. Since 2008, the European Directive 2008/50/EC imposes concentration thresholds for some pollutants as BTEX gas (Benzene, Toluene, Ethylbenzene and Xylene) on a european scale. Many researches have been done to meet the requirements of new european standards for regulating the air quality and many research teams have looked at sensors based on metal oxide for the detection of volatile organic compounds (VOCs) as BTEX.

Several metal oxides have been reported in the literature for the detection of Benzene and Toluene. D. Ghaddab and Al. [3] studied the zinc oxide for its responsiveness to BTEX gas. Tungsten oxide also shows interesting performances in the presence of BTEX and this oxide is less affected by humidity variations. K. Kanda and Al. [4] were able to detect concentrations of 100 ppb of BTEX to 693 K with a sensitive layer of WO_3 of 35 µm of thickness. Moreover, W. Tang and Al. [5] showed a clear improvement in performance of sensitive layers by a "surface doping" with noble metals such as platinum and gold. The addition of noble metals on the surface would reduce the potential barriers and modify areas of space charge of the metal oxides according to a study by F. Li and Al. [6]. Furthermore, it would decrease the activation energies required for the creation of bonds to the surface of the sensitive material. Thus, it facilitates physical and

IOP Conf. Series: Journal of Physics: Conf. Series **939** (2017) 012002 doi:10.1088/1742-6596/939/1/012002

chemical reactions between the compounds studied and metal oxides and improves the performances of the sensors.

In this paper, we will demonstrate that our sensors can detect BTEX concentrations of 5 ppb at temperatures below 593 K, using metal oxide layers produced by reactive magnetron sputtering.

2. Experimental

WO_3 and ZnO thin films were prepared by reactive radio frequency (13.56 MHz) magnetron sputtering using 99.99% pure tungsten and zinc targets. The vacuum chamber was evacuated to 5×10^{-10} bar by a turbo molecular pump. The films were sputtered on SiO_2/Si substrates with platinum (Pt) electrodes, in a reactive atmosphere under different oxygen–argon mixtures. The films thickness are about 50 nm. Both argon and oxygen flow were controlled by mass flow controllers. The total gas flow was maintained constant at 10 sccm, keeping the total pressure in the deposition chamber at 3 mbar. Oxygen content in the gas mixture during the layers' realization, defined as the ratio of oxygen flow to the total flow, is fixed at 50% for WO_3 and at 40% for ZnO. As WO_3 and ZnO layers are highly resistive, interdigitated electrodes were used in order to reduce the sensor resistance. The distance between the electrodes was 50 µm. They were obtained from a sputtered Platine (Pt) film, using photolithography and lift off processes. After the sensitive layer's deposition on the Pt electrodes, the films were annealed at 743 K for 90 minutes in air in order to stabilize the chemical composition and the crystalline structure. The addition of gold particles was performed by thermal evaporation technique.

The samples were kept in dry air and no conditioning step was carried out before the sensor characterizations.

Bragg–Brentano (θ–2θ) X-ray diffraction (XRD Philips X'PERT MPD) was applied to check the crystallinity and the structure of the films. CuKα radiations (λ = 0.154 Å) with a scanning rate of 1°/min were used for the XRD analysis.

To investigate the BTEX sensing properties of WO_3 and ZnO thin films, we designed a measurement system for obtaining gas concentrations from 1 to 500 ppb with control of the relative humidity inside the test chamber. The measuring bench consists of several elements:

- A system of gas dilution and humidification that generates an output mix at very low concentrations of 1 to 500 ppb with a variable humidity (0 to 90%) and an adjustable output rate between 10 and 500 sccm.
- An integrated test cell made in our laboratory to characterize the electrical responses of the sensors in the presence of gas mixture in terms of sensitivity, speed, repeatability.
- An acquisition system that measures the electrical responses of three sensors simultaneously. It consists of three sourcemeters (Keithley 2450) and a programmable power supply (Rhodes & Schwarz HMP4040). The data were collected on a PC through GPIB interface. The programs interface allowing the control of the measuring devices and the signal acquisition have been developed under "LabVIEW Software".

3. Results and discussion

3.1. ZnO and WO_3 thin film structure

The results obtained by X-ray diffraction show diffraction peaks corresponding to WO_3 with a monoclinic structure P21/n and the polycrystalline ZnO films have a preferential growth parallel to the plane (002), which corresponds to the Wurtzite structure. We used Scherrer formula to calculate the size of crystallites (D). The crystallite size in the sensitive layers of ZnO and WO_3 are 15 nm and 10 nm respectively.

3.2. Sensing properties

One of important parameters of the MOX sensor sensibility is the temperature of the sensitive layer. This behavior can be explained by considering the temperature dependence of the surface coverage of chemisorbed species. At low temperature, the desorption phenomenon is weak and a total coverage of the sites can be obtained easily by ambient oxygen making any reversible detection impossible. Conversely, at high temperature the efficiency of the desorption mechanism increases and the system will be less sensitive. There is thus necessarily an optimum temperature at which the sensor has a maximum of sensitivity.

The sensor response was defined as (R_{air}/R_{gas}), where R_{air} and R_{gas} are the resistance of the sensor in air and in tested gas, respectively. In order to check the effect of working temperature on the sensors response, sensors were maintained at various temperatures from 493 to 593 K.

The figure 1 illustrates the response to 500 ppb BTEX versus working temperature. In the case of WO_3, the best response R_{air}/R_{gas}= 2.45 is obtained for an optimum temperature of 533 K. For ZnO, the best response R_{air}/R_{gas}= 2.95 is reached at 573 K. The addition of gold nanoparticles on the surface of the sensitive layers have very significant effects on BTEX detection properties.

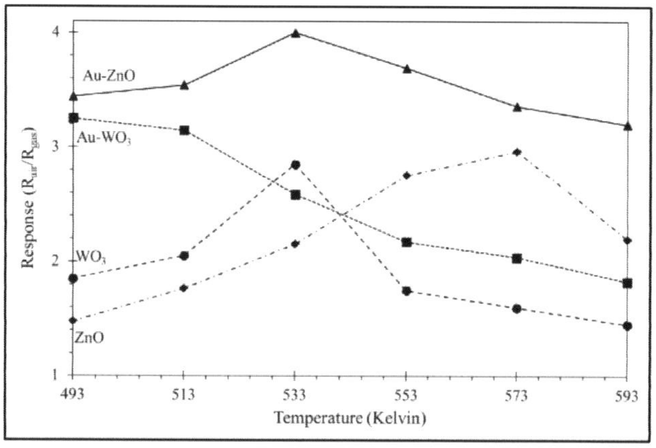

Figure 1. Sensors response vs temperature in presence of 500 ppb of BTEX.

Indeed, the temperature decreases for both metal oxides films studied with the addition of gold nanoparticles and the response is greatly improved. For the case of the sensitive layer of WO_3, it improves the response to BTEX from 2.45 to 3.25 with a decreasing of the optimum temperature from 533 K to 493 K. For ZnO, we can observe the same phenomenon. The response of the modified ZnO with gold is 4 at 533 K against 2.95 at 573 K for the pristine ZnO.

This study also allowed us to determine the response times and recovery times of the studied metal oxides. The addition of gold nanoparticles also decreases the response times and the recovery times. The addition of gold nanoparticles on the surface of metal oxide improve of the sensitivity of the sensor by means of selective reaction of a desired molecule in a selected site.

P. Montmeat and Al. [7] explain this mechanism of improvement of the sensitivity. The reaction of molecules of target gas takes place on the surface of the additives and not the surface of the substrate. These additives change their state of charge, which results in a variation in the height of the potential barrier, which leads to a conductance change on the metal oxide. Furthermore, the addition of gold nanoparticles as active catalyst creates active sites that are considered essential for the improvement of sensitivity obtained on the sensitive layers tested. These hypotheses are in agreement with our results presented in table 1.

In order to investigate the detection limit of our sensors, we submitted our sensitive layers to different concentrations of BTEX between 5 and 50 ppb. The results obtained are shown in figure 2.

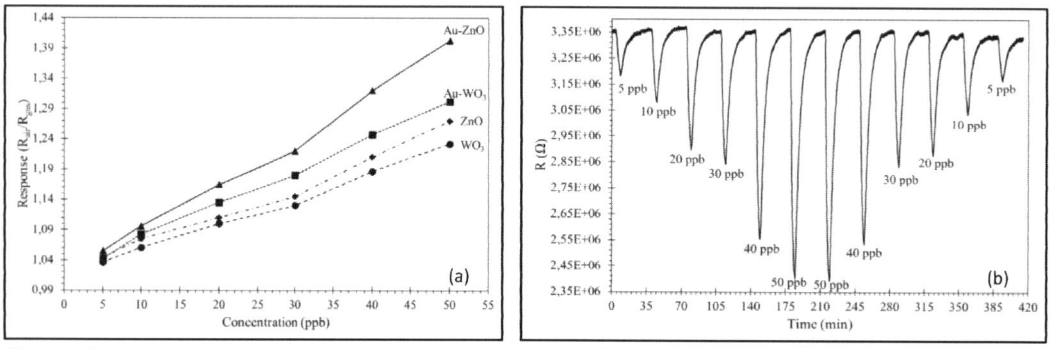

Figure 2. (a) Response of the four studied sensors at different temperature in presence of BTEX, (b) Resistance of Au-ZnO at 533 K for different concentration of BTEX.

Material	Response (R_{air}/R_{gas})	T (K)	Response time (s)	Recovery time (s)
WO_3	2.45	533	225	910
ZnO	2.95	573	120	700
Au-WO_3	3.25	493	160	625
Au-ZnO	4	533	90	425

Table 1. Performances of the four studied sensors in presence of 500 ppb of BTEX

The ZnO and WO_3 based sensors responses respectively at 573 and 533 K in the presence of 5 ppb are 1.075 and 1.02. The addition of gold nanoparticles greatly improves detection performance and the stability of the sensor. Indeed, Au-ZnO and Au-WO_3 based sensors have a response to 5 ppb BTEX of 1.28 and 1.16 respectively. These results are very promising in comparison of the results found in the literature (Table 2).

Material	Concentration (ppb)	Response (R_{air}/R_{gas})	Temperature (K)	Reference
WO_3	500	2.65	693	[5]
ZnO	500	1.2	493	[3]
SnO_2/V_2O_5	80	1.58	543	[9]
TiO_2	100	1.25	368	[9]
WO_3	5	1.02	533	This work
ZnO	5	1.075	573	This work
Au-WO_3	5	1.16	493	This work
Au-ZnO	5	1.28	533	This work

Table 2. Summary of results reported in the literature

4. Conclusion

We realized BTEX sensors by reactive magnetron sputtering. The sensitive layers have been optimized to detect low concentration of BTEX gas. The thin film of ZnO and WO_3 are capable of detecting concentrations of 5 ppb of BTEX vapors. The addition of gold nanoparticles on the surface of the thin film improves the response to BTEX and the response/recovery time. Besides, it decreases the operating temperature of the sensing layer. The best results have been obtained with a thin layer of ZnO doped with gold nanoparticles on the surface and they are very promising for the realization of sensor for the control of the air quality.

Acknowledgement

We wish to thank the Bank for Investment of France (BPI) for their financial support in the SMARTY project as part of the “Programme d’investissements d’avenir”. We also thank Mr. Alain COMBES for his technical support throughout this work.

References

[1] M. Jantunen, E. O. Fernandes, P. Carrer and S. Kephalopoulos, 2011, European Commission Directorate General for Health and Consumers.
[2] A review of human carcinogens: Chemical agents and related occupations, 2012, IARC Monogr Eval Carcinogen Risks Hum.
[3] D. Acharyya and P. Bhattacharyya, 2015, Solid-State Electronics **106**, 18–26.
[4] K. Kanda and T. Maekawa, 2005, Sensors and Actuators B **108**, 97–101.
[5] W. Tang and J. Wang, 2015, Sensors and Actuators B **207**, 66–73.
[6] F. Li, C. Li, L. Zhu, W. Guo, L. Shen, S. Wen and S. Ruan, 2016, Sensors and Actuators B **223**, 761–767.
[7] P. Montmeat, J-C. Marchand, R. Lalauze, J-P. Viricelle, G. Tournier and C. Pijolat, 2003, Sensors and Actuators B **95**, 83–89.
[8] F. Zhang, 2013, Sensors and Actuators B **186**, 126–131.
[9] K. Dutta, B. Bhowmik, A. Hazra, P.P. Chattopadhyay and P. Bhattacharyya, 2015, Microelectronics Reliability **55**, 558-564.

IOP Conf. Series: Journal of Physics: Conf. Series **939** (2017) 012003 doi:10.1088/1742-6596/939/1/012003

Correlating elastic and plastic deformation with magnetic permeability values

S Papadopoulou [1]
Dpt. of Electrical and Computer Engineering,
National Technical University of Athens, Zografou Campus, Athens, 15780, Greece

E-mail: sofiapapadopoulou.ece@gmail.com

Abstract. This paper investigates the utilization of magnetic permeability method in determining elastic and plastic deformation state of ferromagnetic steels. The results have shown a strong degradation of the magnetic values on plastically region due to the irreversible movements of the magnetic domain walls.

1. Introduction

Destructive testing (DT) includes methods where a material is broken down in order to determine its mechanical properties, such as strength, toughness and hardness. for this reason laboratories or industries focus on Non-destructive testing (NDT) techniques, in order to evaluate the properties of a material, without causing any damage. Common NDT methods include eddy currents, magnetic particles, magnetic leakage (MDL), hysteresis loops (B, M – H loops), magnetic permeability (μ – H loops), magnetoacoustic (MAE), Barkhausen noise (MBN) [1-9].

It is well known that the magnetic non-destructive methods are influenced by the microstructural features of the examined ferromagnetic materials [10-23]. It has been also verified that, when a ferromagnetic material is subjected to uniaxial tensile or compressive tests, it undergoes a reconfiguration of its structure [24-27]. A high-precision magnetic sensor was used to measure slight changes in the magnetic field strength. Thus, the deformation influence the final microstructure of the material, resulting in variations of the output magnetic signals.

In this paper, the effect of the elastic and plastic deformation on the magnetic properties of electrical steels has been investigated. The magnetic properties of non-grain-oriented (NGO) electrical steels are highly susceptible to mechanical stresses. It is known that during the tensile loading of a polycrystalline material, each grain experiences a different strain depending on its orientation and constraints inflicted by its neighboring grains [28-33]. The utilization of magnetic permeability method has verified such behavior. The resulting magnetic properties were further evaluated by examining the samples' microstructure by using scanning electron microscopy[34]. Microstructural features can influence these domain processes to modify the energy balance and ease of domain re-alignment and hence affect the magnetic properties of ferromagnetic materials. More details on the domain theory and the effects of the microstructure in ferromagnetic materials on their magnetic domains and properties can be found elsewhere.

2. Experimental procedure

[1] Phone number : +30 210 772 2183, E-mail : xvourna@metal.ntua.gr

The studied alloy was Non-Oriented Electrical Steel samples. The chemical composition, as given by the manufacturer, is given in Table 1.

Table 1. Chemical composition, in %wt, of the as-received NOES.

Si	Mn	Al	P	S	C	Fe
2.18	0.12	0.35	0.00009	0.00009	0.0018	Balanced

Dog-boned samples were cut from the as-received samples, according to the ASTM E8 Standard. These samples were subjected to uniaxial tensile strain at preselected deformation steps. The elastic region was divided into 6 deformation states, while the plastic one into 20 deformation states. Two strain values were defined around the yield point.

The magnetic permeability during the tensile tests were recorded. The permeability magnetic sensor consisted of a double U- shape electromagnet. In this apparatus the receiving coil was wound around the examined sample. The output voltage was proportional to the magnetic permeability of the examined sample.

3. Results

The optimization of NGO when the application includes applications rotating electrical machines, requires comprehensive considerations of spatial variations of magnetic properties in the sheet plane and careful awareness of occurring anisotropic effects. The magnetic permeability values were increased in the elastic deformation region (Fig. 1). During the uniaxial tensile test, the initial easy magnetization axis rotated, in order to become parallel to the direction of the applied stress, resulting in the increment of the 180° domain walls. Thus, the tensile stresses increase the magnetic responses [28-29].

However, in the plastic deformation region it is evident a strong decrease of the magnetic permeability. Thus, the demarcation between elastic and plastic region was evidenced by the drop of both magnetic output signals. Within the elastic region, mechanical stresses up to yield strength cause homogeneous elongation across the entire specimen by deviating atoms of the crystal lattice from their equilibrium position.

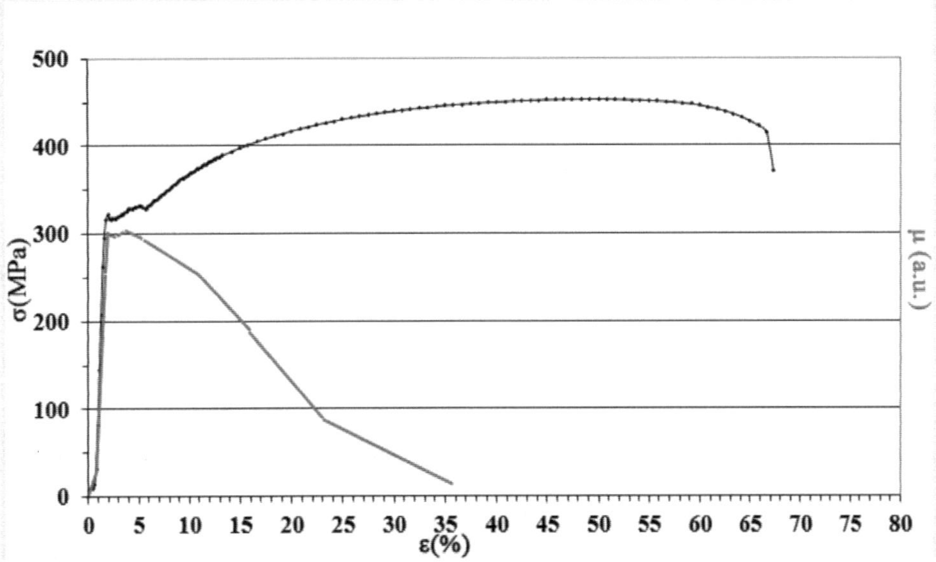

Figure 1. Variations of the magnetic permeability during the uniaxial tensile test.

As soon as the stress passes the yield strength, plastic deformation will start to occur inside the grains. In the polycrystalline material occurred an increment of the dislocation density, at the beginning of the plastic region[30-32]. After the slip system reaches the required critical resolved shear stress value, the dislocation start to slip[33]. Thus, the formation and movement of dislocations will play a crucial role in absorbing the plastic deformation. Moreover, the high density of dislocations formed as tangles for higher plastic strain rates. These tangles acted as strong pinning sites during the magnetization procedure and led to a slight but progressive deterioration of the magnetic behaviour. The magnetocrystalline anisotropy, its change as well as the effect of dislocation interactions and internal stresses are directly linked to the structure of the material and thus, primarily affect the hysteresis term of the losses, which can be affirmed from the results. The excess loss term is nearly not evident in the observed cases.

Study on microstructure revealed that electrical steel consisted of polygonal and equiaxed ferrite grains in the elastic region (Fig. 2). In the plastic region, the ferrite grains were elongated along the direction of the applied tensile stress (Fig. 3). The high dislocation density and the increment in the gain boundary area are pinning sites of higher energy, resulting in the pronounced reduction of the permeability responses.

Figure 2. SEM micrograph of NOES in elastic region (backscatter electrons).

Figure 3. SEM micrograph of NOES in plastic region(backscatter electrons).

4. Conclusions

Broadly, the examined samples of Non-Oriented Electrical Steels results verify the strong dependence between the magnetic properties and the elastic and plastic deformation behaviour, due to the variations of the output magnetic signal. Yet in the plastic deformation the formation and movement of dislocations are crucial, because the high dislocation density and the development of tangles deteriorated the magnetic properties. There is remarkable consistency in magnetic behaviours and properties such as initial/incremental permeability values between the measurements by different techniques. This behaviour has been ascribed to the similar underlying domain processes and hence similar selected microstructural features that are affecting the domain processes.

5. References

[1] Hubert O, 2017 *Journal of Magnetism and Magnetic Materials* **424** 421–442
[2] Piotrowskia L, Augustyniaka B, Chmielewskia M, Labanowskib J and Lech-Gregac 2012 *NDT & E International* **47** 157–162
[3] Hristoforou E, Reilly R E and Niarchos D 1993 *IEEE Trans. Magn.* **29** 3171-3173
[4] Hristoforou E and Reilly RE, 1991 *J. Appl. Phys.* **69** 5008-5010
[5] Hristoforou E, Chiriac H, Neagu M and Darie I 1994 *J. Phys. D: Applied Physics* **27** 1595-1600

[6] Hristoforou E, Chiriac H and Nagacevschi V 1999 *Sensors & Actuators A***76** 442 – 447
[7] Hristoforou E and Chiriac H 2000 *Sensors & Actuators A* **81** 158 – 161
[8] Youroudi I, Orfanidou C and Hristoforou E 2003 *Sensors and Actuators A***106** 179-182
[9] Kepaptsoglou DM, Paluga M, Deanko M, Muller D, Conde CF, Hristoforou E, Janickovic D and Svec P 2006 *J. of Microscopy – Oxford* **223** 288-291
[10] Grabiasa A and Oleszakc D 2012 *J. of Magn. and Magn. Mat.* **324** 2501–2505
[11] Piotrowski L, Augustyniak A, Chmielewski M, Hristoforou E and Kosmas K 2010 *IEEE Trans. Magn.* **46** 239
[12] Makar JM, Tanner BK, 1998 *Journal of Magnetism and Magnetic Materials* **187** 353–365
[13] Ktena A, Hristoforou E, Gerhardt GJL, Missell FP, Landgraf FJG, Rodrigues Jr DL and Alberteris-Campos M 2014 *Physica B: Coned Matter* **435** 109
[14] Piotrowski L, Chmielewski M and Augustyniak B 2012 *J. of Magn. and Magn. Mat.* **324** 2496–2500
[15] Javorskyja I, Kravetsa I, Matskoa I and Yuzefovycha R 2017 *Mechanical Systems and Signal Processing* **83** 406–438
[16] Ondraa V and Severb IA 2017 *Mechanical Systems and Signal Processing* **83** 210–227
[17] Yuning Qiana and Ruqiang Yana 2017 *Mechanical Systems and Signal Processing* **83** 549–567
[18] Hu B and Runqiao Yu 2016 *NDT & E International* **80** 1–5
[19] Kepaptsoglou DM, Svec P, Janickovic D and Hristoforou E 2007 *J. of All. and Comp.* **434** 211-214
[20] Giouroudi I, Ktena A and Hristoforou E 2004 *J. Opt. Adv. Mat.* **401** 45-50
[21] Jiles D 1998 *Introduction to Magnetism and Magnetic Materials*
[22] Luo KY, Liu B, Wu LJ, Yan Z and Lu JZ 2016 *Applied Surface Science* **369** 366–376
[23] Petch NJ 1953 *The cleavage strength of polycrystals* J. *Iron Steel Inst* **174** 25–28
[24] Vourna P, Ktena A, Tsakiridis P E and Hristoforou E 2015 *NDT & E Int.* **71** 33
[25] Vourna P, Hervoches C, Vrána M, Ktena A and Hristoforou E 2015 *IEEE Trans* **51**
[26] Vourna P, Ktena A, Tsakiridis P E and Hristoforou E 2015 *Measurements* **71**
[27] Vourna P, Ktena A and Hristoforou E 2014 *IEEE Trans. Mag.* **50** 1
[28] Weissa HA Leuningb N, Steentjesb S, Hameyerb K, Andorfera T, Jennera S and Volka W 2017 *J. of Magn. and Magn. Mat.* **421** 250–259
[29] Gallaughera M and R Chromik *J. of Magn. and Magn. Mat.* ***382*** *124–133*
[30] Hristoforou E and Reilly R E 1990 *Tran Mag* **26** 1563-1565
[31] Hristoforou E and Niarchos D 1992 *IEEE Trans Mag* **28** 2190-2192
[32] Dimitropoulos P D, Avaritsiotis J N, Hristoforou E 2003 *Sensors & Actuators A* **107** 238-247
[33] Deanko M, Kepaptsoglou DM, Muller D, Janickovic D, Korvanek IS, Hristoforou E and Svec P 2006 *J. of Microscopy – Oxford* **223** 260-263
[34] Deanko M, Paluga M, Kepaptsoglou DM, Muller D, Mrafko P, Janickovic D, Hristoforou E, Skorvanek I and Svec P 2007*J. of Alloys and Compounds* **434** 248-251

IOP Conf. Series: Journal of Physics: Conf. Series **939** (2017) 012004 doi:10.1088/1742-6596/939/1/012004

The effect of noble metal additives on the optimum operating temperature of SnO_2 gas sensors

S. Mohammad-Yousefi[1], S. Rahbarpour[1*], H. Ghafoorifard[2]
[1]Electrical Engineering Department, Shahed University, Tehran, Iran.
[2]Electrical Engineering Department, Amirkabir University of Technology, Tehran, Iran
*s.rahbarpour@shahed.ac.ir

Abstract. The effect of Pd and Au additives on gas sensing properties of SnO_2 was investigated. SnO_2 pallets were fabricated and sintered at 900 °C for 90 minutes. Several nanometer layers of Pd and Au were deposited on separate SnO_2 pallets and were intentionally dispersed into the SnO_2 pallets by long heat treatment (400 °C for 1 Day). All metal loaded samples showed significant enhancement in response level and optimum operating temperature compare to pure SnO_2 gas sensors. The amount of enhancement was strongly dependent on the material and the thickness of deposited metal layer. Studying butanol response showed that increasing the thickness of metal causes the response level to increase. Further thickness increase caused contrary effect and decreased the performance of sensors. Best results were achieved at 10 nm-thick Au and 7 nm-thick Pd. Generally, Pd-SnO_2 samples demonstrated better performance than Au-SnO_2 ones, however, Au-SnO_2 samples were proved to be good candidate to sense reducing gases with lower hydrogen atoms in their formula. Given experimental results were also good evidence of chemical activity of gold and simply confirms the relation between chemical activity and gold particle size. Results were qualitatively described by gas diffusion theory and surface reactions take place on metal particles.The first section in your paper

Keywords : Palladium, Gold, Noble metal catalyst, Temperature behaviour, Gas sensor, Low operating gas sensor.

1. Introduction

Metal oxide semiconductor sensors especially tin oxide based sensors are widely used in detection reducing gases such as toxic or flammable gases [1]. The benefits of tin oxide-based sensors, high sensitivity, simply design, low weight and price. Pure tin oxide sensors are less sensitive than are sensors contain noble metals. As the noble metals are catalysts in many reactions, modifying the surface of metal oxides with these materials improves their sensing properties. Many researchers have reported increasing the sensitivity of sensors modified with noble metal such as Pd [2-4], Pt [5-6] and Ag [7-8] noble metals. In 1983, Yamazoe observed that the addition of noble metals on the surface of tin oxide sensors improved its sensing features. They proposed two mechanisms: Ag and Pd affect the tin oxide work function while Pt acts as a catalyst. The work function of palladium and silver increase 0.7 and 0.3 respectively, which can these reduce to zero when put in the reducing atmosphere [9]. The origin of metal work function increases. Oxygen adsorption on the surface of metal particles which consequently increases the barrier height at connection points of metal and metal oxide particles. The resistance of the sensitive layer changes via this barrier alteration. In the case of platinum, gas molecules breakup the metal surface and active protons flow to sensitive surface. In other words, the sensitivity to gas is due to increased rate of chemical reactions on the surface [9].

Both above mentioned mechanisms change the depletion layer at the metal oxide grain boundaries. Although gold is structurally similar to platinum and silver [10] but less attention paid to its gas sensing properties because of the lack of its nobility and lack of chemical activity [11-13]. Major study in the field of chemistry of gold are focused on catalytic behavior of gold nanoparticles for oxidation of carbon monoxide [14,15]. Thus researchers in the field of gas sensors have focused their investigation on CO sensing properties of gold loaded metal oxides, and others gold sensing properties are missed in gas sensing literature. Studies the gold loaded metal oxides are limited to works of Wang [16], Goto [17], Ramgir [19] and several other [20-28] reporters. Adding gold nanoparticles on the surface of metal oxides can improve the sensitivity of these sensors to carbon monoxide. Different chemical properties of gold nanoparticles compared to bulk gold, it necessitates doing more research to investigate gas sensing properties of gold nanoparticles sensors is the subject of many researchers in the field of chemistry and physics [14].

Here we have reported the effect of adding gold and palladium nanoparticles on sensing properties of tin oxide in response to acetone and butanol. To understand working mechanism of fabricated sensor and find the optimum fabrication and working different amount of gold and palladium were loaded to samples and the sensors were investigated. It was shown that sensitivity of the sensor is highly dependent on the amount of metal nanoparticles on the surface of the sensitive layer. It was also shown that the optimum operating temperature, where the sensor shows the highest gas sensitivity can be controlled by altering the fabrication process.

2. Experimental

Tin oxide sensitive layer created on the ceramic substrate and was sintered at 900 °C for 90 minutes. Sintering temperature and time were respectively varied in the range from 400°C to 1200 °C and 30 to 120 minutes. All fabricated samples were tested and samples with best performance were selected for further examination and reporting their features. Fabrication conditions were 900 °C and 90 minutes sintering condition. By using sputtering system, thin layers of gold and thin layers of palladium were deposited on different samples on top of sensitive layer tin oxide. Although the reason of using this method was controllable depositing process other conventional methods such as wet chemical method [10], are simple and inexpensive, they suffer from controllability. Sputtering systems provide process control about one nanometer. Gold and palladium layers were created in 3, 7, 10, 15 and 20 nm thicknesses. In order to avoid redundancy, the abbreviated name in accordance with table 1 is used for sensors. To penetrate metal nanoparticles into sensitive layer, samples were placed at a temperature of 400 °C for 24 h. Electrical connections established by two platinum electrodes. After placing sensors on micro-heaters, all of the fabricated sensors were placed under saturation concentration of 1-butanol gas for two hours. This stage was done in order to stabilize and increase the response of the sensor.

IOP Conf. Series: Journal of Physics: Conf. Series **939** (2017) 012004 doi:10.1088/1742-6596/939/1/012004

Table 1: Abbreviated names of modifiedSnO_2 sensors with gold and palladium nanoparticles.

Abbreviation Name	Metal loaded	Thickness of metal layer (nm)
S_0	----	0
SA_3	Au	3
SA_7		7
SA_{10}		10
SA_{15}		15
SA_{20}		20
SP_3	Pd	3
SP_7		7
SP_{10}		10
SP_{15}		15

The thickness of deposited layers was varied in the range of 3 to 20 nm. As metal deposited between 2 electrodes, it was considered to be continues to prevent short circuiting the sample beyond this layer, metal particles deposited separately on the surface and their cluster size depend on their layer thickness.

Gold reformer layer discolor the surface of the sensor, but due to the thin layer of gold, its color wasn't gold. To prevent short-circuiting the electrodes of the sensor, the metal layer will not be Continuous. It is expected that be made discrete layer of metal for thickness less than 25 nm. By reducing the thickness of metal layers, reduced the number of particles per unit area, an electron reactions decreases and the density of charge carriers at the interface between grains decreases.

Measurement system consisted of a metal cage attached to the earth to eliminate static charges, a glass chamber to create an area certain concentration of contaminating gas, temperature control circuit and resistance measuring circuit. At each test, sensor was first in clean air and rapidly was positioned in a chamber with predetermined conditions. Sensor resistance variations were recorded during the best. Response of the sensor was calculated by equation (1):

$$S = \frac{R_a}{R_g} \qquad (1)$$

In which R_a is sensor resistance in the air and R_g is sensor resistance in the chamber of target gas [5].

3. Results and Discussion

Fabricated samples were examined in different concentrations of 1-butanol and acetone gas (70ppm to 1000ppm) at different operating temperatures between 100-400 °C with 10 °C temperature steps.As shown in Fig.1, pure tin oxide sensor showed maximum sensitivity at 250°C. The amount of sensor response to 1000ppm of 1-butanol and acetone gas vapor was 23 and 7/2 respectively based on equation (1), and is consistent with response levels for pure metal oxide resistive sensors reported in the literature [39]. According to Fig. 2, in this type of sensors, operating temperature (Tp) in which maximum response occurred is independent on concentration.

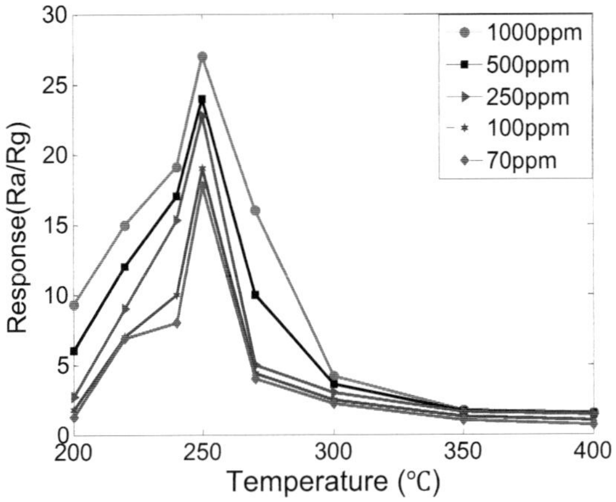

Fig.1: Responses curves of pure tin oxide gas sensor versus operating temperature at various concentrations of 1-butanol.

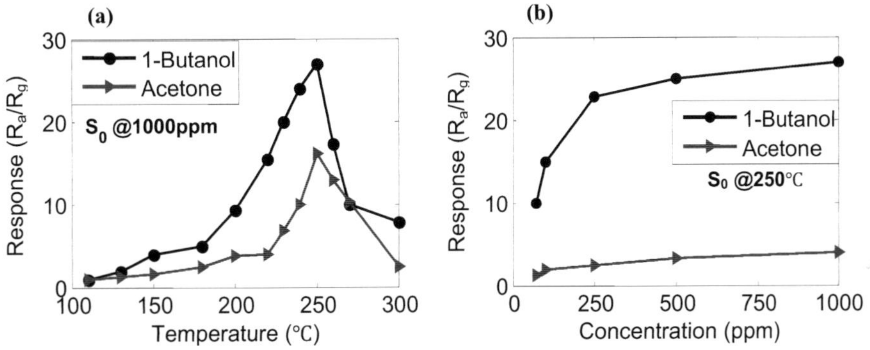

Fig.2: Responses of pure SnO_2 gas sensor vs. (a) operating temperature and (b) different level of 1-butanol and acetone.

Gas concentration of Fig.2.a was 1000 ppm. Fig. 1 and 2 depict that Tp is equal to ~ 250°C.
Responses of all Au loaded samples were studied at different concentrations of 1-butanol and at different operating temperatures. Results are show in Fig.3. As can be seen in Fig.3 Tp is decreased up to 50°C. A notable effect was that the amount of decrease of Tp was related to the thicknesses of the samples. Data related to the SA_3 is not given in Fig.3 because it didn't have significant changes from pure ones. Results are summarized in Table 2.According to data given in Table 2, if the lowest operating temperature is important one may select SA_7, otherwise SA_{10} demonstrate higher response level.

IOP Conf. Series: Journal of Physics: Conf. Series **939** (2017) 012004 doi:10.1088/1742-6596/939/1/012004

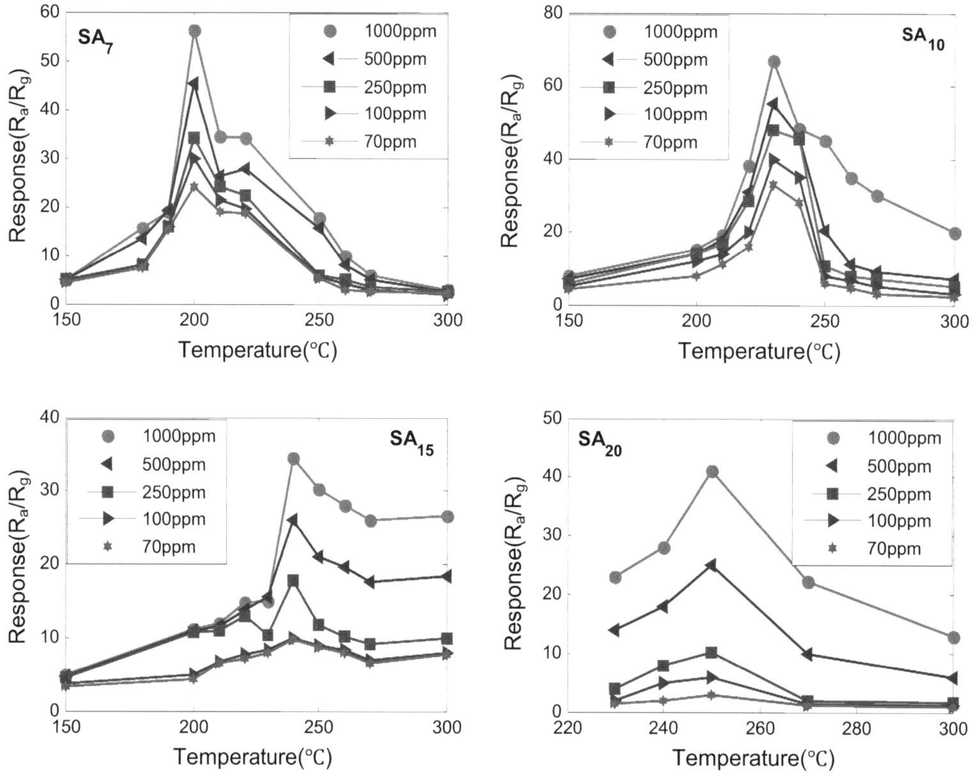

Fig.3: Responses of SnO_2-Au gas sensors versus operating temperature to various concentration of 1-butanol.

Table 2: Responses and operating temperature of Au-SnO_2 sensors in response to 1000 ppm 1-butanol.

Sample name	Pure SnO_2	SA_3	SA_7	SA_{10}	SA_{15}	SA_{20}
Tp (°C)	250	250	200	230	240	250
Maximum response	27	35	60	70	35	41

Some of the samples were modified by palladium nanoparticles. Some experiments were done to Pd-SnO_2 samples. Results are shown in Fig.4. Similar to SnO_2-Au sensors, adding Pd to SnO_2 body caused reduction in sensors operating temperature. However, the amount of reduction of Tp was more considerable in the case of Pd additive and Tp as low as 130°C was achieved. The summary of results is given in table 3.As Tp was ppm independent, 1000ppm of examined target gases was selected to report in Fig.5. Another difference of Pd and Au was that Pd-SnO_2 samples were sensitive to target gas and generally showed different Tp at each target gas, thus, the results of acetone were also given in Fig.4.

IOP Conf. Series: Journal of Physics: Conf. Series **939** (2017) 012004 doi:10.1088/1742-6596/939/1/012004

Table 3: Responses and operating temperature of Pd-SnO_2 sensors in response to 100 ppm 1-butanol.

Sample	Pure SnO_2	SP_3	SP_7	SP_{10}	SP_{15}
Tp (°C)	250	200	150	180	200
Maximum response	27	220	260	100	145

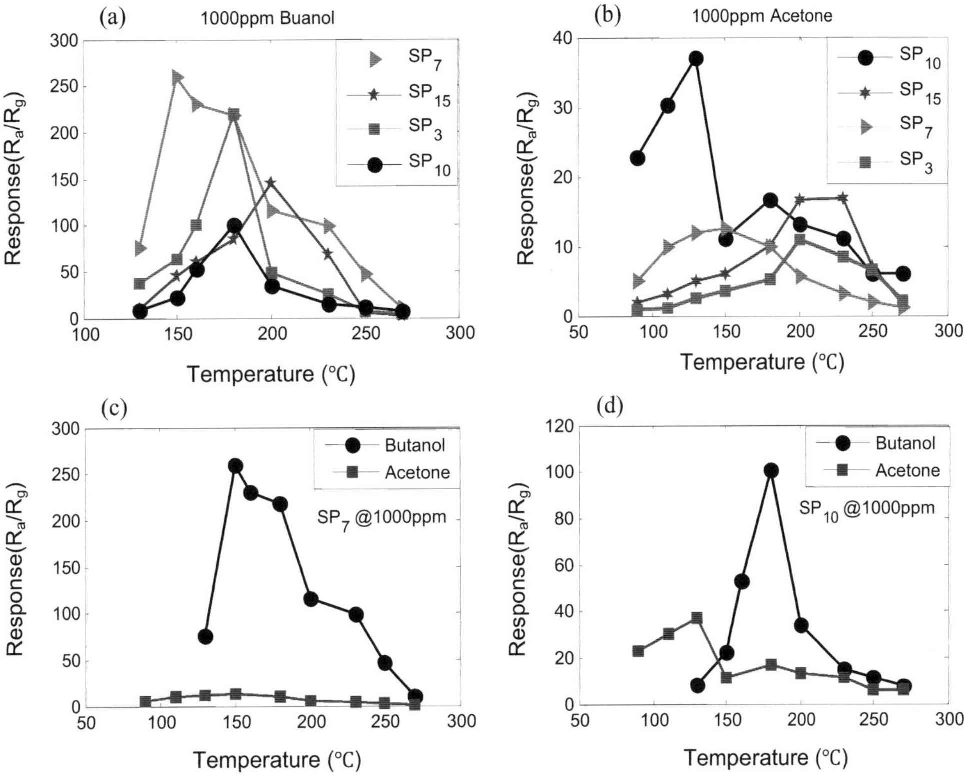

Fig.4: Responses of Pd-SnO_2 gas sensors versus operating temperature to (a) 1000ppm of 1-butanol. (b) 1000ppm of acetone. (c) Comparing responses of SP_7 to 1000ppm of 1-butanol and acetone. (d) Comparing responses of SP_{10} to 1000ppm of 1-butanol and acetone. (d) Comparing responses of SP_7 in operating temperature.

Fig.4. also depicts higher 1-butanol response, that considering high reaction of hydrogen with Pd, it seems causes in the case of hydrogen atmosphere in 1-butanol.
The obvious fact is that deposition of noble metal layers on SnO_2 sensing layer improves the response level and decreases the operating temperature. However these parameters are not related directly. For example, increasing the thicknesses from few nanometer to distinct value (7nm for Pd and 10 nm for Au). Increase shows reverse effect and decreases response levels at sensors. We described this by considering two opposite phenomena: 1- the size of nanopaticle and 2-The amounts of them. First one

increasing with thickness, increase and lead to chemical activity reaction. The second one increase with thicknesses increase and makes layer more sensitive.

It means that although thicknesses increase, higher is the population of the dimensions of particles become larger and the intensity of their chemical activity is reduced. As these two properties affect exactly sensors in opposite directions, an extermum point is observed.

Increase of response level of sensors is a sign of surface reaction increase. This drastically affects temperature of maximum response (Tp). Sakaeiet.al. has described this by gas diffusion theory [30-32] and defined effective parameters on Tp. Although this model considers linear assumption between gas concentration and sensor response, the model is still one of the best models to describe thermal response in resistive gas sensors. Main affecting parameters were mentioned in the model were the thickness of sensing layer (which was constant in our study) and the first order reaction activation energy (E_k). More adsorbed oxygen species on sensing layer fall E_k and consequently decreases Tp. This can happen when metal nanoparticles such as Pd exist on the surface of sensitive layer of sensors. As oxygen adsorption on Au nanoparticles is proved in the literature [30, 31], lower Tp can be achieved even in Au loaded samples. Increasing the population of gold species (thicker deposited layers of Au) leads to further reduction in E_k, if only chemical activity of gold particles remains constant.

Size dependence of gold chemical activity is frequently mentioned in the background literature [33] and it is proved by many theoretical and experimental reports [2]. Here, different fabricated samples had gold particle of different size in their surfaces. The mechanism of gold deposition is as follow: the first layer is an aggregation center which its grain growth by continuing the deposition process. At~ 25 nanometer deposited layer, grains bind together and eventually become a continuous layer [33,34]. Therefore increasing the thickness of the gold layer increases not only the number of gold particles per unit area but also the size of gold particles. Hence, E_k is increased and Tp would rise. The result of both events is a peak in sensor quality Vs. additive thickness.

4. Conclusion

Adding noble metal nanoparticles on the surface of metal oxides can improve the sensitivity of these sensors and reduce their operating temperature. It was shown that reducing the thickness of the noble metal layer reduced the operating temperature of the sample. This phenomenon was related to different chemical activity seen in metal particles of different size. Among Au-SnO_2 sensors exposed to 1000ppm of 1-butanol, sample with 10 nm gold layer showed the highest response (~70at 230 ° C) and sample with 7 nm gold layer showed the lowest operating temperature (200 ° C with highest response of ~60). Samples with metal thicknesses less than 5 nm of gold have demonstrated an inverse performance and increased their operating temperature (same as pure SnO_2 sensors at 250 ° C). It was shown that changing profile of operation temperature is similar in both acetone and butanol and no difference was observed.

In Pd-SnO_2 sensors, sample with 7 nm palladium layer showed the highest response (~260) and lowest operating temperature (equal to 150°C) when exposed to 1-butanol. But the result of acetone was different. The best thickness for acetone sensing was 10nm which its highest response was~ 50 at 130°C.

5. References

[1] L.A.Patil and D.R.Patil: Department of Physics, Pratap College Vol. 93 (2006), p. 316.
[2] N.Yamazoe, Y.Kurokawa and T.Seiyama: Sensors and Actuators 4 Vol. 4 (1983), p. 283.
[3] N.Van Toan, N.Viet Chien and N.Van Duya: Hazardous Materials (2016), p. 433.
[4] Lu, Y.Li, J.Han and C.Meyyappan: *Chem.* Phys. Lett (2004), p. 344.
[5] I.Fasaki, M.Suchea and G.Mousdis: Thin Solid Films Vol. 518 (2015), p. 1109.
[6] D.Haridas, V.Gupta and K.Sreenivas: Bull. Mater. Sci Vol. 31 (2009), p. 397.

IOP Conf. Series: Journal of Physics: Conf. Series **939** (2017) 012004 doi:10.1088/1742-6596/939/1/012004

[7] S.Wang, Z.Ma, S.Dai, J.Liu, Z.Nie and M.H.Engelhard: Phys. Chem. C Vol. 112 (2008), p. 13499.
[8] A.M.Ruiz, A.Cornet, K.Shimanoe, J.R.Morante and N.Yamazoe: Sensors and Actuator B Vol. 108 (2005), p. 34.
[9] N.Yamazoe: Sens. Actuators B: Chem Vol.5 (1991), p. 7.
[10] G.Korotcenkov, V.Brinzari, K.Cho and L.B. Gulina: Applied Surface Science Vol. 353 (2015), p. 793.
[11] T. Hyodo, T. Yamashita and Y. Shimizu: Sensors and Actuators B Vol. 207 (2015), p. 105.
[12] A. Cabot, J. Arbiol, J.R. Morante, U. Weimar, N. Barsan and W. Gopel: Sens. Actuators B Vol. 70 (2000), p. 87.
[13] D.Y. Cha and G. Parravano: J. Catal Vol. 18 (1970), p.200.
[14] M. Haruta, N. Yamada, T. Kobayashi and S. Iijima: J. Catal Vol.115 (1989), p. 301.
[15] S. Wang, Y. Zhao and J. Huang: Applied Surface Science Vol. 140 (2007), p. 3057.
[16] T. Goto and T. Itoh: Sensors and Actuators B Vol. 223 (2016), p. 774.
[17] D. Kohl: Sens. Actuators B Vol. 1 (1990), p. 158.
[18] N. Ramgir, Y.K. Hwang, S.H. Jhung., H.K. Kim, J.S. Hwang and I.S. Mulla: Appl Surf Sci Vol. 252 (2006), p. 4298.
[19] S. Wang, J. Huang, Y. Zhao, S. Wang, X. Wang, T. Zhang, S. Wu, S. Zhang and W. Huang: J. Mol. Catal. A Chem Vol. 259 (2006), p. 245.
[20] M. Haruta and N. Yamada: Journal of catalysis Vol. 115 (1989), p. 301.
[21] L.H. Qian, K. Wang, H.T. Fang and Y. Li and X.L. Ma: Mater. Chem. Phys Vol. 103 (2007), p. 132.
[22] D.Y. Cha and G. Parravano: J. Catal Vol. 18 (1970), p. 200.
[23] S.T. Daniells, A.R. Overweg, M. Makkee and J.A. Moulijn: J Catal Vol. 230 (2005), p. 52.
[24] F. Moreau and G.C. Bon: Top Catal Vol. 44 (2007), p. 95.
[25] R. Grisel and B. Nieuwenhuys: J Catal Vol. 16 (2001), p. 48.
[26] S. Nicoletti, L. Dori, G.C. Cardinali and A. Parisini: Sensors and Actuators B Vol. 60 (1999), p. 90.
[27] N.S. Ramgir, Y.K. Hwang, S.H. Jhung, H.K. Kim, J.S. Hwang, I.S. Mulla and J.S. Chang: Appl. Surf: Sci Vol. 252 (2006), p. 4298.
[28] S. Rahbarpour, S. Sajed and H. Ghafoorifar: Key Engineering Materials Vol. 644 (2015), p. 181.
[29] G. Sakai,G. Matsunaga, N. Shimanoe and K. Yamazoe: Sensors and Actuators B Chemical Vol. 80 (2001), p. 125.
[30] S. Rahbarpour and F. Hossein-Babaei: Key Engineering Materials (2012), p. 289.
[31] S. Rahbarpour and S.M. Hosseini-Golgoo: Gas Sensing Mechanism and Modeling. 14th International Meeting on Chemical Sensors – IMCS.
[32] W.E. Thiele: Industrial & Engineering Chemistry (1939), p. 916.
[33] G. Korotcenkov, V. Brinzari and B.K. Cho: A focused review. Materials Letters Vol. 147 (2015), p. 101.

IOP Conf. Series: Journal of Physics: Conf. Series **939** (2017) 012005 doi:10.1088/1742-6596/939/1/012005

Comparison of information content of temporal response of chemoresistive gas sensor under three different temperature modulation regimes for gas detection of different feature reduction methods

S M Hosseini-Golgoo[1*], F Salimi[2], A Saberkari[1] and S Rahbarpour[3]
[1] Electrical Engineering Department, University of Guilan, Rasht 41635-3756, Iran
[2] Guilan Science and Research Branch, Islamic Azad University, Rasht, Iran
[3] Electrical Engineering Department, Shahed University, Tehran, Iran
*E-mail: smhosseini@guilan.ac.ir

Abstract. In the present work the feature extraction of transient response of a resistive gas sensor under temperature cycling, temperature transient, and temperature combination methods were compared. So, the heater were stimulated by three pulse (cycling), ramp (transient) and staircase (combination) waveforms. The period or duration of all waves was equal to 40 s. Methanol, ethanol, 1-propanol, 1-butanol, toluene and acetone each at 11 different concentration levels in the range of 100 to 2000 ppm were used as the target gases. The utilized sensor was TGS-813 that made by Figaro Company. Recorded results were studied and heuristic features such as peak, rise time, slope and curvature of recorded responses were extracted for each heater waveform. Results showed that although application of this feature extraction method to all waveforms led to gas diagnoses, best results were achieved in the case of staircase waveform. The combination waveform had enough information to separate all examined target gases.

1. Introduction

Resistive gas sensors (RGS) are widely used for gas detection. RGS is cheap, small, durable and has high sensitivity. Such sensors work at elevated temperatures which are provided by a tiny heater embedded in the vicinity of sensitive surface. Despite their great benefits, the lack of selectivity is one of the great weaknesses of RGSs. The use of multiple sensors in array form is the most common method used to increase selectivity of an RGS. However, some efforts have also been made to establish the selectivity of just one sensor [1-2]. Among the various techniques provided to create selectivity in a single sensor [3-4], modulation of operating-temperature of a sensor can be cited. In this method, the heater element is activated by a time variant voltage [5-6]. Temperature modulation approaches of RGSs can be divided in two main categories: Thermal transient (TT) and Thermal cycling (TC). Also, there are combined methods which benefit from properties of both methods [7].

In the work done by Hiranaka et al, the TT method was applied on the heater of tin oxide based Taguchi Gas Sensor (TGS) [8]. In this method, first the sensor warmed by applying 5V voltage and, then exposed to a polluting gas. By doing this, in response to sensor's transitions, some peaks were created and the timing of their occurrence depends on the type of the target gas and was not depended so much on the target gas concentration. Thus, simply provides the possibility to distinguish between analytical gases which were tested. Y. Kato and et. al [9] applied a thermal pulse to a print tin

IOP Conf. Series: Journal of Physics: Conf. Series **939** (2017) 012005 doi:10.1088/1742-6596/939/1/012005

oxide sensor and observed changes of behavioral disturbance resistance at the sensor's output. S. Bukowiecki and et. al provided TC method for RGS sensors. These researchers applied a sinusoidal heating voltage along with a decline to TGS812 heating gas sensor and observed that the number of the position and shape of peaks is related to the gas identity and the gas density is in relation with domain and frequency of the sinusoidal voltage heater [10]. Hosseini and et. al in line with previous experiments applied a staircase voltage, a combination of two methods of thermal transitions and thermal cycles, to a TGS813 resistive gas sensor. Used gases in this experiment were: Methanol, ethanol, 1-Propanol, 2-Propanol, 1-Butanol, 2-Butanol, iso-butanol, tert-butanol, Neon, pentanol, acetone, and hydrogen, each in 11 concentration levels. They could distinguish the 11 mentioned target gases by feature extraction methods [11]. Also in another essay, Hosseini et. al [12], using this thermal modulation method, conducted a comparison among feature extraction methods and obtained acceptable results.

Considering applied methods for producing fingerprint from sensor transient responses, transient compression methods can be categorized in three: One is sampling method: This method, by sampling of response of sensor, provides dynamic information of the sensor's response in various times in actions or recovery phase. Second is feature extraction methods: these methods compress the transient response using some features of transient response such as ascent time, maximum/minimum slope, and integral time. Third is system identification methods: These methods fit a theoretical model to experimental data and use model parameters as property.

In this work, feature extraction method was used and Innovative features such as peak, rise time, slope and curvature of the recorded response were extracted for each shape of heater waveforms and one can find obtained results in the following.

In the second section, the way of conducting experiments, applying waveforms of sensor's heater and recording their response will be provided. In the third section, the way of extracting feature vector from modeled responses and categorizing obtained features using educational data will be evaluated and expressed. Finally, the conclusion will be presented at section 4.

2. Experiment Method

In this work, six alcohol including methanol, ethanol, 2-propanol, 1-butanol, acetone and toluene were used as target gas and TGS-813 produced by Figaro Company was used as RGS. These sensors possess a tin oxide sensitive layer deposited on a tubular alumina substrate and a tiny heater provides the required operating temperature. Variable voltage level leads to creation of varying operating temperature with time. Resistance of RGS depends to operating temperature of sensor and combination of gas polluted air. The way of changing temperature of sensor's sensitive level through applying three different waveforms is illustrated in Fig.1(a)-(c). These waveforms were selected from three different temperature modulation methods: ramp from TC, pulse from TT and staircase waveform from CT. the main parameters of these waveforms are depicted in fig.1. First, maximum amplitude (MA) voltage is considered rated voltage recommended by manufacturer, i.e. 5V. First time delay (TD0) includes a period of minimum 5 minutes in clean air; the sensor is warmed in TD0 interval. This is because of cleaning sensor's surface from former pollution. After placing sensor into the chamber, sensor's voltage was kept constant to 5V for second time delay (TD). This time includes a 2-minute period in order to reach sensor response to its stable condition. At TD, the heater voltage is declined from 5V to 1V. This is maximum amplitude range (AR) of applied waveforms. Each waveform in pulse duration (PD) of 40 seconds will change its condition and the obtained response will be transferred to computer and this process will be repeated for all gases.

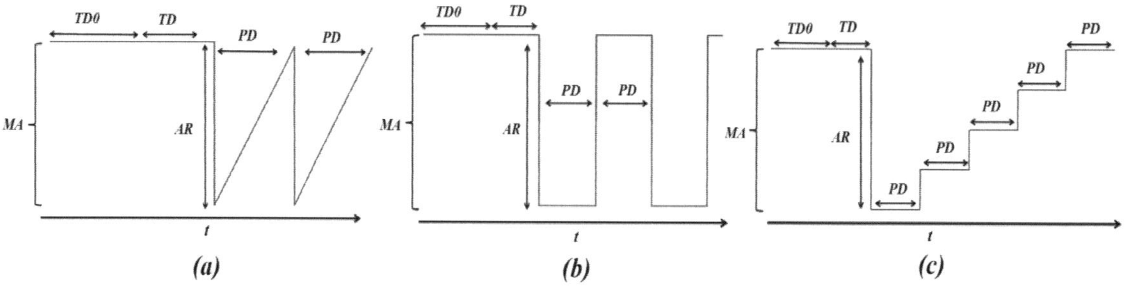

Fig1. The configuration of waveforms of applied voltages to sensor's heater (a) Ramp (b) Pulse (c) Staircase

3. Analysis Method

Fig. 2 illustrates the response of TGS-813 in wide concentration range of six target gases when sensor heater is excited by ramp, pulse, and staircase waveforms. The intended features will be produced in dynamic response of sensor using analysis of multilevel waves. Feature selection is one of the most important topics in pattern categorization and data extraction application. In this regard, innovative features such as peak, rise time, slope, and curvature of recorded response for each waveform were examined. In these figures, three features are considered for each six target gases. For ramp and pulse waveforms in the second peak and for staircase waveform a combination of previous two figures is considered. The first feature at the beginning of the second peak (F1), the second one at the slope of descending curve when heater voltage changes from 5 to 1V (F2) and the third one at the slope of ascending curve when heater voltage varies from 1 to 5V (F3). The results are depicted in Fig.3.

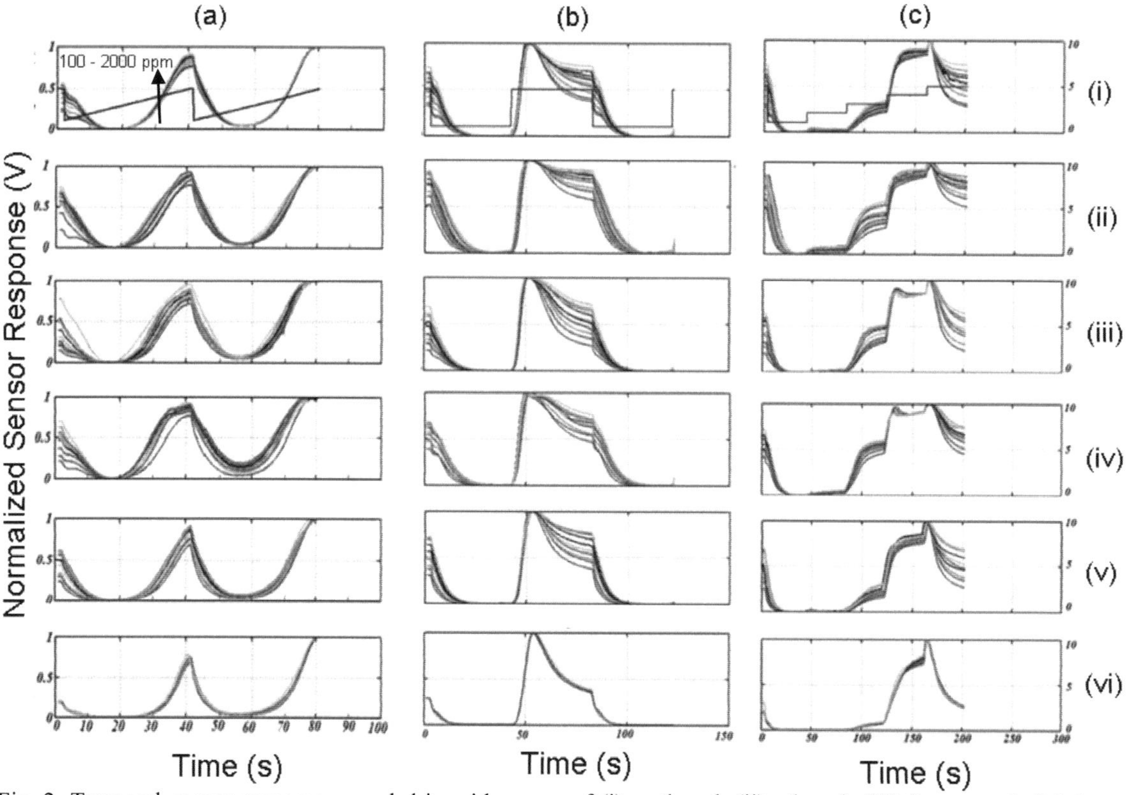

Fig. 2. Temporal sensor response recorded in wide range of (i) methanol, (ii) ethanol, (iii) 2-propanol, (iv) 1-butanol, (v) acetone, and (vi) toluene when sensor heater voltage is excited by (a) Ramp, (b) Pulse, (c) Staircase waveforms. PD of applied waveforms is equal to 40 second.

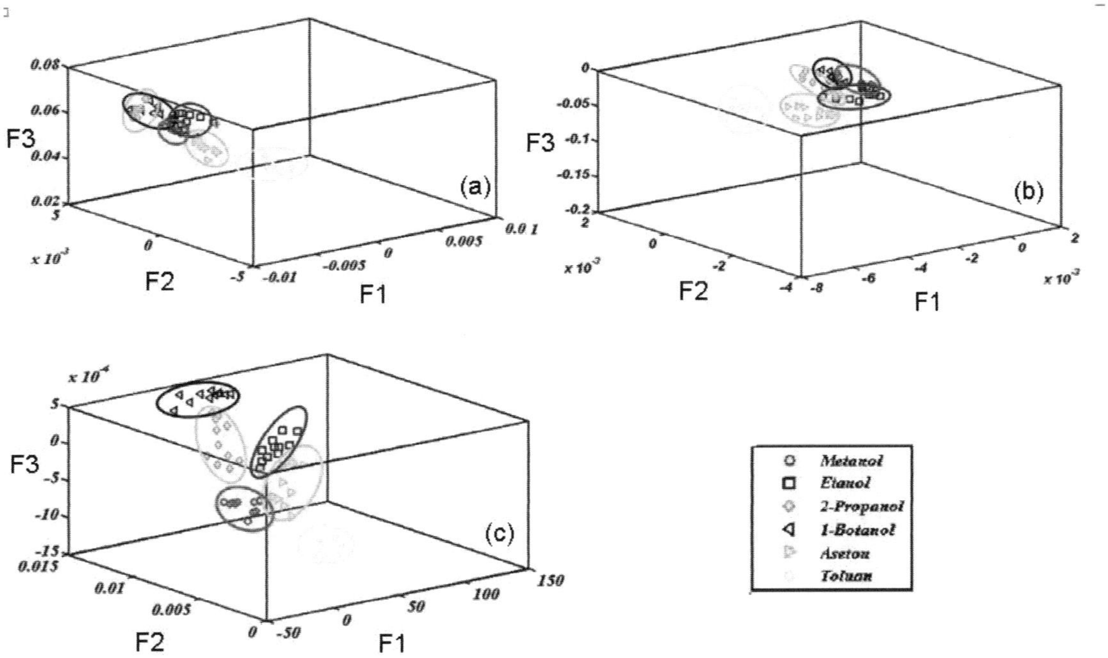

Fig. 3: The representation of feature vectors extracted from sensor responses in exposure to polluted air by methanol, ethanol, 2-propanol, 1-botanol, acetone, and toluene in 3 dimensions for heater waveforms of (a) slope (b) pulse (c) staircase

As can be seen, application of this feature extraction method to all waveforms led to gas diagnoses, best results were achieved in the case of staircase waveform. The combination waveform had enough information to separate all examined target gases.

4. Conclusion

According to fig. 3 it can be expressed that by comparing three waveforms of ramp, pulse, and staircase in thermal modulation, staircase waveform provides the best difference for all 6 gases and using it in this experiment, the 6 gases were distinguished but the other two waveforms distinguished 4 gases of methanol, 2-propanol, acetone, and toluene and the gases of ethanol and 1-bonatol were not separated. Consequently, considering this experiment, the best results are obtained for staircase waveform.

5. References

[1] H.T. Nagle, S.S. Schiffman and R. Gutierrez-Osuna: IEEE Spect. J, Vol. 38 (1998), p. 22.
[2] N. Barsan, D. Koziej, and U. Weimar: Sens. Actuators B VOl. 121 (2007), p.18.
[3] K. Persaud and G.H. Dodd: Nature Vol. 299 (1982), p. 352.
[4] F. Hossein-Babaei and M. Orvatinia: IEEE Sensors J Vol. 4 (2004), p. 802.
[5] F. Hossein-Babaei and S.M. Hosseini-Golgoo: IEEE Sensors J Vol. 8 (2008), p. 1837.
[6] S.M. Hosseini-Golgoo and F. Hossein-Babaei: Meas. Sci. Technol Vol. 22 (2011), P. 35201.
[7] A.P. Lee and B.J. Reedy: Sens. Actuators B Vol. 60 (1999), p. 35.
[8] T. Amamoto, T. Yamaguchi, Y. Matsuura and Y. Kajiyama: Sens. Actuators B Vol. 13-14 (1993), p. 587.
[9] Y. Kato, K. Yoshikawa and M. Kitora: Sens. Actuators B Vol. 40 (1997), p. 33.
[10] W.M. Sears, K. Colbow and F. Consadori: Semicond. Sci. Technol Vol. 4 (1989), p. 351.
[11] S.M. Hosseini-Golgoo, H. Bozorgi and A. Saberkari: Meas. Sci. Technol Vol. 26 (2015), P. 65103.

[12] S.M. Hosseini-Golgoo and N. Ebrahimpour: Materials Science and Engineering Vol. 108 (2016), P. 12001.

IOP Conf. Series: Journal of Physics: Conf. Series **939** (2017) 012006 doi:10.1088/1742-6596/939/1/012006

Wearable strain sensors based on thin graphite films for human activity monitoring

Takanari Saito, Yusuke Kihara and Jun-ichi Shirakashi

Department of Electrical and Electronic Engineering, Tokyo University of Agriculture and Technology, 2-24-16 Nakacho, Koganei, Tokyo 184-8588, Japan

shrakash@cc.tuat.ac.jp

Abstract. Wearable health-monitoring devices have attracted increasing attention in disease diagnosis and health assessment. In many cases, such devices have been prepared by complicated multistep procedures which result in the waste of materials and require expensive facilities. In this study, we focused on pyrolytic graphite sheet (PGS), which is a low-cost, simple, and flexible material, used as wearable devices for monitoring human activity. We investigated wearable devices based on PGSs for the observation of elbow and finger motions. The thin graphite films were fabricated by cutting small films from PGSs. The wearable devices were then made from the thin graphite films assembled on a commercially available rubber glove. The human motions could be observed using the wearable devices. Therefore, these results suggested that the wearable devices based on thin graphite films may broaden their application in cost-effective wearable electronics for the observation of human activity.

1. Introduction

Recently, human activity monitoring using wearable devices have been widely applied in disease diagnosis and health assessment. In general, wearable health-monitoring devices for measuring human motions and human physiological signals, apply various flexible materials, including polymer nanofibers [1, 2], nanowires [3-5], carbon nanotubes [6-8], and graphene [9-11]. Despite the high flexibility and sensitivity, the wearable devices still present challenges because the fabrication of wearable devices is, in many cases, a complicated multistep procedure which results in the waste of materials and requires expensive facilities. The application and markets for wearable devices are limited by high costs and complex processes. Therefore, new low-cost materials and cost-effective fabrication processes for wearable health-monitoring devices are strongly required.

A commercially available pyrolytic graphite sheet (PGS) [12] is an inexpensive flexible sheet, and is easy to cut into any desired shapes and sizes using knives or scissors. Previously, it has been reported that the PGS, which possesses high thermal conductivity has been adopted as a heat spreader [13], and the diamagnetic levitation characteristics of the PGS has been studied for actuators and transport systems [14]. In addition, we have reported the use of thin graphite films as strain sensors, which were fabricated by cutting PGSs [15]. The unique properties of the thin graphite films make it suitable for the application in flexible devices. In this paper, we focused on PGS as a low-cost and simple flexible material, we have investigated wearable devices based on PGS for monitoring human activity.

2. Experimental methods

Figure 1(a) shows the schematic of the simple fabrication procedure of strain sensors using the thin graphite films. First, the thin graphite films were produced by cutting small films from 17-µm-thick PGSs. The thin graphite films were then attached to a flexible plastic substrate. Next, the thin graphite films were wired using silver conducting paste for electrical measurements. Finally, adhesives were used to cover the silver electrodes in order to prevent mechanical failure in the junctions connecting the thin graphite films and the silver electrodes. In case of need, the thin graphite film can be then thinned down using the mechanical exfoliation method. In addition, the thin graphite films attached to the adhesive tapes, which are usually the waste of the mechanical exfoliation method, may also be used in wearable strain sensors, as show in figure 1(b). Previously, the sensitivity of the sensors has been reported to increase when the thickness of the thin graphite films is reduced [15]. The relative change in the resistance of a 5-µm-thick sample was much larger than that of thicker samples; these relative changes in resistance were +3089% under +1.8% tensile bending strains, and −49% under −1.9% compressive bending strains [15]. To demonstrate the potential of the thin graphite films for monitoring human activity, we fabricated wearable devices using the thin graphite films for the observation of the elbow and finger motions, as shown in Figures 2(a), (b), and the performances of these wearable devices were investigated.

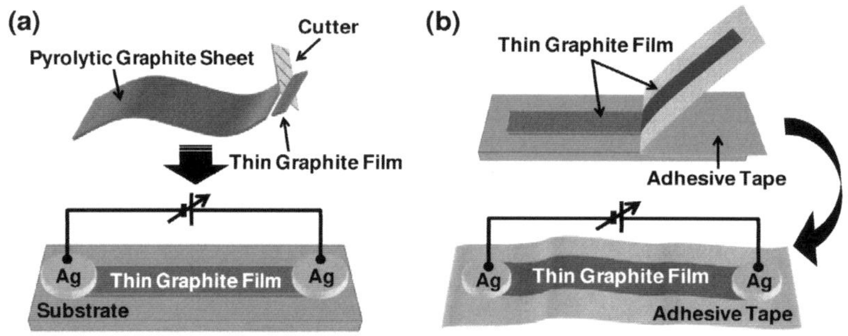

Figure 1. Schematic of the fabrication of a thin graphite film. (a) The thin graphite films as strain sensors fabricated by cutting small films from a 17-µm-thick PGS. (b) A strain sensor based on a thin graphite film attached to an adhesive tape.

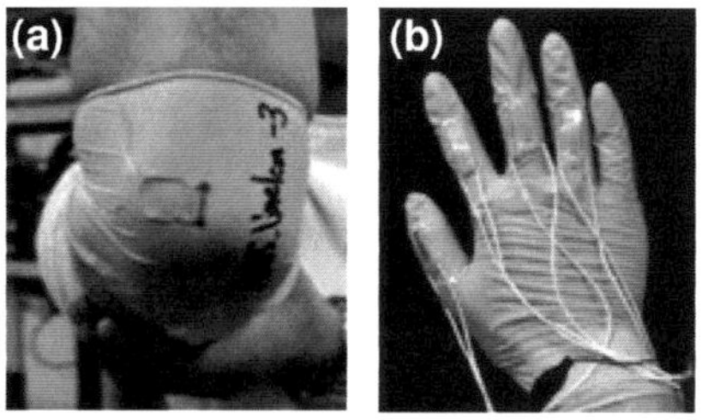

Figure 2. Photographs of the wearable devices based on thin graphite films for monitoring of the (a) elbow and (b) finger.

3. Results and discussions

3.1. Application for elbow motion monitoring

To detect large-scale human motions, the thin graphite film was assembled on a commercially available rubber glove with some modifications over the elbow joint. Figure 3 shows that the resistance had changed during the folding and extending motions of the elbow. In accordance with the movement of the human arm, the resistance of the thin graphite film changed. First, the resistance increased from 3.5 Ω to 13 Ω during the flexion of the elbow, in the time range from 3.1 to 3.8 s, which was attributed to tensile bending strain. Then, during the extending elbow motion, the tensile bending strain decreased, and the resistance recovered to the initial state (R = 3.5 Ω) at a time of 5.8 s. The results indicate that the elbow motions, which are relatively larger joint motions, could be measured using the wearable devices using the thin graphite films.

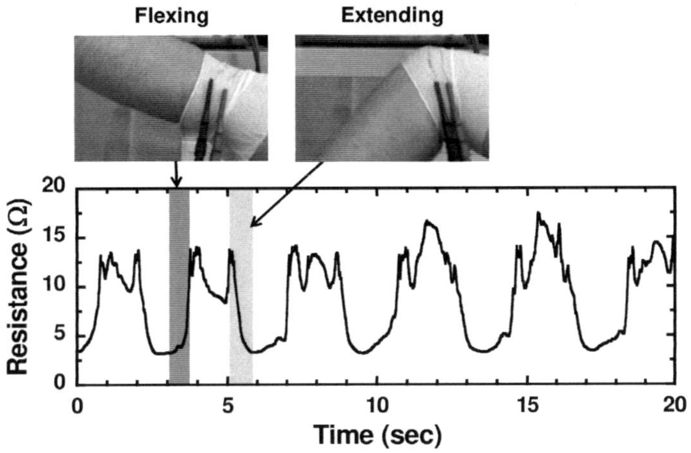

Figure 3. Resistance as a function of time during the folding and extending of elbow motions.

3.2. Fabrication of data glove for finger motion monitoring

We have demonstrated the potential of the thin graphite films on wearable data gloves. The thin graphite films were assembled on an ultrathin glove over the middle finger joint. Figure 4 shows the middle finger's motions of subject A, at 3 types of hand gestures, patterns α, β, and γ. The motions of the middle finger could be monitored using the data glove, and the output of the sensor was measured. In these results, the resistance change of the middle finger has peaks between pattern β and γ, as indicated with the dashed circles in Figure 4. We have concluded that the cause of the peaks is the motion specific to the person or the habit of the person. Insets of Figure 4 show the resistance change of the middle finger between pattern β and γ of two other people (subject B and C). In all cases, similar peaks could be measured, as indicated with the dashed circles in insets of Figure 4. Therefore, the cause of the peaks is possibly characteristics specific to human motions. These results indicated that our data glove could be applied as health care devices and even as controllers for virtual games.

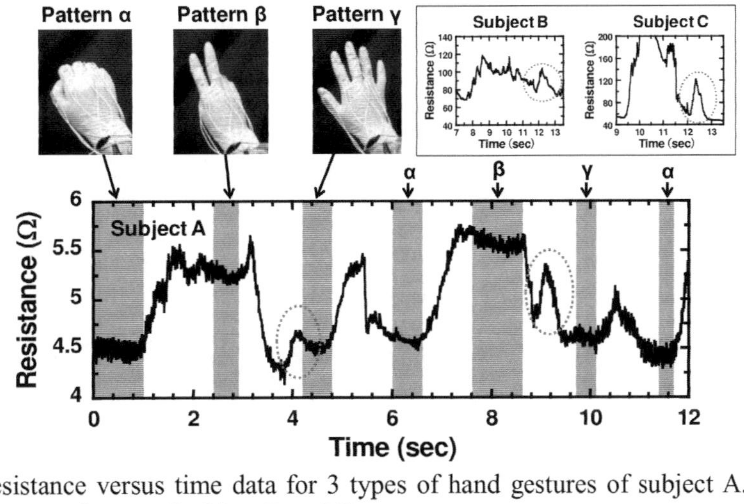

Figure 4. Resistance versus time data for 3 types of hand gestures of subject A. The insets show the resistance change of subject B and C.

IOP Conf. Series: Journal of Physics: Conf. Series **939** (2017) 012006 doi:10.1088/1742-6596/939/1/012006

4. Summary and conclusions

We have investigated wearable devices simply and easily fabricated from PGSs for monitoring human activities. To demonstrate the potential of the wearable devices using the thin graphite films, human motions including elbow and finger motions were observed. The resistance of the thin graphite films were measured in accordance with the movement of the elbow and finger. In these results in of finger motion monitoring, typical resistance change in the middle finger were also measured, which are possibly characteristics specific to human motions. These results suggested that the wearable devices based on thin graphite films hold great potential for the detection of human motion, and for monitoring personal health and therapeutics.

References

[1] Pang C, Lee G Y, Kim T i, Kim S M, Kim H N, Ahn S H, and Suh K Y 2012 *Nat. Mater.* **11** 795-801.
[2] Yang J, Chen J, Su Y, Jing Q, Li Z, Yi F, Wen X, Wang Z, and Wang Z L 2015 *Adv. Mater.* **27** 1316-1326.
[3] Lee M, Chen C Y, Wang S, Cha S N, Park Y J, Kim J M, Chou L J, and Wang Z L 2012 *Adv. Mater.* **24** 1759-1764.
[4] Gong S, Schwalb W, Wang Y, Chen Y, Tang Y, Si J, Shirinzadeh B, and Cheng W 2014 *Nat. Commun.* **5** 3132-1-8.
[5] Amjadi M, Pichitpajongkit A, Lee S, Ryu S, and Park I 2014 *ACS Nano* **8** 5154-5163.
[6] Yamada T, Hayamizu Y, Yamamoto Y, Yomogida Y, Izadi-Najafabadi A, Futaba D N, and Hata K 2011 *Nat. Nanotechnol.* **6** 296-301.
[7] Cai L, Song L, Luan P, Zhang Q, Zhang N, Gao Q, Zhao D, Zhang X, Tu M, Yang F, Zhou W, Fan Q, Luo J, Zhou W, Ajayan P M, and Xie S 2013 *Sci. Rep.* **3** 3048-1-8.
[8] Wang X, Gu Y, Xiong Z, Cui Z, and Zhang T 2014 *Adv. Mater.* **26** 1336-1342.
[9] Bae S H, Lee Y, Sharma B K, Lee H J, Kim J H, Ahn J H 2013 *Carbon* **51** 236-242.
[10] Zhu B, Niu Z, Wang H, Leow W R, Wang H, Li Y, Zheng L, Wei J, Huo F, and Chen X 2014 *Small* **18** 3625-3631.
[11] Wang Y, Wang L, Yang T, Li X, Zang X, Zhu M, Wang K, Wu D, and Zhu H 2014 *Adv. Funct. Mater.* **24** 4666-4670.
[12] Pyrolytic Graphite Sheet Product Datasheet, Panasonic Industrial Devices, Japan.
[13] Wen C Y, and Huang G W 2008 *J. Power Sources* **178** 132-140.
[14] Kobayashi M, and Abe J 2012 *J. Am. Chem. Soc.* **134** 20593-20596.
[15] Saito T, Shimoda H, and Shirakashi J 2015 *J. Vac. Sci. Technol.* **33** 042002-1-5.

IOP Conf. Series: Journal of Physics: Conf. Series **939** (2017) 012007 doi:10.1088/1742-6596/939/1/012007

Field-assisted diffusion of silver in SnO_2 thin films

Mohsen Gharesi[1(c)], Maryam Moalaghi[2], Niloofar Dehghani[3], and Mohammad Ansari[4]

Electronic Materials Laboratory, Electrical Engineering Department, K. N. Toosi University of Technology, Tehran 16317-14191, Iran.

[1(c)]m.gharesi@gmail.com [2]maryam.moalaghi@gmail.com
[3]niloofardehghani91@gmail.com [4]mohammad.ansari1371@gmail.com

Abstract. Tin oxide films are known for their wide spread applications in the field of optoelectronics as transparent conductors and as atmosphere sensitive ceramic pellets for chemical sensor fabrication. Forming durable ohmic contacts to these films is vitally necessary for the effective performance and long life of almost all such devices. Silver electrodes are commonly considered as low resistance ohmic contacts to many metal-oxide semiconductors. Particularly in devices operating at elevated temperatures, silver migration can considerably decrease useful device lifetime. Here, we study the performance of silver electrodes paste printed on SnO_2 thin films and show that the I-V characteristics of the silver/SnO_2 contact changes with operating time at elevated temperatures. SnO_2 films are grown on alumina substrates by ultrasonic spray pyrolysis technique. A pair of Ag electrodes are paste-printed on the surface of the grown films followed by a heat-treatment step to stabilize the deposits. The as-fabricated Ag/SnO_2 contacts present ohmic characteristics. Continuous application of DC voltages to the samples operating at 350 °C results in a significant drop in the devices' resistance. The results originate from field-assisted migration of silver ions from the anode to the cathode electrode and formation of Ag micro-filaments in the polycrystalline SnO_2 layer.

1. Introduction

Metal oxides, known in the art for their thermal stability, resistance to harsh environs, and unique combination of electrical and optical characteristics, are widely used in sensor technology [1-6]. Furthermore, their promising efficacy in energy harvesting and optoelectronic devices has led to increasing research and investment in the field [7-9]. SnO_2 is an n-type wide bandgap semiconductor and the dominant material for the fabrication of chemical sensing pellets [10-13]. Dense polycrystalline tin oxide films have exhibited great potential for utilization as transparent electrodes in optoelectronic devices and as thin film high power resistors in electric heaters [14-16].

Metal oxide devices mostly operate at elevated temperatures and under harsh environmental conditions. Therefore, noble metals with their excellent corrosion resistance are preferred for utilization as the electrode materials. Ohmic contacts are of crucial importance for establishing low-resistance connections to the external circuitry [17-19]. On the other hand, Schottky contacts are advantageous in many device structures including chemical gas sensors, high frequency transistors, and resistive memories [20-23]. Gold and platinum have work functions of 5.1 and 5.65 eV, respectively, which are higher than the SnO_2 electron affinity (4.5 eV) [24, 25]. Theoretically, Au and Pt fail to form ohmic connections to tin oxide films. The fabrication method is an important parameter influencing the films' crystallinity and stoichiometry [26]. These parameters, in turn, control the quality of the contacts. It has been reported that Au/ and Pt/SnO_2 contacts may exhibit ohmic characteristics and the ohmicity is attributed to the films polycrystalline nature and the oxygen deficiency at the surface of the tin oxide grains [27, 28].

Silver, with the least price and resistivity among all noble metals, is a favourable material for electrode fabrication on metal oxide semiconductors. With the work function of 4.26 eV, the metal forms a low-resistance ohmic contact with tin oxide layers [17]. In certain cases, silver migration is a prominent factor affecting the device functionality. The phenomenon is observed in many oxides and is considered to be beneficial for attaining switching behaviour in resistive memories [29, 30]. In the case of forming ohmic contacts though, the effect can induce instabilities in the electrical and structural properties of the film.

Here, the field-assisted diffusion of silver in SnO_2 layers is studied at 350 °C. Tin oxide films are grown on alumina substrates by ultrasonic spray pyrolysis (USP). Silver electrodes are paste printed on the deposits. By applying constant external electric fields to the samples operating at elevated temperatures, instabilities in the devices' current are observed. The recorded phenomena are shown to stem from the electric field-induced movement of the mobile Ag^+ ions, which result in the formation of conductive paths in the cross-section of the grown oxide layer.

2. Experimental

Standard polycrystalline alumina ceramics with a thickness of 0.5 mm are laser cut into 3×3 mm^2 squares, washed with acetone and distilled water, and dried on a hot plate at 150 °C. The utilized USP deposition system is a refined form of the system previously described [31]. 0.2 M solution of the precursor ($SnCl_2.2H_2O$, Merck 107815) in absolute ethanol is poured into a polymer container with an average rate of 30 mL/h. Two piezoelectric transducers working at 1.7 MHz produce the aerosol droplets and a continuous flow of air with 3 L/min average rate transfer them to the deposition chamber. The substrate is placed on a heater and heated to 325 °C. Tin oxide thin films with ~50 kΩ/□ sheet resistance are grown on the substrates in less than 30 seconds. The deposits are annealed at 600 °C for 1 hour in air.

As shown in figure 1, the contacts are asymmetrically configured so that the effective area of the counter electrode is ~10 times more than that of the other. Micron-thick silver layers are deposited on the SnO_2 films using a commercially available silver paste. Ag wires of 0.1 mm diameter are attached to the silver electrodes and provide the connections to the external circuitry. Silver deposits are dried at 110 °C to remove the organic solvent and heat-treated at 400 °C for 15 minutes to ensure the mechanical stability of the connections [17].

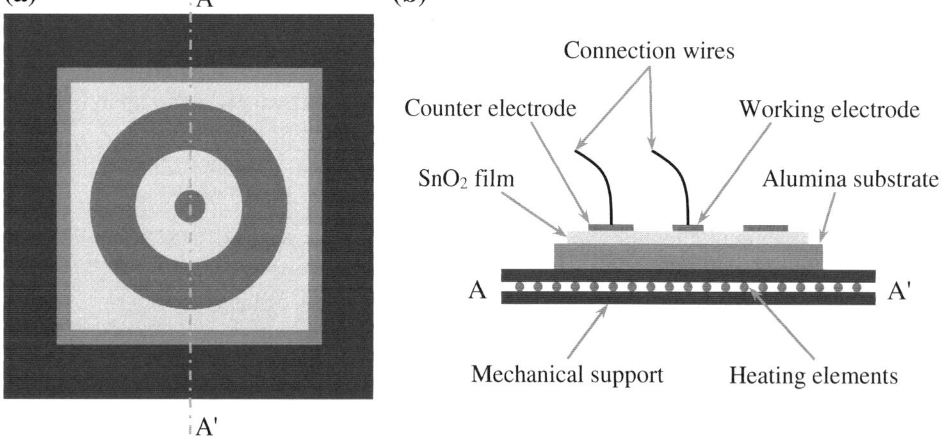

Figure 1. Plan (a) and cross-sectional (b) schematic views of the fabricated Ag/SnO_2 samples.

IOP Conf. Series: Journal of Physics: Conf. Series **939** (2017) 012007 doi:10.1088/1742-6596/939/1/012007

During the I-V tests, device's voltage is directly monitored; the current is measured by reading the voltage drop on a series resistance. The data is read using a high input impedance operational amplifier and recorded by a 12-bit data acquisition card (Advantech, PCI-1711) with a sampling rate of 50 kilosamples per second. The contact durability tests are performed under continuous application of DC biasings to the samples operating at elevated temperatures. Devices' temperature is monitored by a fine S-type thermocouple and controlled by changing the voltage applied to an underlying resistive heater. The utilized experimental setup is schematically illustrated in figure 2.

3. Results

A biasing of 2.0 V is constantly applied to the samples operating at 350 °C. The variations of the device current are recorded (figure 3). The current is limited to 0.1 A so that the wirings cause negligible error in the measured resistance values. The samples' resistance has a decreasing trend with time which is attributed to the emergence of conductive pathways in the oxide film. The results are similar to those reported in the silver-based resistive switching devices [29, 30]. There are 2 main mechanisms involved in this process: (a) the production of Ag^+ ions, and (b) their drift toward the cathode. Owing to high mobility of charged silver species, formation of Ag^+ ions is believed to be the rate limiting factor [32]. Performing the same experiments under argon atmosphere causes no such change in the device resistances indicating the prominent effect of the atmospheric oxygen on the filament formation process.

The recorded I-V characteristics before and after the forming process at 350 °C are shown in figure 4. The results are obtained using 0.05 Hz triangular voltage waveforms at 25-300 °C. In spite of asymmetric contacts, the I-V curves have an obvious symmetry, indicating ohmic quality of the contacts. At temperatures above 250 °C, hysteresic I-V characteristics emerge, which demonstrate resistive switching. Consecutive formation and collapse cycles of conductive Ag microfilaments has given rise to the observed hysteresic property. Owing to the ionic nature of the phenomenon, the hysteresis loop disappears when the biasing waveform frequency goes above 1 Hz.

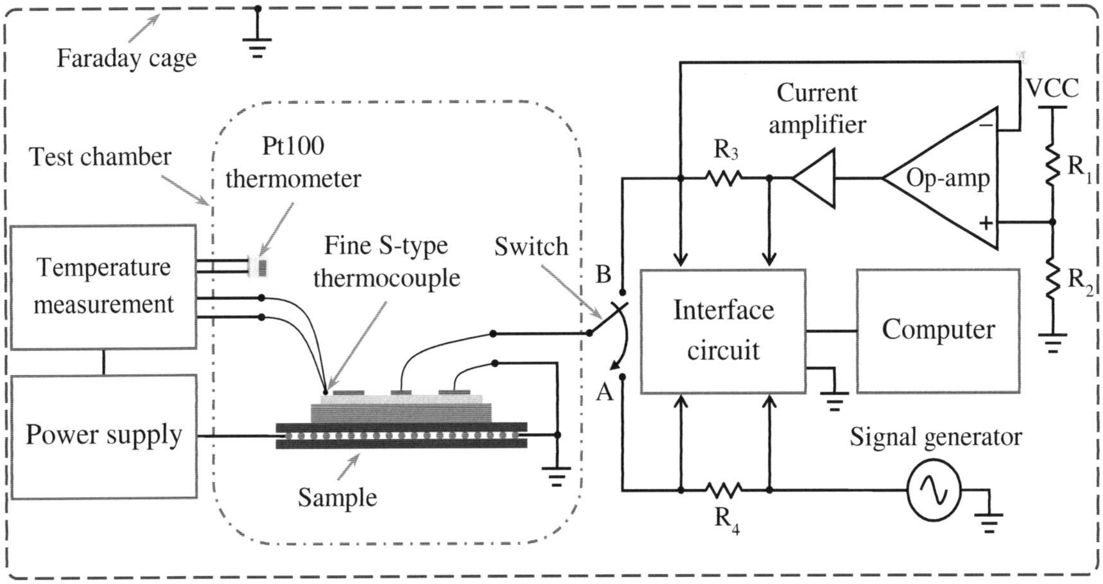

Figure 2. The schematic illustration of the utilized experimental setup. The switch is in the A and B positions for the I-V recording and durability tests, respectively.

IOP Conf. Series: Journal of Physics: Conf. Series **939** (2017) 012007 doi:10.1088/1742-6596/939/1/012007

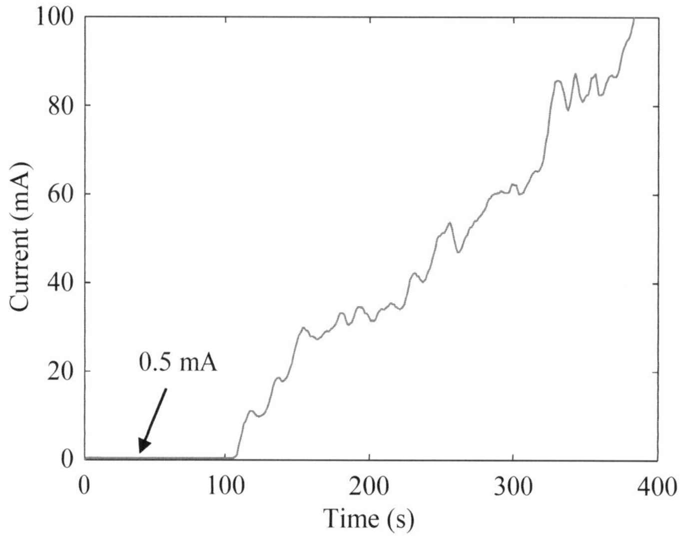

Figure 3. Current variation in a tin oxide sample operating at 350 °C under a 2.0 V DC bias.

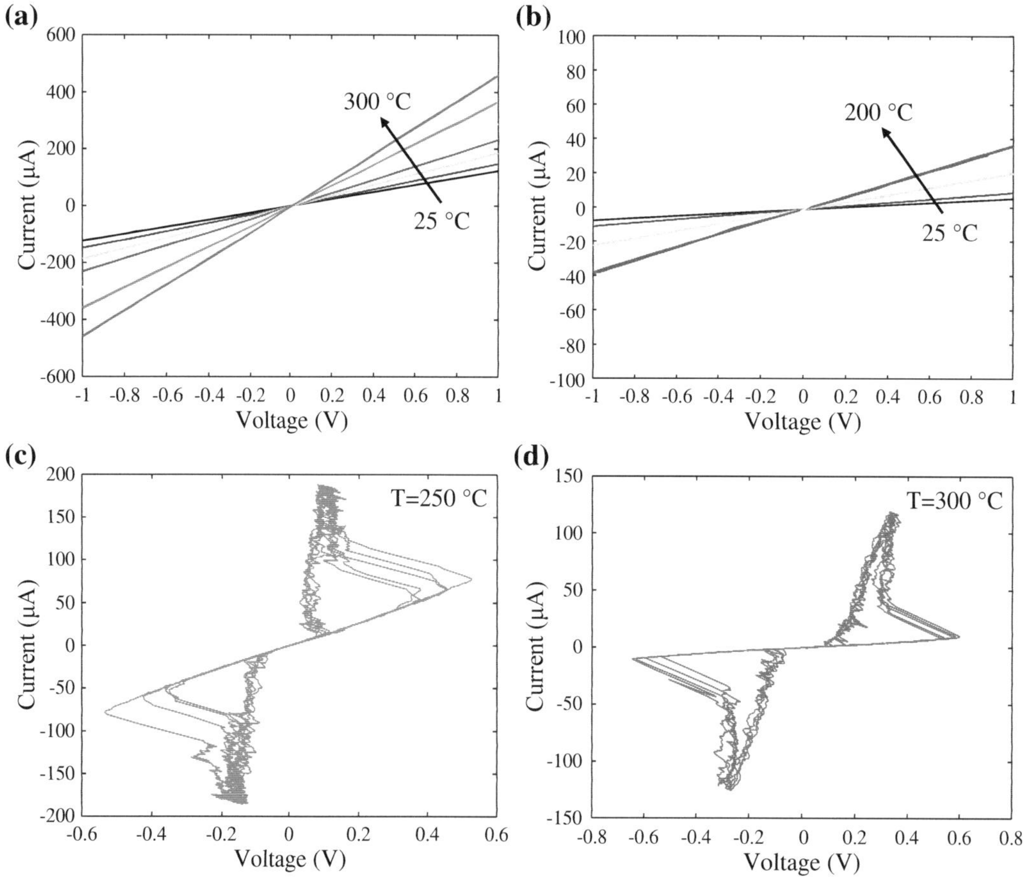

Figure 4. I-V characteristics of a tin oxide sample in response to a triangular voltage biasing at 0.05 Hz frequency before (a) and after (b, c, and d) the 350 °C-forming process.

IOP Conf. Series: Journal of Physics: Conf. Series **939** (2017) 012007 doi:10.1088/1742-6596/939/1/012007

4. Conclusions

Ohmic silver contacts were formed on the tin oxide films deposited by spray pyrolysis. The contacts reliability and durability were examined by the continuous application of DC biasings to the samples operating at elevated temperatures. The electric field-induced motion of mobile Ag^+ ions caused the formation of silver filaments in the oxide layer cross-section, which caused a significant drop in the samples resistance. It was found that oxygen adsorption plays a major role in the ionic migration process. According to the recorded I-V characteristics, the Ag/SnO_2 devices have a potential for utilization as resistance switching devices.

References

[1] Hossein-Babaei F and Rahbarpour S 2014 Alteration of pore size distribution by sol-gel impregnation for dynamic range and sensitivity adjustment in Kelvin condensation-based humidity sensors *Sens. Actuators, B* **191** 572-8.

[2] Wang J, Yang P and Wei X 2015 High-performance, room-temperature, and no-humidity-impact ammonia sensor based on heterogeneous nickel oxide and zinc oxide nanocrystals *ACS Appl. Mater. Interfaces* **7** 3816-24.

[3] Hossein-Babaei F, Paknahad M and Ghafarinia V 2012 A miniature gas analyzer made by integrating a chemoresistor with a microchannel *Lab Chip* **12** 1874-80.

[4] Alaei-Sheini N and Rohani M 2016 Ag-doped titanium dioxide gas sensor *IOP Conf. Ser.: Mater. Sci. Eng.* **108** 012033.

[5] Hossein-Babaei F, Zare A H and Ghafarinia V 2016 Transient molecular diffusion in microfluidic channels: Modeling and experimental verification of the results *Sens. Actuators, B* **233** 646-53.

[6] Ghalamboran M and Saedi Y 2016 TiO_2-TiO_2 composite resistive humidity sensor: ethanol crosssensitivity *IOP Conf. Ser.: Mater. Sci. Eng.* **108** 012039.

[7] Walia S, Balendhran S, Nili H, Zhuiykov S, Rosengarten G, Wang Q H, Bhaskaran M, Sriram S, Strano M S and Kalantar-zadeh K 2013 Transition metal oxides-Thermoelectric properties. *Prog. Mater. Sci.* **58** 1443-89.

[8] Tian W, Lu H and Li L 2015 Nanoscale ultraviolet photodetectors based on onedimensional metal oxide nanostructures *Nano Res.* **8** 382-405.

[9] Choy W C and Zhang D 2016 Solution-processed metal oxides as efficient carrier transport layers for organic photovoltaics *Small* **12** 416-31.

[10] Hossein-Babaei F and Amini A 2014 Recognition of complex odors with a single generic tin oxide gas sensor *Sens. Actuators, B* **194** 156-63.

[11] Zhang D, Liu A, Chang H and Xia B 2015 Room-temperature high-performance acetone gas sensor based on hydrothermal synthesized SnO_2-reduced graphene oxide hybrid composite *RSC Adv.* **5** 3016-22.

[12] Suman P H, Felix A A, Tuller H L, Varela J A and Orlandi M O 2015 Comparative gas sensor response of SnO_2, SnO and Sn_3O_4 nanobelts to NO_2 and potential interferents *Sens. Actuators, B* **208** 122-7.

[13] Hossein-Babaei F and Orvatinia M 2003 Thickness dependence of sensitivity in thin film tin oxide gas sensors deposited by vapor pyrolysis *Int. J. Eng., Trans. B* **16** 33-40.

[14] Liu X Y, Gao Y Q and Yang G W 2016 A flexible, transparent and super-long-life supercapacitor based on ultrafine Co_3O_4 nanocrystal electrodes *Nanoscale* **8** 4227-35.

[15] Morales-Masis M, Dauzou F, Jeangros Q, Dabirian A, Lifka H, Gierth R, Ruske M, Moet D, Hessler-Wyser A and Ballif C 2016 An Indium-Free Anode for Large-Area Flexible OLEDs: Defect-Free Transparent Conductive Zinc Tin Oxide *Adv. Funct. Mater.* **26** 384-92.

[16] Gupta R, Rao K D, Kiruthika S and Kulkarni G U 2016 Visibly transparent heaters *ACS Appl. Mater. Interfaces* **8** 12559-75.

[17] Hossein-Babaei F, Moghadam S and Masoumi S 2015 Forming ohmic Ag/SnO_2 contacts *Mater. Lett.* **141** 141-4.

[18] Zhang H, Shallcross R C, Li N, Stubhan T, Hou Y, Chen W, Ameri T, Turbiez M, Armstrong N R and Brabec C J 2016 Overcoming electrode-induced losses in organic solar cells by tailoring a quasi-ohmic contact to fullerenes via solution-processed alkali hydroxide layers *Adv. Energy Mater.* **6** 1502195.
[19] Jahromi K E, Ara M H M, Mousavi S S and Efafi B 2016 Investigation of a reliable ohmic contact to n-type ZnO thin films prepared by sol-gel method *IEEE Electron Device Lett.* **37** 43-5.
[20] Hossein-Babaei F and Rahbarpour S 2011 Separate assessment of chemoresistivity and Schottky-type gas sensitivity in M-metal oxide-M′ structures *Sens. Actuators, B* **160** 174-80.
[21] Fan D, Kang N, Ghalamestani S G, Dick K A and Xu H Q 2016 Schottky barrier and contact resistance of InSb nanowire field-effect transistors *Nanotechnology* **27** 275204.
[22] Li Y, Gong Q and Jiang X 2014 Feasibility of Schottky diode as selector for bipolar-type resistive random access memory applications *Appl. Phys. Lett.* **104** 132105.
[23] Hossein-Babaei F, Lajvardi M M and Alaei-Sheini N 2015 The energy barrier at noble metal/TiO_2 junctions *Appl. Phys. Lett.* **106** 083503.
[24] Michaelson H B 1977 The work function of the elements and its periodicity *J. Appl. Phys.* **48** 4729-33.
[25] Zhao H, Fan Z, Liang H, Selopal G S, Gonfa B A, Jin L, Soudi A, Cui D, Enrichi F, Natile M M and Concina I 2014 Controlling photoinduced electron transfer from PbS@CdS core@shell quantum dots to metal oxide nanostructured thin films *Nanoscale* **6** 7004-11.
[26] Hossein-Babaei F and Alaei-Sheini N 2016 Electronic conduction in Ti/Poly-TiO_2/Ti structures *Sci. Rep.* **6** 29624.
[27] Körber C, Harvey S P, Mason T O and Klein A 2008 Barrier heights at the SnO_2/Pt interface: in situ photoemission and electrical properties *Surf. Sci.* **602** 3246-52.
[28] Nagata T, Bierwagen O, White M E, Tsai M Y and Speck J S 2010 Study of the Au Schottky contact formation on oxygen plasma treated n-type SnO_2 (101) thin films *J. Appl. Phys.* **107** 033707.
[29] Lee C B, Kang B S, Benayad A, Lee M J, Ahn S E, Kim K H, Stefanovich G, Park Y and Yoo I K 2008 Effects of metal electrodes on the resistive memory switching property of NiO thin films *Appl. Phys. Lett.* **93** 042115.
[30] Yang Y, Gao P, Gaba S, Chang T, Pan X and Lu W 2012 Observation of conducting filament growth in nanoscale resistive memories *Nat. Commun.* **3** 732.
[31] Gharesi M and Ansari M 2016 Tin oxide microheater for chemical sensors *IOP Conf. Ser.: Mater. Sci. Eng.* **108** 012018.
[32] Sears W M and Love D A 1993 Measurements of the electrical mobility of silver over a hot tin oxide surface *Phys. Rev. B* **47** 12972.

IOP Conf. Series: Journal of Physics: Conf. Series **939** (2017) 012008 doi:10.1088/1742-6596/939/1/012008

Flexible and highly sensitive humidity sensors using screen-printed TiO2 nanoparticles as sensitive layer

G Dubourg[1], J Katona[2], M Rodović[1], S Savić[1], G Kitic[1], G Niarchos[1] N Jancović and V Crnojević-Bengin

[1] Biosense Institute, University of Novi Sad, Novi Sad, Serbia
[2] Faculty of Technology, University of Novi Sad, Novi Sad, Serbia

Abstract. In this work, we present an original example of flexible and miniaturized humidity sensor using screen-printed TiO_2 nanoparticles as sensitive layer. The proposed sensors operate under room-temperature conditions and exhibit high sensitivity to changes in relative humidity. The humidity sensor consists of a TiO_2-based sensitive nanomaterial printed on the top of interdigitated electrodes (IDEs) patterned by laser ablation on flexible substrate. A TiO_2 screen-printing paste from commercially-available nanopowders was developed in order to allow large-scale fabrication of humidity sensors at low temperature. The I-V characteristic of the sensor shows good linearity and resistance measurements confirm the water vapour's absorption on the sensitive layer, showing high sensitivity in the range of 24 to 90% relative humidity, while exhibiting relatively fast response and recovery times, without the need of any refreshing method.

1. Introduction

Humidity sensors find applications from semiconductor manufacturing to soil moisture monitoring or in food industry and are generally fabricated on ceramic and silicon substrates. Nevertheless, recent advancements in the field of printed electronics show increased potential in substitution of rigid substrates by flexible substrates, due to their low cost and easy fabrication processing.

In parallel, semiconducting metal oxide nanostructures such as TiO_2 nanoparticles, SnO_2 nanowires, or ZnO nanorods have been reported to be good candidates for highly sensitive and stable humidity sensors due to their high surface-to-volume ratio [1][2][3]. Nevertheless, their integration in printed sensors fabricated on flexible substrates is still non trivial, since plastic substrates deform or melt at temperatures of only 100–200 °C [4], which makes them incompatible with the standard silicon and ceramic technologies. Lots of studies have focused on the fabrication of flexible electronic devices using metal oxide materials with good electrical and optical properties [5]. However, their integration as highly-sensitive materials into flexible chemical sensors is still lacking, since they are generally deposited by drop-casting or spin-coating methods [6] which cannot be consider for the mass-production of sensors. One of the main challenges is then the development of cost-effective and low temperature fabrication process allowing the large-scale production of flexible sensors with high performances.

In this work, we demonstrate the feasibility of screen-printed TiO_2 sensing material at low-temperature for the large-scale fabrication of humidity sensors on flexible substrate. The humidity sensors proposed and realized in this work show a high sensitivity in the range of 24 to 90% relative humidity and fast response and recovery times.

2. Materials and methods

A resistive transducer was chosen for the conception of the humidity sensor because it has a simple structure, a low fabrication cost, and it does not require complex read-out circuitry. Typically, it consists of a TiO_2-based sensitive nanomaterial deposited on top of gold interdigitated electrodes (IDEs) patterned on a flexible substrate.

As a principle, the absorption of water vapour by the sensitive film creates an increase of film electrical conductance due to the dissociation of water molecules to ionic functional hydroxyl groups resulting to a resistance change between interdigitated electrodes which can be easily measured.

2.1. Materials

TiO2 nanoparticles-based paste was used as sensitive layer and was developed for the screen-printing process. Water was chosen as the main solvent of the paste because it is cheap, environmental friendly and it can be evaporated at low temperature.

First, HPMC (hydroxypropylmethyl cellulose) was dissolved in water to increase its viscosity. Next, the resulting water/HPMC solution was mixed with a mixture of propylene glycol/propanol. In parallel, dispersant (Solspers 40000) was dissolved in a mixed solvent of water/propylene glycol/propanol, and TiO_2 powder (anatase, sigma Aldrich) was added to the solution. Afterwards, the dispersion was performed by ultrasonication for 20 min. The suspended TiO_2 nanoparticles suspension wereas added to the solution of HPMC in a ratio of 1:1 under vigorous stirring at RW20 in 2000 rpm during 20min, to give a final formulation.

2.2. Fabrication process

Keeping the intended final application closely in mind, we aimed to for the development of a simple and economic process for the fabrication of humidity sensors, in order to be able to preserve their attractiveness as low-cost, potentially mass-produced devices. The fabrication process of the sensors proposed in this work is fast, compatible with rolls-to-rolls technologies and does not require the high-cost semiconductor manufacturing equipment and high temperature steps normally used for silicon or ceramic microfabrication. It combines a laser ablation technique for the fabrication of the micro-scale electrodes and a screen-printing process for the patterning of the TiO_2 nanoparticles-based sensing material. The process sequence is illustrated in Figure 1.

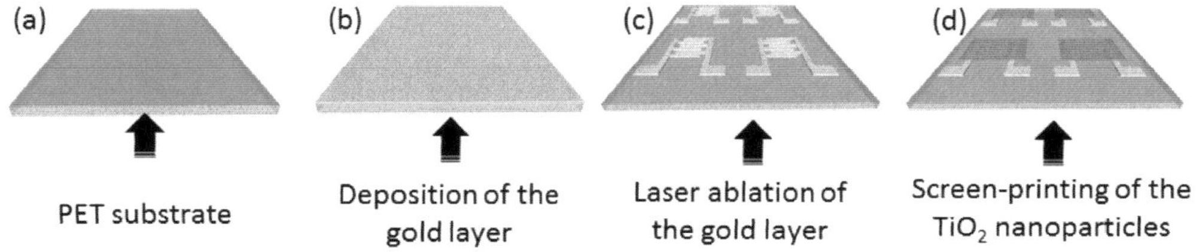

Figure 1: Process sequence of the humidity sensors: (a) Cleaning of the PET substrate. (b) Deposition of the gold layer by electron beam evaporation. (c) Laser ablation of the gold layer. (d) Screen-printing of the TiO_2 nanoparticles.

Initially, a commercial PET flexible substrate of 50 μm thick was cleaned with iso-propanol and dried at 80°C for 15 min. Next, a gold layer was deposited on the substrate by electron-beam evaporation. Afterwards, the resulting layer was directly patterned by laser ablation (Nd:YAG-1064 nm, Rofin) in order to create micro-scale interdigitated electrodes using a process inspired from [7]. The TiO_2 nanoparticles-based paste was then patterned on the surface of electrodes by screen-printing (EKRA 2H screen printer) which is a widespread industrially-applied method. A screen made of photosensitive film of 50 μm thick and a mesh of 40 μm thick were used to successfully achieve the deposition of the sensing material. Finally, the resulting devices were dried at 80°C for 30 min in an oven to accelerate the solvent evaporation. Note that the solvent evaporation can be also performed at room temperature during 1 day.

This process has the advantage to be fast and cost-effective; it allows also the mass-production of humidity sensors at low-temperature. In addition, due to their small size, the resulting devices can be easily integrated in a microchip for industrial use. The characterisation of the resulting humidity sensors is described in the following parts.

3. Results

3.1. Characterization of the device geometry

Using the process described above, the large scale fabrication of humidity sensors has been successfully achieved. Figure 2a shows a matrix of 3x3 sensors. The interdigitated electrodes are 600 µm long and 60 µm wide, and they are separated by a gap of 50 µm (Figure 2b). Here, small electrode geometry was obtained compared to the standard printed sensors generally above 100 µm which is very important for restive-type sensor performances [8].

The screen-printed TiO_2 nanoparticles layer forms a rectangle of 1 mm wide and 1.5 mm long covering perfectly the surface of electrodes (Figure 2c)

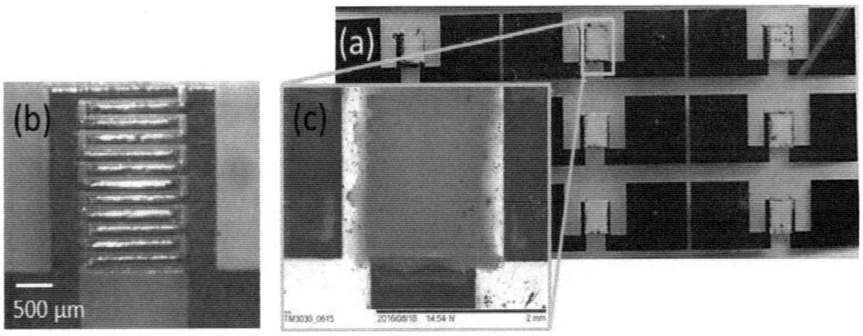

Figure 2 : (a) Matrix of humidity sensors fabricated on PET substrate. (b) Optical image of the intedigitated electrodes before the deposition of the TiO_2 nanoparticles. (c) SEM (scanning electronic microscope) picture of the TiO_2 layer screen-printed on the surface of electrodes.

3.2. Electrical characterization

Figure 2a shows the I–V characteristics of the humidity sensor and the patterned electrodes without coating of the sensitive layer measured at humidity levels of 12% and 40% using semiconductor parameter analyser. In Figure 3a, we can observe that the current did not flow on the unprocessed substrate implying that the PET substrate is an insulator. The I–V characteristics of the device with TiO_2 nanoparticles show Ohmic behaviour in both conditions, at 12% RH and 40% RH (Figure 3b). The resistance is constant over the supply voltages; the sensitivity is then identical regardless of the operation bias. In other words, a low voltage operation does not hinder the sensitivity which is essential for low power operation.

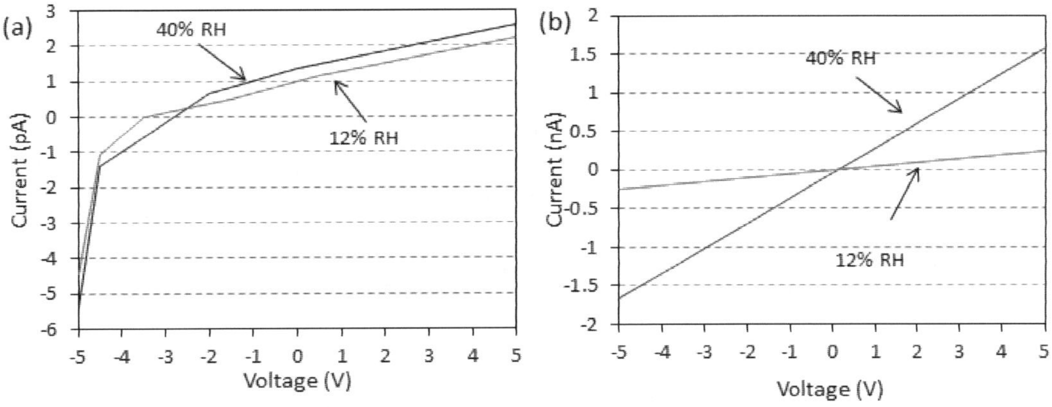

Figure 3: I–V characteristics measured under humidity levels of 12% and 40%, (a) for a device without TiO2 deposited on the electrodes, and (b) for a device with TiO_2 nanoparticles.

3.3. Humidity sensing properties

The humidity-sensing performances of the TiO_2-based sensors were investigated under various levels of relative humidity (RH) in a climatic test chamber with a control module (Hygrometer testo 608-H1) and at room temperature conditions (26°C). Figure 4a shows the sensor response when the relative humidity in the chamber was sequentially changing from 12% to 52% in periods of 10 min. The response curves are almost identical at a given humidity indicating repeatability of the results with a fast response time and almost immediate recovery. Figure 4b shows the resistance changes when the sensor was placed in different humidity conditions. In this figure, we can see that low concentration of humidity (24%) can be measured proving the ability of the sensors to be used for accurate humidity detection.

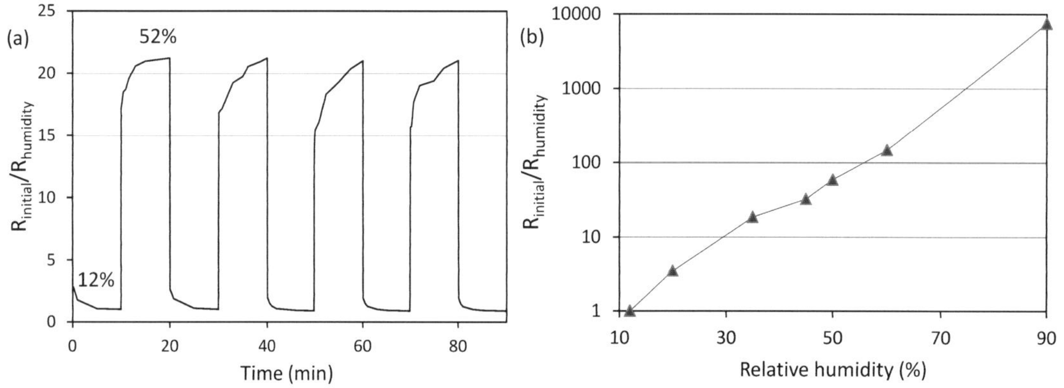

Figure 4: (a) Repeatability curves of the humidity sensor for humidity level variations from 12% to 52%. (b) Resistance changes of the sensor as function of the relative humidity.

4. Conclusion

The large-scale production of humidity sensors on flexible substrate was successfully achieved using a laser ablation method for the fabrication of micros-scale electrodes and a screen-printing technique for the patterning of the TiO2-based sensing material. Both methods are fast, cost-effective, and compatible with flexible substrate. This approach addresses the technological barriers that can hamper the development of chemical sensors on flexible substrates. In addition, the resulting humidity sensors exhibit high sensitivities in the range of 24 to 90% relative humidity and fast response and recovery times, which paves the way to large-scale fabrication of printed sensors on flexible substrate.

References

[1] Hu, Dai CL, Hsu CC 2014 *Sensors* **14** 4177-4188
[2] Zhang Y, Yu K, Jiang D, Zhu Z, Geng H, Luo L 2005 *Appl. Surf. Sci.* **242** 212–217
[3] Kuang Q , Lao C , Wang ZL, Xie Z , Zheng L 2007, *J. Am. Chem. Soc.* **129** 6070–6071
[4] McAlpine MC, Ahmad H, Wang D, Heath JR 2007 *Nat. Mater.* **6** 379-384
[5] Lai WH, Su YH, Teoh LG, Hon MH 2008 *J. Photochem. Photobiol. A* **195** 307–313
[6] Zheng ZQ, Yao JD, Wang B, Yang GW 2008 *Sci. Rep.* **5** 11070
[7] Dubourg G, Zlebic C, Matovic J, Bengin VC 2015 *Proc. Int. Conf on NEMS* 553-556
[8] Gerwen PV, Laureyn W, Laureys W, Huyberechts G, Op De Beeck M, Baert K, Suls J, Sansen W, Jacobs P, Hermans L, Mertens R 1998 *Sensor Actuat. B-Chem* **49** 1-2

Thermal stability of γ-Fe_2O_3 nanoparticles and their employment for sensing of acetone vapours

Š Luby[1], J Ivančo[1,4], M Jergel[1], P Švec, Jr.[1], M Kotlár[1,2], D Kostiuk[1], J Halahovets[1], J Kollár[3], J Mosnáček[3] and E Majková[1]

[1] Institute of Physics of Slovak Academy of Sciences, Dúbravská cesta 9, SK-84511 Bratislava, Slovakia
[2] Faculty of Electrical Engineering and Information Technology, Slovak University of Technology, Ilkovičova 3, SK-812 19 Bratislava, Slovakia
[3] Polymer Institute of Slovak Academy of Sciences, Dúbravská cesta 9, SK-84511 Bratislava, Slovakia
[4] Author to whom any correspondence should be addressed.

E-mail: jan.ivanco@savba.sk

Abstract. Stability of γ-Fe_2O_3 nanoparticles-based films upon an isochronal annealing in air was investigated by x-ray diffraction, differential scanning calorimetry, and thermogravimetry. The $\gamma \rightarrow \alpha$ transformation temperature increased owing to the nanoscaling of Fe_2O_3; the higher stability of the γ phase was explained on the ground of the surface free energy of nanoparticles (with the size of about 6.4 nm). Further, chemiresistors based on the Fe_2O_3 nanoparticle bilayer prepared by the Langmuir-Schaefer method were fabricated and examined in terms of their sensitivity to acetone vapours down to 500 ppb concentration in air.

1. Introduction

Among hundreds of volatile organic compounds (VOCs) involved in a human exhaled breath, various of them–referred to as biomarkers–were associated with a specific disease [1]. This makes the detection of breath biomarkers applicable in non-invasive screening and real-time monitoring of particular diseases. For example, diabetes patients exhibit a higher acetone concentration in their breath: whilst the values lower than 0.76 ppmv were associated with healthy persons, acetone in diabetics` breath exceeds about 1.7 ppmv [2].

Contrary to an expensive and complicated instrumentation, solid-state chemiresistor sensors of acetone vapours may dispose of a manufacture potential for fabrication of simple and affordable and thereby personal monitoring device providing an alternative to standard blood analysis. Considering the mentioned concentration range, the sensitivity down to sub ppm range is required. Various semiconducting –often nano-scaled–metal oxides have been used in the role of sensing elements, such as Fe_2O_3 [3-8], WO_3 [9-11], SnO_2 [3,12], TiO_2 [13], CeO_2 [14], and ZnO [15], or related compounds [5,16] and/or activated by impurities [3] being the most popular ones. An operating temperature of a few hundred degrees is typically necessary by virtue of the adsorption/desorption kinetics of detected molecules, and/or technology and application issues; optimal temperatures, *i.e.* corresponding to the highest response, were reported to range from 175°C to 500°C [3-8,13].

Fe_2O_3 can occur in the α- and γ-phase, whilst the phases display reverse conductivity behaviours [17]. The γ-Fe_2O_3 undergoes a transformation to α-Fe_2O_3 above a transformation temperature (t_t) (reported to be of 481°C for γ-Fe_2O_3 NPs (8-20 nm) [18]) impairing the sensitivity and selectivity to reducing gases [19]. Thereby, the necessity of the elevated working temperatures up to t_t of γ-Fe_2O_3-based sensors arises a question on stability of γ-Fe_2O_3. In recent works, t_t was increased by doping of Fe_2O_3 by *e.g.* Pt [18] or Mn [20]. In this study, the achievement of higher t_t with pure Fe_2O_3 NP*s* will be demonstrated.

2. Experimental

The Fe_2O_3 nanoparticles (NP*s*) have been synthesized via reaction between Fe acetylacetonates with 1,2-hexadecanediol, and oleic acid and oleyl amine in the role of the surfactant precursor in phenyl ether. The average NP diameter has been of 6.4±0.6 nm including the surfactant thickness being of about 1 nm. Mixtures of Fe_2O_3/Pd nanoparticles with ratios of 100/1 and 100/5 were prepared. Whereby the Pd NPs (6-7 nm) covered by oleyl amine were purchased from PlasmaChem.

The NP films intended for characterization of sensing properties were prepared by the modified Langmuir-Schaffer (LS) deposition [21] of colloid NP*s* from the water subphase onto encased alumina substrates (2×2 mm) equipped with interdigitated electrodes with the channel length of 50 μm allowing the film resistivity determination. The investigated films consisted of arrays of two sequentially deposited NP monolayers. The deposited films were annealed at 400°C in dry air for 30 min to stripe off the insulating organic surfactant.

Resistance NP layers was examined at voltage of 1 V upon varying concentration of acetone vapors in dry air. This was achieved by mass flow controller-controlled mixing of dry air and a certified mixture of 20 ppm acetone vapors in dry air. The total medium flow of 10 or 100 sccm was kept constant for any acetone concentration. The elevated working temperature of the sensing layer was attained by means of planar resistive heaters incorporated on the back side of the alumina substrates.

The NP films intended for structural characterization were drop-casted onto polished silicon substrates. The x-ray diffraction (XRD) characterization was performed using D8 Discover SSS (Bruker AXS) facility with rotating Cu anode.

The differential scanning calorimetry (DSC) and thermogravimetry (TG) measurements were performed using TA Instruments Q 200 and SDT Q 600 facilities, respectively, in Ar ambient with heat ramps of 1, 10, and 40 K/min.

3. Results and discussion

3.1. Structural properties of Fe2O3 nanoparticles

The γ-α transformation is the phase change from cubic γ-Fe_2O_3 (maghemite) to rhombohedral α-Fe_2O_3 (hematite). The phase transformation of NPs is generally affected by the NP preparation, supporting substrate, size, doping, and heating rate. Randrianantoandro *et al.* [22] reported that γ-Fe_2O_3 NPs with size from 5 to 30 nm are stabilized owing to the high surface free energy, and even a reverse α→γ transition was achieved by ball milling. 8-20 nm γ-Fe_2O_3 NP*s* transformed to α phase at 481°C and t_t increased to 547°C upon the doping of NP*s* with 0.25 wt % of Pt [18].

Here, the γ→α transformation was investigated by means of XRD and DSC-TG methods. The probed NPs did undergo isochronal annealing durative one hour in air at temperatures ranging from 500°C to 770°C. Figure 1 contrasts the XRD of the Fe_2O_3 NPs annealed at 550°C and 600°C. Apparently, the γ-Fe_2O_3 phase preserves up to at least 550°C, the annealing at 600°C reveals the mixed phase, as features pertaining to the α-Fe_2O_3 appear. It means that 550°C $< t_t <$ 600°C. On the other side, the features indicative α-Fe_2O_3 appeared already at 500°C for both the 100/1 and 100/5 Fe_2O_3/Pd mixtures (not shown here) suggesting that the presence of Pd decreases the transformation temperature by at least 50°C. Accordingly, Pd shows an adverse effect in comparison to Pt [18].

IOP Conf. Series: Journal of Physics: Conf. Series **939** (2017) 012009 doi:10.1088/1742-6596/939/1/012009

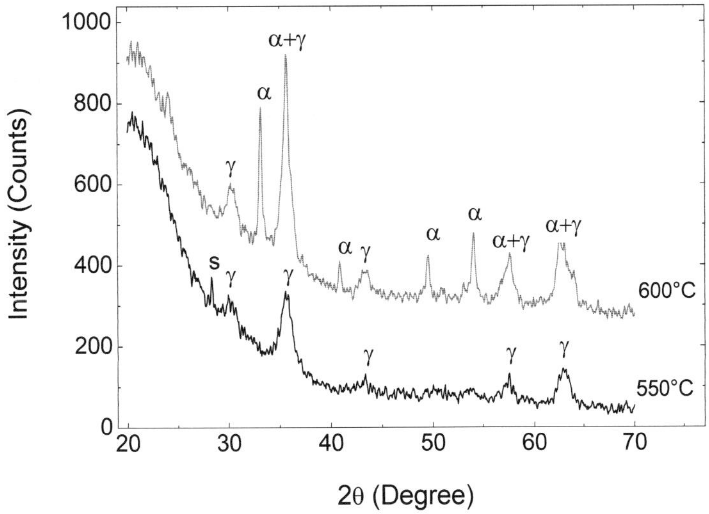

Figure 1. X-ray diffractogram of Fe_2O_3 nanoparticles annealed one hour at 550°C or 600°C with labelled features indicative of α- and γ-Fe_2O_3 phases, and the substrate (s).

In parallel, the γ-α transformation was examined by the DSC-TG method. Figure 2a presents DSC dependences taken on couples of the pure and Pd-doped Fe_2O_3 samples. The dependences reveal peaks at about 560 K, which is associated by dramatic weight loss in TG (Figure 2b). This suggests that the surfactant is removed in two steps.

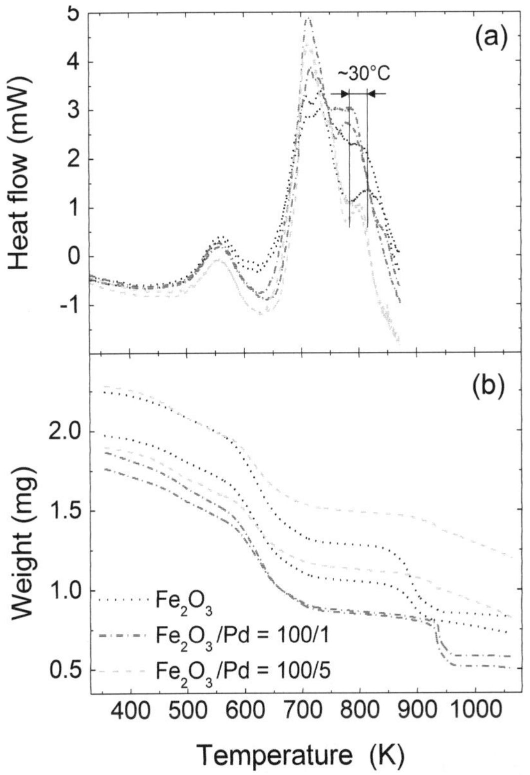

Figure 2. The differential scanning calorimetry (a) and the thermogravimetry (b) dependences taken with the ramp rate of 10 K/min for three couples of samples: undoped Fe_2O_3 nanoparticles, and Fe_2O_3/Pd nanoparticle mixtures.

Similar two steps TG weight loss was reported for γ-Fe_2O_3 NPs covered by oleic acid surfactant earlier [20]. Authors assumed that the first and second weight loss correspond to weakly bound secondary and to the strongly bound primary surfactant layer, respectively. Double or even triple peaks are present at higher temperatures (Figure 2a). We presume that the peaks between 700 and 760 K are related to the crystallization, which was supported by XRD characterization (not shown here), while the features around 800 K correspond to the $\gamma \rightarrow \alpha$ transformation. This–in terms of the pure Fe_2O_3–is finished at about 870 K, thus complying to about 600°C arising from the XRD measurements. In terms of the Fe_2O_3/Pd samples, the transformation is shifted down by about 30°C. Since these peaks are related to the phase transition, no weight losses are expected, as indicated by plateaus of the TG dependences in the same temperature region (Figure 2b). The further weight loss observed above 900 K (Figure 2b) is not understood at this moment.

3.2. Sensing performance of Fe2O3 nanoparticle films

Polycrystalline films Fe_2O_3 can manifest electrical resistance of hundreds MΩ [3,18]. Even though resistance decreases upon reducing acetone exposure, chemiresistors with the extremely low conductivity are rather precluded from standard laboratory characterization. Resistance of the layers examined here was about 6 MΩ at the operating temperature, which allowed the characterization at voltage of 1 V. The sensor response, *i.e.*, the resistivity measured in dry air, *R*air, ratioed to resistivity detected upon precisely controlled trace concentrations of acetone vapours, *R*acetone, was examined in dependence on acetone concentration, operating temperature, and total gas flow.

Figure 3 shows the response to 1 and 0.5 ppm acetone vapors in air upon the constant total gas mixture flow of 100 sccm. The sensor temperature was kept at 420°C, since the highest response was observed in the temperature range from about 370 to 450°C (not shown here).The concentration change in the testing cell is not immediate upon the switching the probed mixture; it depends on the total flow of the medium and the volume of the testing chamber. Given the filling time of the employed testing cell (with volume of about 25 cm^3) being about 15 seconds upon the flow of 100 sccm, the onsets of the responses suggest that rate of the response is virtually determined by the time necessary to establish the requested acetone concentration in the testing cell. The response saturates at about 7.5 % upon acetone concentration of 500 ppb.

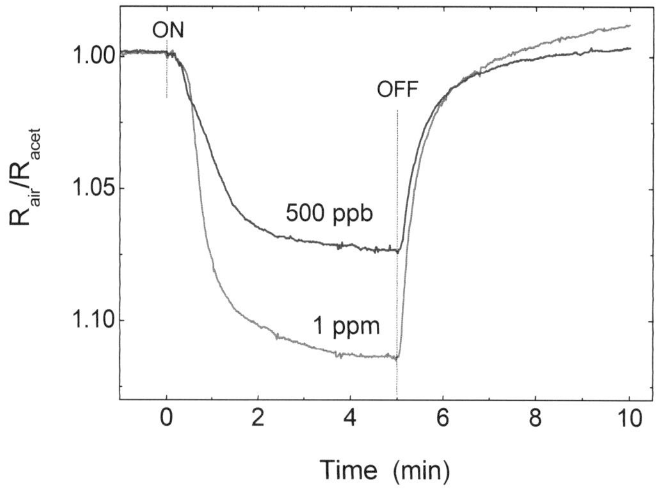

Figure 3. Response R_{air}/R_{acet} of two arranged monolayers of Fe_2O_3 nanoparticles upon 5 minute exposures of 0.5 and 1 ppm acetone vapours with the preserved total gas flow of 100 sccm. Vertical dash bars indicate the switching ON and OFF of the acetone vapours. The sensor temperature was kept at 420°C.

The time necessary for the response saturation upon the probed mixture makes the sensor rather slow. The response time can be shortened by employment of a higher gas flow, yet, that augments the cooling of the sensor, which affects the response. Therefore, the operation at a lower total gas flow would be preferable. Figure 4 shows the responses upon one-minute durative exposure of various concentrations of acetone vapours with the preserved total gas flow of 10 sccm. Such flow results in the filling time of about 2:30 minutes. The dependences suggest that employment of such stringent conditions, *i.e.* both the short exposure period and the very low total gas flow, allows to obtain response near to the diabetes diagnosis threshold.

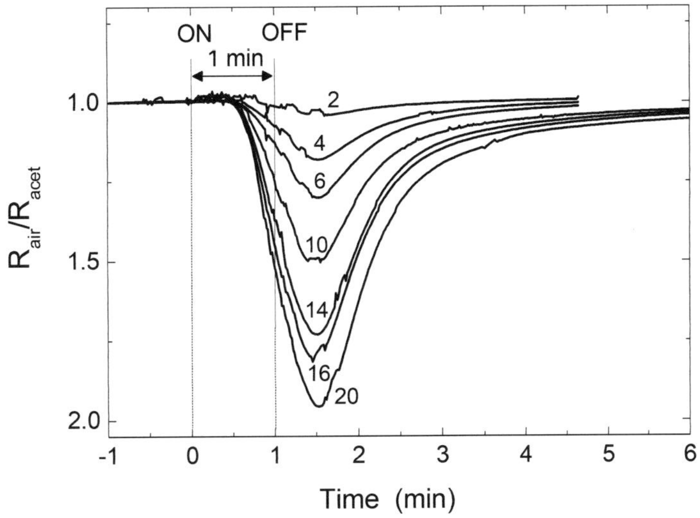

Figure 4. Response R_{air}/R_{acet} of two monolayers of Fe_2O_3 nanoparticles upon various concentrations of acetone vapours allowed flowing for one minute with the constant overall gas flow of 10 sccm. Numbers next to the dependences indicate the acetone concentrations in ppm. The switching ON and OFF of the acetone vapours are indicated by vertical bars.

4. Summary

The $\gamma \rightarrow \alpha$ transformation temperature of Fe_2O_3 nanoparticles was investigated by combination of the x-ray diffraction, the differential scanning calorimetry, and the thermogravimetry methods. It was demonstrated that the γ-Fe_2O_3 phase preserves up to at least 550°C, the magnitude sufficient for the operating temperature of a gas sensor. The doping of Fe_2O_3 nanoparticles by Pd ones decreases the $\gamma \rightarrow \alpha$ transformation temperature of Fe_2O_3 NP*s* by about 50°C. The Fe_2O_3 films formed by two monolayers of arranged nanoparticles proved to detect acetone vapours down to 500 ppb. The sensor appear sufficient swift to detect acetone concentration near to the diabetes diagnosis threshold within one minute.

Acknowledgments
We acknowledge the support via APVV-14-0891, VEGA 2/0010/15 and CNR-SAS 2016-2018 grants.

References

[1] Wang C and Sahay P 2009 *Sensors* **9** 8230
[2] Deng C, Zhang J, Yu X, Zhang W and Zhang X 2004 *J. Chromatogr.* B **810** 269
[3] Ryabtsev S V, Shaposhnick A V, Lukin A N and Domashevskaya E P, 1999 *Sens. Actuators* B **59** 26
[4] Chakraborty S, Banerjee D, Ray I and Sen A 2008 *Curr. Sci.* **94** 4794

[5] Ivanco J, Luby S, Jergel M, Siffalovic P, Benkovicova M, Halahovets Y, Majková E, Rella R and Manera M G 2013 *Sens. Lett.* **11** 2322
[6] Kim D H, Shim Y-S, Jeon J-M, Jeong H Y, Park S S, Kim Y-W, Kim J-S, Lee J-H and Jang H W 2014 *ACS Applied Mater. Interfaces* **6** 14779.
[7] Capone S, Manera M G, Taurino A, Siciliano P, Rella R, Luby S, Benkovicova M, Siffalovic P and Majkova E 2014 *Langmuir* **30** 1190
[8] Liang S, Li J, Wang F, Qin J, Lai X and Jiang X 2017 *Sens. Actuators* B **238** 923
[9] Wang L, Teleki A, Pratsinis S E and Gouma P I 2008 *Chem. Mater.* **20** 4794
[10] Righettoni M, Tricoli A and Pratsinis S E 2010 *Anal. Chem.* **82** 3581
[11] Chen D, Hou X, Li T, Yin L, Fan B, Wang H, Li X, Xu H, Lu H, Zhang R and Sun J 2011 *Sens. Actuators* B **153** 373
[12] Bamsaoud S F, Rane S B, Karekar R N and Aiyer R C 2011 *Sens. Actuators* B **153** 382
[13] Rella R, Spadavecchia J, Manera M G, Capone S, Taurino A, Martino M, Caricato A P and T. Tunno T 2007 *Sens. Actuators* B **127** 426
[14] Bene R, Perczel I V, Réti F, Meyer F A, Fleisher M and Meixner H 2000 *Sens. Actuators* B **71** 36
[15] Liu L, Li S, Zhuang J, Wang L, Zhang J, Li H, Liu Z, Han Y, Jiang X and Zhang P 2011 *Sens. Actuators* B **155** 782
[16] Zhou X, Li X, Sun H, Sun P, Liang X, Liu F, Hu X and Lu G 2015 *ACS Appl. Mater. Interfaces* **7** 15414
[17] Flak D, Braun A, Mun B S, Döbeli M, Graule T and Rekas M 2012 *Procedia Engineering* **47** 257
[18] Biswal R C 2011 *Sens. Actuators* B **157** 183
[19] Tianshu Z, Hongmei L, Huanxing Z, Ruifang Z and Yusheng S 1996 *Sens. Actuators* B **32** 181
[20] Gnanaprakash G, Ayyappan S, Jayakumar T, Phillip J and Raj B 2006 *Nanotechnology* **17** 5851
[21] Chitu L, Siffalovic P, Majkova E, Jergel M and Luby S 2014 Industrial Property Office of SR Patent No. 288234
[22] Randrianantoandro N, Mercier A M, Hervieu M and Grenèche J M 2001 *Mater. Lett.* **47** 150

IOP Conf. Series: Journal of Physics: Conf. Series **939** (2017) 012010 doi:10.1088/1742-6596/939/1/012010

A model for the electric conduction in metal/poly-TiO_2/metal structure

Faramarz Hossein-Babaei[1,a], Navid- Alaei-Sheini[1,b]

[1]Electronic Materials Laboratory, Electrical Engineering Department, K. N. Toosi University of Technology, Tehran 16317-14191, Iran

E_email: [a]fhbabaei@kntu.ac.ir, [b]navid_alaei@ee.kntu.ac.ir

Abstract. Intensely studied memristive devices have M'/MO/M″ structures, wherein MO is a nanometer-sized metal oxide crystallite sandwiched between the M' and M″ metal electrodes. The most widely used oxide for this purpose is TiO_2 and the electrodes are of noble metals such as platinum, silver, and gold. The memristive features of the device is believed to originate from the motion of the ionized oxygen vacancies within the oxide crystal. The operation of the device is further complicated by the motion of the mobile cations originating from the metal electrodes. The complexity of the device performance increases further when the noble metal electrodes form Schottky barriers at their junctions with TiO_2, as the conduction takes place through these energy barriers. In a recent publication, the authors have shown that, owing to the ohmicity of the Ti/TiO_2 junctions, electronic observations on the devices with Ti/TiO_2/Ti structure can be easier to model. The presented model clarified that in a Ti/poly-TiO_2/Ti structure, the ionic motion and the electronic conduction take place on the TiO_2 grain surfaces and grain boundaries rather than the grain interiors. Here, we show that the suggested model has important implications for chemical sensor design and fabrication.

Keywords: TiO_2, Polycrystalline, Oxygen vacancy, Chemical sensor.

1. Introduction

Profound resistance switching, observed in nanostructured metal oxide semiconductors, originates from the alteration of the conduction mechanism or route caused by the motion and reformatting of the distribution patterns of the present mobile ionic species [1]. The resistive memory devices (memristors) [2] operating based on such resistance switching are candidate memory devices for the future nanoelectronic circuits [3]. The memristor structures studied are generally of M'/MO/M'' type, where MO is a nanometer sized metal oxide crystallite sandwiched between the M' and M'' metal electrodes [4]. The most widely studied oxide for memristor fabrication is TiO_2 which is, in most cases, connected to noble metal electrodes such as platinum [5], silver [6], or gold [7], which add to the complexity of the device performance by forming Schottky energy barriers at the junctions with TiO_2 [8-10]. The motion, patterning and relaxation of the mobile ionic species at varying electric field levels form or eliminate different electron conduction routes and result in complex I-V characteristics generally difficult to predict. While the migration of the ionized oxygen vacancy (IOV) has been established as the most effective cause of resistance variations in these devices [11-16], it has been shown that the motions of all cations, including those originating from the electrodes, can determine the main conduction mechanism and promote hysteretic behaviors [3, 17, 18]. It appears that better understanding of the device performance would be afforded using samples with Ti/TiO_2/Ti structure in which the varieties of the candidate mobile species is limited. Moreover, unlike many other transition metals, no energy barrier is formed at the junction between titanium and TiO_2 [19- 21] and the electronic structure of the device is simpler to model.

Here, we study the I-V characteristics of the Ti/poly-TiO_2/Ti structures made of 400 nm-thick oxide layers with asymmetric cross-sectional oxygen vacancy distributions and describe the obtained results based on the two different conduction mechanisms prevailing respectively in the low and high biasing fields. It is shown that the I-V diagram recorded for such a sample strongly depends on the biasing voltage sweeping frequency and the relative humidity level in the surrounding atmosphere.

2. Experimental

TiO_2 layers are grown on 10 mm × 10 mm Ti chips cut from 1 mm thick titanium foil by thermal oxidation at 650 ºC for 60 minutes in air. Samples are air quenched from the soaking temperature. The temperature vs. time profiles used for the oxidation and quenching are given in figure 1a. Owing to the heat capacity of the Ti substrate and the refractory underneath, the samples cool asymmetrically. The thickness of the oxide layers is in the 350 to 400 nm range and the speed of cooling is higher at the top of the layer. According to the obtained XRD patterns, given in figure 1b, the grown oxide layers are of rutile phase. The oxide layers grown at lower temperatures, e.g. 500 ºC, in similar conditions, are rutile and anatase mixtures [8, 22]. The Ti/TiO_2/Ti devices are made by selected area titanium deposition by thermal evaporation of the metal on the grown oxide layer. Each chip accommodates 16 samples as shown in figure 2a. The schematic diagram of the sample structure is given in figure 2b wherein the micrographs of the oxide layer are given as insets indicating a polycrystalline structure with average grain size of 50 nm.

The titanium chip acts as the backside electrode and is connected to the measurement circuitry using conductive paste and silver wire. The top Ti electrode is connected to the measurement system via a pressure-connected titanium probe which can move from one device to another using a manually controlled x-y-z micromanipulator. This connection methods prevents contamination of the oxide layer with foreign metallic species. All samples are briefly annealed at 50 ºC in air, which proved to improve the evenness in the electrical features among similar samples. All samples are primarily tested by the application of an AC voltage of 1 V peak-to-peak to the sample connected in series with a 100 Ω resistor; samples allowing current levels above 1 mA/mm^2 are rejected. The experimental setup is depicted in figure 3.

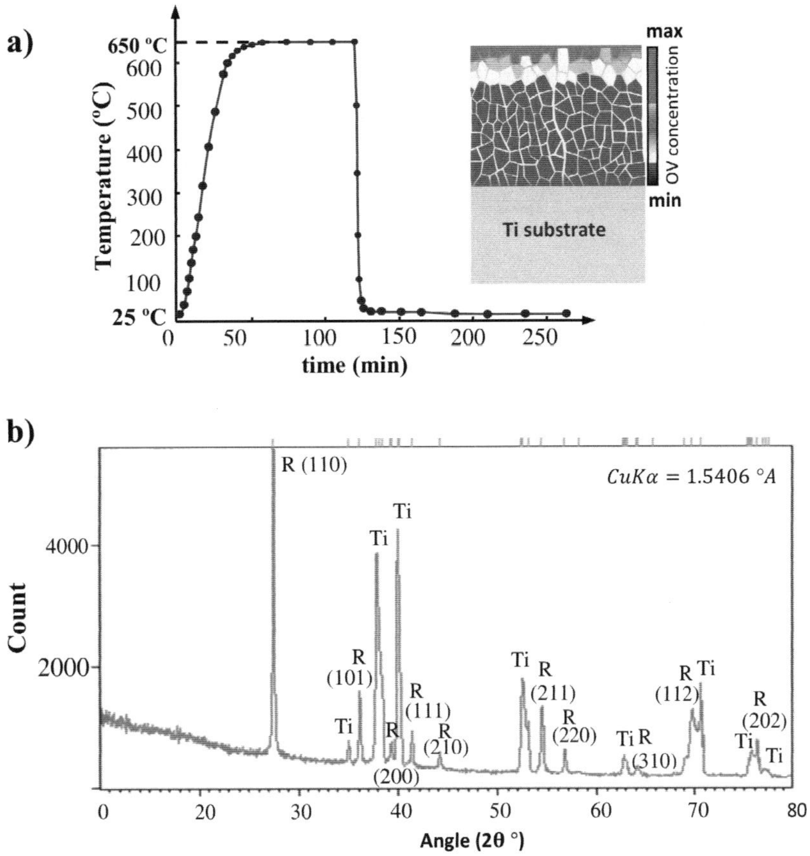

Figure 1. (a) The temperature vs. time diagrams of the oxidation and quenching processes, and (b) the XRD patterns of the grown oxide layers. The inset in (a) schematically presents the uneven distribution of the oxygen vacancies.

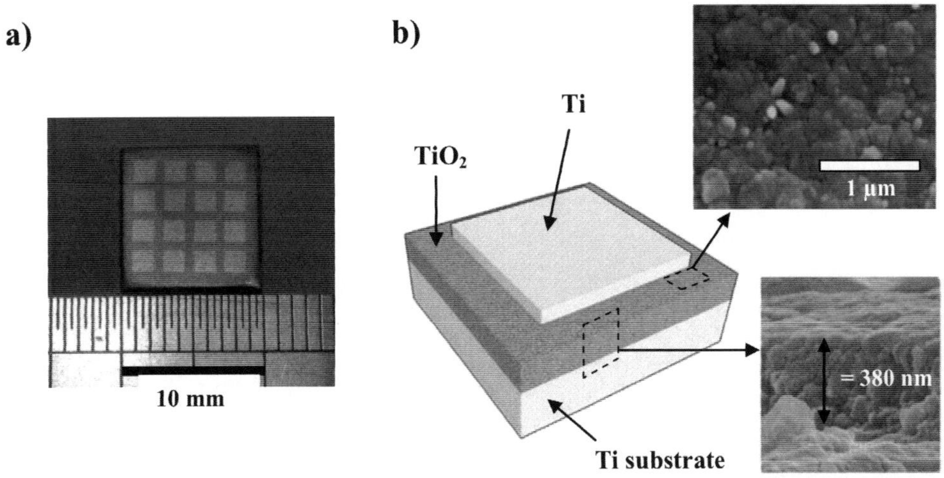

Figure 2. (a) The photograph of the Ti/poly-TiO_2/Ti samples fabricated on a thermally oxidized titanium chip. (b) The schematic diagram of the device structure. The insets in (b) are the plan and cross-sectional view micrographs of the TiO_2 layer.

IOP Conf. Series: Journal of Physics: Conf. Series **939** (2017) 012010 doi:10.1088/1742-6596/939/1/012010

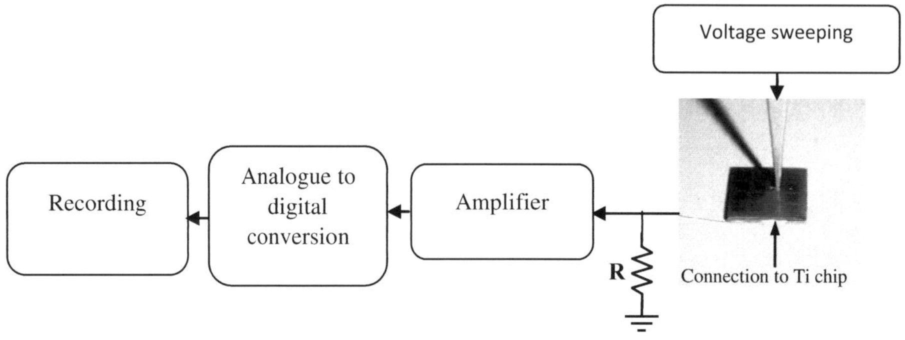

Figure 3. The experimental setup used for the I-V measurements on the Ti/poly-TiO_2/Ti samples.

3. Results and Discussion

The I-V diagram of the sample at five different voltage sweeping frequencies of 0.01, 0.1, 1.0, 10, and 100 Hz are shown in figure 4. The applied field is considered positive when the connection probe is positively biased with respect to the Ti substrate. These diagrams are plotted after the reduction of the current contribution of the parallel capacitor of the sample devices, which is considerable particularly at 100 Hz. In figure 4, the configuration of the I-V diagram respective to 100 Hz is in agreement with the previous reports on the Ti/TiO_2 contacts [8, 10, 19-22]. This diagram presents characteristics consistent with the symmetric structure of the device, the zero junction energy barrier, ohmic connection between the titanium electrodes and TiO_2, and normal electronic conduction through an n-type space-charge-limited medium. The resistance of the oxide layer calculated based on the slope of the profile at around zero biasing is 0.3 MΩ. Considering the geometry of the device, the calculated conductivity for the rutile layer is 8.3×10^{-9} S/cm. The I-V diagrams recorded by utilizing low voltage sweeping frequencies are strikingly different. At 0.01 Hz, for instance, the device presents a rectifying diode-like I-V profile and a "soft breakdown" at around biasing voltage of +1.5 V. This is encountered with a profound reduction in the sample resistance and hysteretic behavior of the device.

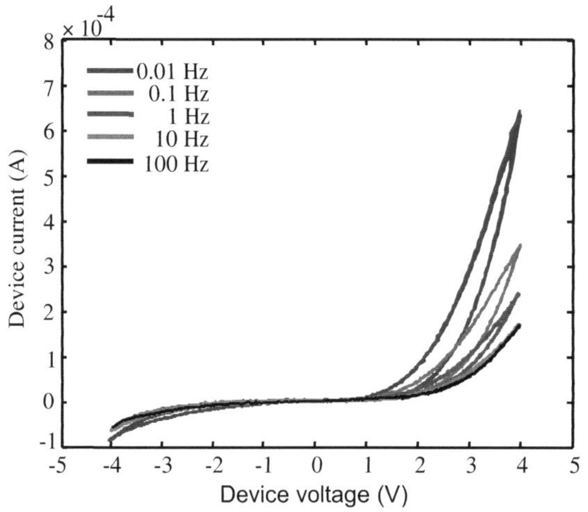

Figure 4. The I-V diagrams of a Ti/poly-TiO_2/Ti diode recorded using different voltage sweeping frequencies.

IOP Conf. Series: Journal of Physics: Conf. Series **939** (2017) 012010 doi:10.1088/1742-6596/939/1/012010

We have proposed a model which describes the observed resistance change, the asymmetric profile, hysteretic behavior, the strong dependence of the I-V profiles of the device on the voltage sweeping frequency, and the I-V characteristics change upon thermal annealing in air [1]. The model assumes ionized oxygen vacancy (IOV) motion on the surface of the grains, rather than the grain interior. The IOVs motion is in the direction of forming conductive filaments within the grain boundary regions; the process is schematically presented in figure 5. This assumption is justified as we prove that the field applied through the whole I-V profile is an order of magnitude lower than the level required for the IOV mobilization within the rutile crystal.

The implications of the proposed model are interesting as the surface conduction would allow influence of the external species on the conduction process. For instance, it was predicted that the water monomers adsorbed to the surface of the rutile grains should affect the IOV motion therein and, hence, alter the conductance through the polycrystalline layer. Our quantitative considerations, based on the results of direct STM observations of the oxygen vacancy migration on rutile {110} facets [23-26], showed that the presence of the H_2O molecules and hydroxyl group decrease the activation energy of the IOV migration. Our experimental results, presented in figure 6, verify the predictions of the model. According to these results, considering a constant forward biasing level, say +1.5 V, the device current increases with the relative humidity level in the surrounding atmosphere.

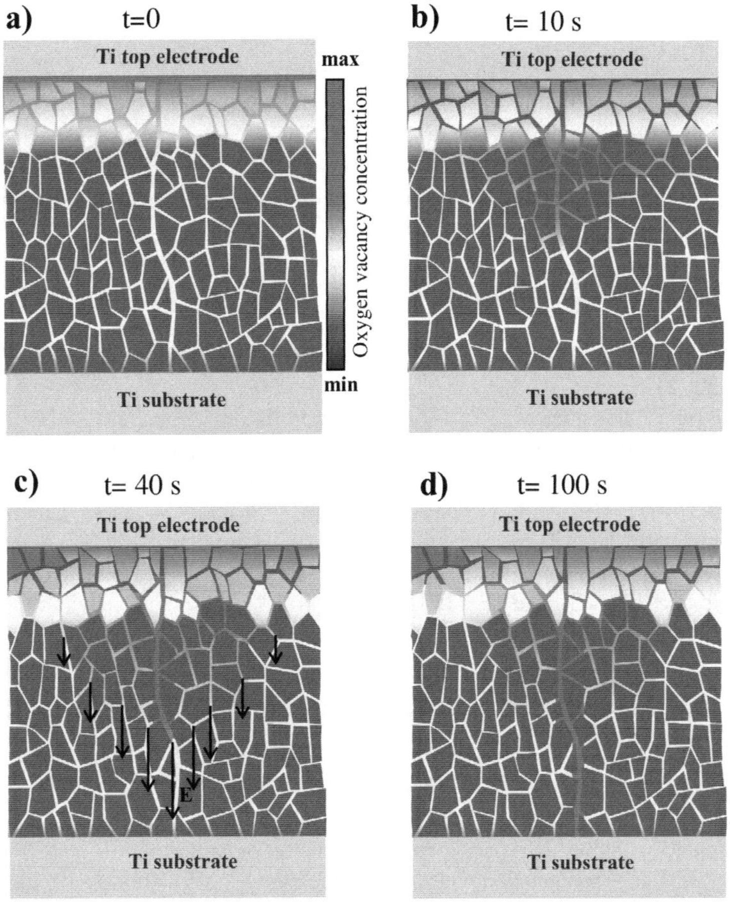

Figure 5. (a-d) The schematic presentation of the IOV filament formation within the grain boundaries after applying step voltage.

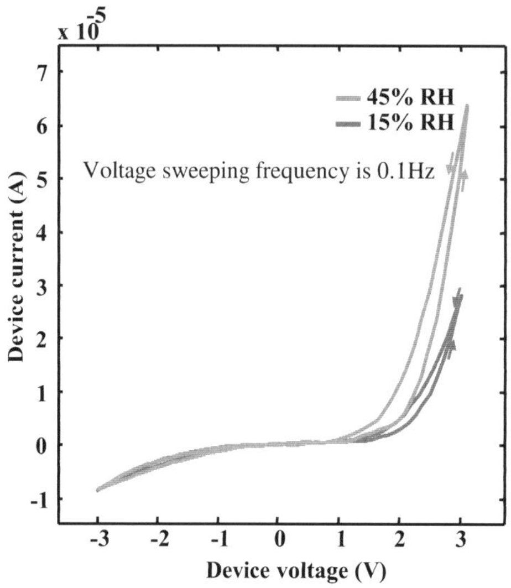

Figure 6. The I-V diagram of the Ti/poly-TiO_2/Ti device plotted at two different relative humidity levels.

4. Conclusions

Due to the uneven distribution of the oxygen vacancy, the I-V characteristics of the diode with Ti/poly-TiO_2/Ti structure is strongly asymmetric. In the forward biased device, the current is conducted via conductive filaments formed by the motion of the IOVs on the surfaces of the grains. Interactions between the grain surface and the surrounding atmosphere is easy and, hence, the described conduction mechanism can easily be influence by environmental parameters.

References

[1] Hossein-Babaei F and Alaei-Sheini N 2016 Electronic conduction in Ti/Poly-TiO_2/Ti Structures *Sci. Rep.* **6** 29624

[2] Chua L O 2011 Resistance switching memories are memristors *Appl. Phys. A* **102** 765-83.

[3] Waser R and Aono M 2007 Nanoionics-based resistive switching memories *Nat. Mater.* **6** 833-40.

[4] Wong H S *et al.* 2012 Metal–oxide RRAM *Procc. IEEE* **100** 1951-70

[5] Rogala M *et al.* 2015 Resistive switching of a quasi homogeneous distribution of filaments generated at heat-treated TiO2 (110) surfaces. *Adv. Funct. Mater.* **25** 6382–9

[6] Prada S *et al.* 2011 Density functional theory study of TiO_2/Ag interfaces and their role in memristor devices *Phys. Rev. B* **83** 245314

[7] Maslova N *et al.* 2015 Features of titanium oxide memristor fabrication by pulsed laser deposition *Phys. Status Solidi C* **12** 242–5

[8] Hossein-Babaei F, Lajvardi M M and Alaei-Sheini, N 2015 The energy barrier at noble metal/TiO_2 junctions *Appl. Phys. Lett.* **106** 083503

[9] Jeong H Y, Lee J Y and Choi S Y 2010 Interface Engineered Amorphous TiO_2-Based Resistive Memory Devices *Adv. Funct. Mater.* **20** 3912–7

[10] Hossein-Babaei F, Lajvardi M M and Boroumand F A 2012 Large area Ag-TiO_2 UV radiation sensor fabricated on a thermally oxidized titanium chip *Sens. Actuators, A* **173** 116–21

[11] Li D *et al.* 2012 Oxygen vacancy filament formation in TiO_2: A kinetic Monte Carlo study *J. Appl. Phys.* **112** 073512

[12] Kwon D H *et al.* 2010 Atomic structure of conducting nanofilaments in TiO_2 resistive switching memory *Nat. Nanotech.* **5** 148–53

[13] Janousch M *et al.* 2007 Role of oxygen vacancies in Cr-doped $SrTiO_3$ for resistance-change memory *Adv. Mater.* **19** 2232–5
[14] Chen J Y *et al.* 2015 Switching kinetic of VCM-based memristor: evolution and positioning of nanofilament *Adv. Mater.* **27** 5028–33
[15] Yang J J *et al.* 2009 The mechanism of electroforming of metal oxide memristive switches *Nanotech.* **20** 215201
[16] Hu C *et al.* (2014) Highly controllable and stable quantized conductance and resistive switching mechanism in single-crystal TiO_2 resistive memory on silicon *Nano Lett.* **14** 4360-7.
[17] Lv H *et al.* 2015 Atomic view of filament growth in electrochemical memristive elements *Sci. Rep.* **5** 13311
[18] Waser R *et al.* 2009 Redox-based resistive switching memories–nanoionic mechanisms, prospects, and challenges *Adv. Mater.* **21** 2632–63
[19] Hossein-Babaei F and Rahbarpour S 2011 Titanium and silver contacts on thermally oxidized titanium chip: electrical and gas sensing properties *Solid-State Electron.* **56** 185–90
[20] Yang J J *et al.* 2011 Metal/TiO2 interfaces for memristive switches *Appl. Phys. A* **102** 785–9
[21] Hossein-Babaei F and Rahbarpour S 2011 Separate assessment of chemoresistivity and Schottky-type gas sensitivity in M–metal oxide–M' structures *Sens. Actuators, B* **160** 174–80
[22] Hossein-Babaei F, Alaei-Sheini N and Lajvardi M M 2016 Oxygen adsorption at noble metal/TiO_2 junctions *IOP Conf. Ser. Mater. Sci. Eng.* **108**
[23] Wendt S *et al.* 2005 Oxygen vacancies on TiO_2(110) and their interaction with H_2O and O_2: A combined high-resolution STM and DFT study *Surf. Sci.* **598** 226–45
[24] Schaub R *et al.* 2003 Oxygen-mediated diffusion of oxygen vacancies on the TiO_2(110) *Surf. Sci* **299** 377–379
[25] Wahlström E *et al.* 2004 Electron transfer-induced dynamics of oxygen molecules on the TiO_2(110) *Surf. Sci.* **303** 511–3
[26] Matthiesen J *et al.* 2009 Formation and diffusion of water dimers on rutile TiO_2(110) *Phys. Rev. Lett.* **102** 226101

IOP Conf. Series: Journal of Physics: Conf. Series **939** (2017) 012011 doi:10.1088/1742-6596/939/1/012011

Transparent heaters made by ultrasonic spray pyrolysis of SnO_2 on soda-lime glass substrates

Mohammad Ansari[1], Mehdi Akbari-Saatloo[2] and Mohsen Gharesi[3(c)]

Electronic Materials Laboratory, Electrical Engineering Department, K. N. Toosi University of Technology, Tehran 16317-14191, Iran.

[1]mohammad.ansari1371@gmail.com [2]akbari@email.kntu.ac.ir [3(c)]m.gharesi@gmail.com

Abstract. Transparent heaters have become important owing to the increasing demand in automotive and display device manufacturing industries. Indium tin oxide (ITO) is the most commonly used material for production of transparent heaters, but the fabrication cost is high as the indium resources are diminishing fast. This has been the driving force behind the intense research for discovering more durable and cost-effective alternatives. Tin oxide, with its high temperature stability and coexisting high levels of conductivity and transparency, can replace expensive ITO in the fabrication of transparent heaters. Here, we propose tin oxide films deposited using ultrasonic spray pyrolysis as the raw material for the fabrication of transparent heaters. Silver contacts are paste printed on the deposited SnO_2 layers, which provide the necessary connections to the external circuitry. Deposition of films having sheet resistance in the 150 Ω/sq range takes only ~5 minutes and the utilized methods are fully scalable to mass production level. Durability tests, carried out for weeks of continuous operation at different elevated temperatures, demonstrated the long load life of the produced heaters.

1. Introduction

Transparent conducting films (TCFs) are extensively employed as the heating element in numerous device structures. Fabrication of heaters for lab-on-a-chip applications and defogger systems for outdoor display panels, vehicle windows, and camera lenses are just a few of many fields where high-power TCFs are required [1-3]. Indium tin oxide (ITO) has been conventionally the major material used for the production of transparent heaters. High fabrication cost due to rare indium resources, brittleness, and poor temperature stability, have raised questions on ITO's suitability for the sustainable use in industrial applications [4, 5]. Carbon nanotubes, graphene layers, and metallic nanowires were recently proposed as ITO alternatives [6-8]. Despite their superior optical properties, long-term stability of these heaters is yet questionable and more has to be done on large-area fabrication methods.

Figure of merit for TCFs is defined as the ratio of electrical conductivity (σ) to visible light absorption coefficient (α) [9]. Assuming equal α values, this definition leads to higher figure of merit for more conducting layers. This is a practical guideline for transparent electrodes fabrication, but a transparent material for on-chip heater applications should be selected more carefully. Doped metal oxides utilized for fabrication of TCFs have low resistivities of $\sim 10^{-4}$ Ω.cm. In order to obtain durable operational characteristics at elevated temperatures, the metal oxide deposits have micrometric thicknesses and thus sheet resistances in the 1-4 Ω/sq range [10, 11]. These values are too low for heater fabrication and cause excessive power consumption in the metal electrodes and wirings. Geometries with high length-to-width aspect ratios are usually regarded as possible solutions, although, the method complicates the fabrication process and results in extra costs.

Tin oxide, an n-type wide-bandgap semiconductor, is mostly known for its widespread application as the sensing element of gas sensors. Exceptional high-temperature stability and remarkable sensitivity to a wide variety of analyte molecules have led to its popularity in the field of chemical sensors [12-16]. Doped tin oxide films with transparency in the whole visible spectrum have been widely suggested for utilization as electrodes in optoelectronic devices [17]. The undoped material is also a promising candidate for TCF fabrication. Low formation energy of oxygen vacancies causes significant nonstoichiometry in the SnO_2 structure even at room temperature. Oxygen vacancies donate electrons to the conduction band resulting in metal-like conductivity in $SnO_{2-\delta}$ films while not changing their optical absorption characteristics in comparison with stoichiometric SnO_2 [18].

Here, transparent tin oxide heaters are fabricated on standard soda-lime glasses using a homemade ultrasonic spray pyrolysis (USP) system. USP technique is chosen due to its simplicity, cost effectiveness and suitability for mass production [19, 20]. The method is optimized to result in dense polycrystalline films at deposition temperatures as low as 300 °C. Ohmic Ag/SnO_2 contacts are formed on the grown films [21]. The load life of the fabricated samples is investigated over weeks of continuous operation at temperatures in the 80-150 °C range.

2. Experimental

Standard 1-mm thick soda-lime glass slides are washed with absolute acetone, rinsed with distilled water, and dried on a hot plate at 150 °C for 10 min. These are laser cut into 5 mm × 5 mm chips and subjected to the same washing procedure. The utilized USP deposition system was previously described in ref. [22]. The precursor solution is prepared by dissolving 0.2 moles of $SnCl_2.2H_2O$ (Merck Millipore, 107815) in 1.0 litre of absolute ethanol (99.5% v/v). Piezoelectric transducers working at 1.7 MHz frequency are embedded in a water filled tank. A polymer container is placed on top of the water tank which prevents the solution-caused corrosion of the vaporization system. The precursor solution is poured into the polymer container through a flow control valve with ~30 mL/h rate. Continuous flow of air with an average rate of 3 L/min carries the generated aerosol droplets to the deposition chamber. The aerosol strikes the substrate surface which is heated to 300 °C by an underlying resistive heater. The substrate temperature is monitored using a fine S-type thermocouple and controlled by the voltage applied to the heater.

Some of the samples were annealed at 400 °C in air for half an hour. Lateral silver electrodes are paste-printed on the films followed by field-assisted heat treatments at 300 °C to ensure ohmicity of the connections [21, 23]. Durability tests are performed by continuously applying AC heating voltages to the fabricated samples. Device resistances are measured with 24 h time intervals while controlling the heating voltages to keep the operating temperatures constant. The utilized experimental setup is schematically illustrated in figure 1.

3. Results

Owing to the lower growth temperature compared to the previously reported USP depositions [9], the method is applicable for production of tin oxide films on soda-lime glasses, as well. The planar scanning electron microscopy (SEM) image of the surface of an as-deposited sample is shown in figure 2.a. The polycrystalline film has a dense morphological structure. The average grain size, measured using the line averaging method, is ~70 nm. The cross-sectional micrograph shown in figure 2.b illustrates a 400 nm-thick homogenously deposited film on the glass substrate. The average sheet resistance of the samples is around ~150 Ω/sq at room temperature. Considering the sample geometry, the resistivity of the deposited material is calculated to be ~6×10^{-3} Ω.cm.

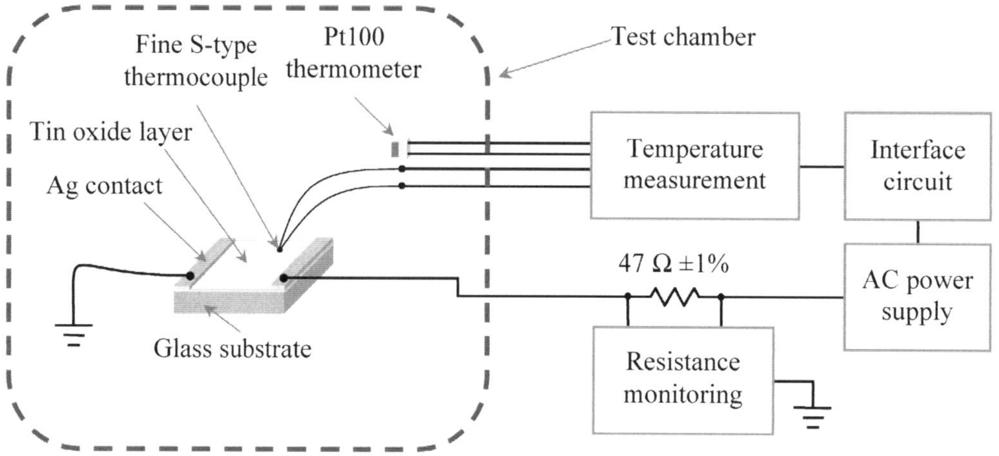

Figure 1. The schematic presentation of the sample structure and the experimental setup utilized for the durability tests.

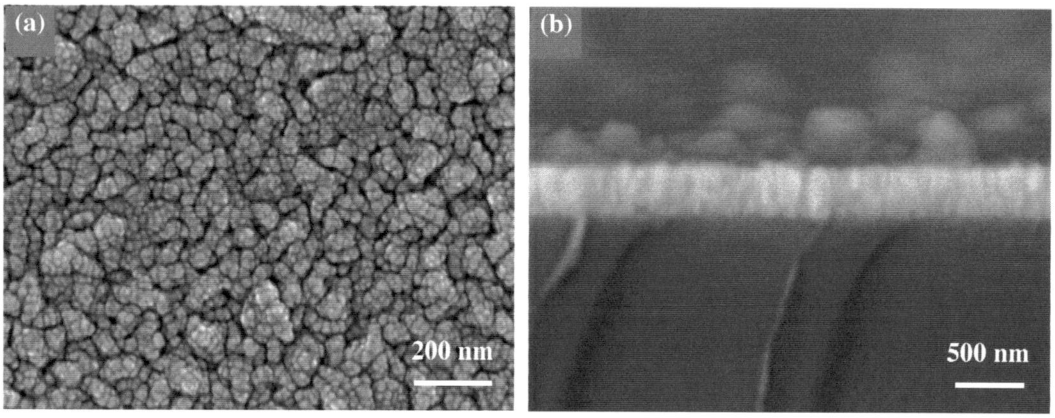

Figure 2. SEM micrographs from the tin oxide layer grown on a soda-lime glass substrate; (a) the plan-view and (b) the cross-section.

Photographs of as-deposited and annealed tin oxide samples are represented in figure 3. The annealed tin oxide films have a clearly enhanced visible transparency which is induced by their improved stoichiometric composition relative to the as-grown ones. This is consistent with a rise of ~20% in layers sheet resistance occurred after the annealing process; the increase in resistance is attributed to the decrease in the oxygen vacancy concentration in the layer.

Resistance-temperature diagram of a fabricated sample is shown in figure 4.a. Clearly, the device shows negative temperature coefficient of resistance (TCR) in the investigated temperature range which originates from both the increased oxygen vacancy concentration and the enhanced ionization mechanism at higher temperatures [24, 25]. Negative TCR heaters are known for their resistance to hot-spot generation and, hence, excellent reliability for long-term operations [26]. As shown in figure 4.b, the resistance values show negligible variation in a 2-week test period; only 0.9, 1.0 and 2.5% increase was recorded for the devices operating continuously at 80, 100 and 150 °C, respectively. The annealed devices offer even better long run stability; no resistance change was recorded during a similar experimental session.

Figure 3. Photograph of the glass substrate before (a) and after (b) tin oxide deposition; (c) is the same sample after thermal annealing at 400 °C in air.

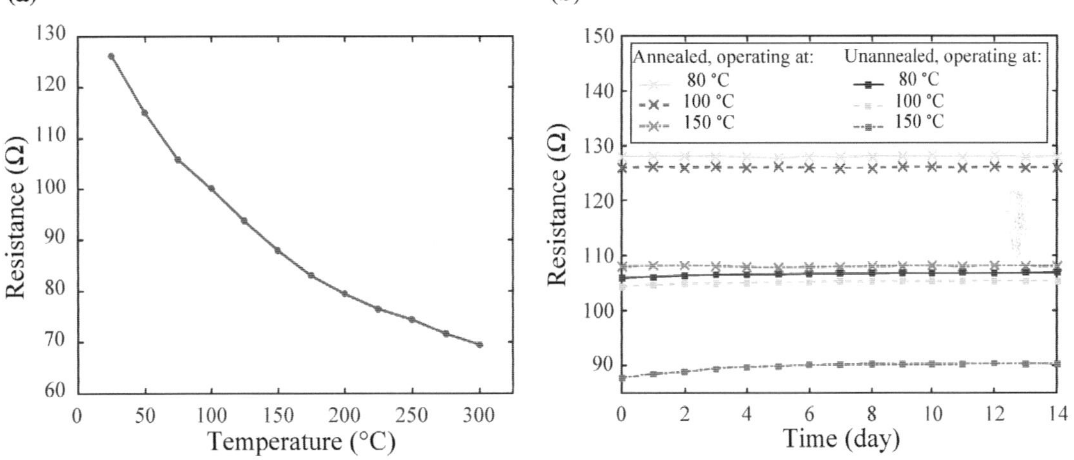

Figure 4. The resistance vs. temperature (a) and the resistance vs. time plotted for the fabricated tin oxide resistors.

4. Conclusions

Tin oxide films were grown on soda-lime glass substrates using a homemade USP system. The deposition temperature is 300 C which is safe for the glass substrates. Layers with sheet resistances of ~150 Ω/sq are grown in deposition times as low as 5 minutes. SEM micrographs revealed the polycrystalline nature of the deposits and a homogenous layer thickness of ~400 nm across the substrate. The samples exhibited fine optical transparency in visible spectrum. Ohmic silver contacts were formed on the films, and the reliability of the resistor samples for long-term use as on-chip transparent heaters was investigated. The devices showed negligible resistance variation after a 2-week of continuous operation at the 80-150 °C temperature range. The results show that undoped tin oxide layers are promising alternatives to the expensive ITO device components.

References

[1] Im K, Cho K, Kim J and Kim S 2010 Transparent heaters based on solution-processed indium tin oxide nanoparticles *Thin Solid Films* **518** 3960-3.

[2] Kang J, Kim H, Kim K S, Lee S K, Bae S, Ahn JH, Kim YJ, Choi JB and Hong BH 2011 High-performance graphene-based transparent flexible heaters *Nano Lett.* **11** 5154-8.

[3] Zhang X, Yan X, Chen J and Zhao J 2014 Large-size graphene microsheets as a protective layer for transparent conductive silver nanowire film heaters *Carbon* **69** 437-43.

[4] Jung D, Kim D, Lee K H, Overzet L J and Lee G S 2013 Transparent film heaters using multi-walled carbon nanotube sheets *Sens. Actuators, A* **199** 176-80.

[5] Rao K D and Kulkarni G U 2014 A highly crystalline single Au wire network as a high temperature transparent heater *Nanoscale* **6** 5645-51.
[6] Yoon Y H, Song J W, Kim D, Kim J, Park J K, Oh S K and Han C S 2007 Transparent Film Heater Using Single-Walled Carbon Nanotubes *Adv. Mater.* **19** 4284-7.
[7] Kang J, Kim H, Kim K S, Lee S K, Bae S, Ahn J H, Kim Y J, Choi J B and Hong B H 2011 High-performance graphene-based transparent flexible heaters *Nano lett.* **11** 5154-8.
[8] Bob B, Machness A, Song T B, Zhou H, Chung C H and Yang Y 2016 Silver nanowires with semiconducting ligands for low-temperature transparent conductors *Nano Res.* **9** 392-400.
[9] Gordon R G 2000 Criteria for choosing transparent conductors *MRS Bull.* **25** 52-7.
[10] Gupta R, Rao K D, Kiruthika S and Kulkarni G U 2016 Visibly Transparent Heaters *ACS Appl. Mater. Interfaces.* **8** 12559-75.
[11] Gao K H, Lin T, Liu X D, Zhang X H, Li X N, Wu J, Liu Y F, Wang X F, Chen Y W, Ni B and Dai N 2013 Low temperature electrical transport properties of F-doped SnO_2 films *Solid State Commun.* **157** 49-53.
[12] Hossein-Babaei F and Amini A 2014 Recognition of complex odors with a single generic tin oxide gas sensor *Sens. Actuators, B* **194** 156-63.
[13] Wang C, Yin L, Zhang L, Xiang D and Gao R 2010 Metal oxide gas sensors: sensitivity and influencing factors *Sensors* **10** 2088-106.
[14] Hossein-Babaei F, Zare A H, Ghafarinia V and Erfantalab S 2015 Identifying volatile organic compounds by determining their diffusion and surface adsorption parameters in microfluidic channels *Sens. Actuators, B* **220** 607-13.
[15] Hossein-Babaei F, Paknahad M and Ghafarinia V 2012 A miniature gas analyzer made by integrating a chemoresistor with a microchannel *Lab Chip* **12** 1874-80.
[16] Hosseini-Golgoo S M and Hossein-Babaei F 2011 Assessing the diagnostic information in the response patterns of a temperature-modulated tin oxide gas sensor *Meas. Sci. Technol.* **22** 035201.
[17] Elangovan E and Ramamurthi K 2003 Optoelectronic properties of spray deposited SnO_2: F thin films for window materials in solar cells *J. Optoelectron. Adv. Mater.* **5** 45-54.
[18] Kılıç Ç and Zunger A 2002 Origins of coexistence of conductivity and transparency in SnO_2 *Phys. Rev. Lett.* **88** 095501.
[19] Jadsadapattarakul D, Euvananont C, Thanachayanont C, Nukeaw J and Sooknoi T 2008 Tin oxide thin films deposited by ultrasonic spray pyrolysis *Ceram. Int.* **34** 1051-4.
[20] Lee S Y and Park B O 2006 Structural, electrical and optical characteristics of SnO_2: Sb thin films by ultrasonic spray pyrolysis *Thin solid films* **510** 154-8.
[21] Hossein-Babaei F, Moghadam S and Masoumi S 2015 Forming ohmic Ag/SnO_2 contacts Mater. Lett. **141** 141-4.
[22] Gharesi M and Ansari M 2016 Tin Oxide Microheater for Chemical Sensors *IOP Conf. Ser.: Mater. Sci. Eng.* Vol. **108** 012018.
[23] Hossein-Babaei F, Lajvardi M M and Alaei-Sheini N 2015 The energy barrier at noble metal/TiO_2 junctions *Appl. Phys. Lett.* **106** 083503.
[24] Gurlo A 2006 Interplay between O_2 and SnO_2: oxygen ionosorption and spectroscopic evidence for adsorbed oxygen *ChemPhysChem* **7** 2041-52.
[25] Hossein-Babaei F and Alaei-Sheini N 2016 Electronic conduction in Ti/Poly-TiO_2/Ti structures *Sci. Rep.* **6** 29624.
[26] Simon I, Bârsan N, Bauer M and Weimar U 2001 Micromachined metal oxide gas sensors: opportunities to improve sensor performance *Sens. Actuators, B* **73** 1-26.

IOP Conf. Series: Journal of Physics: Conf. Series **939** (2017) 012012 doi:10.1088/1742-6596/939/1/012012

Structural and Galvanomagnetic properties in Mn-Bi_2Te_3 thin films

K. Bidinakis and Th. Speliotis
Institute of Nanoscience and Nanotechnology, NCSR Demokritos 15310 Aghia Paraskevi, Greece

t.speliotis@inn.demokritos.gr

Abstract. Bismuth-based binary chalcogenide compounds such as Bi_2Te_3 and Bi_2Se_3 are well known materials for their excellent thermoelectric properties due to their near-gap electronic structure. In the last few years these materials have received attention for exhibiting new physics of 3D topological insulators (TI). Possible applications of TI based devices range from quantum computing, spin based logic and memory to electrodynamics. The 3D TIs present spin-momentum-locked surface states by time reversal symmetry (TRS). Introducing magnetic doping in a TI, brakes the TRS and is predicted to open the gap at Dirac point, resulting in exotic quantum phenomena. This interaction between magnetism and topologically protected states is of potential attention for applications in modern spintronics. Quantum phenomena such as weak antilocalization observed in these nanostructures are described. In this work, granular Mn-Bi_2Te_3 thin films were grown by DC magnetron sputtering on Si(111) substrates and were submitted to ex situ annealing. We present results for the crystal structure of sputtered and annealed films characterized with X-ray diffraction and high-resolution scanning electron microscopy (HRSEM). The surface analysis was studied with atomic force microscopy (AFM). Magnetotransport measurements were performed using standard four probe technique with Hall and MR configurations, with perpendicular magnetic fields up to 9T and temperatures from 300 to 3K.

1. Introduction

Topological insulators (TI) are exotic materials which denote new physics of condensed matter and unusual quantum phenomena[1,2]. They are characterised by an insulating bulk and conducting surface states of massless helical Dirac fermions as a consequence of strong spin-orbit coupling[3].
Among the numerous three-dimensional (3D) topological insulators are the group V-VI chalcogenide materials, such as Bi_2Se_3 and Bi_2Te_3 binary compounds. These materials have been known for a long time to be excellent thermoelectric materials[4] and are indeed the most studied materials. The large band gap in the bulk is 300meV and 150meV respectively. The distinct properties of TI surface electron states appear attractive for fundamental research, as well as for applications in the spintronic area[5], quantum information[6] and the emerging field of spin caloritronics[7]. Among the various Bi-based TIs that have been studied, Bi_2Te_3 is one of the promising candidates for future electronic device applications. Bi_2Te_3 in crystalline form has a rhombohedral structure and upon magnetic doping the TRS may be broken and that would enable us to observe exotic phenomena such as the quantized anomalous Hall (QAH) effect, extraordinary galvanomagnetic effects or other emerging phenomena[8-9]. The possibility to control the topological properties and the low-dimensional electron transport by introduced magnetic spin, concerned significantly the interest of materials scientists to develop new TI materials and to implement magnetic atoms in 3D TIs. Furthermore, a number of groups have also experimentally prepared Mn/Fe/Cr doped Bi_2Se_3 or Bi_2Te_3 using several techniques such as molecular

IOP Conf. Series: Journal of Physics: Conf. Series **939** (2017) 012012 doi:10.1088/1742-6596/939/1/012012

beam epitaxial (MBE) growth, laser ablation or growth in the bulk form[10]. Wide transport measurements were showed in these systems, for example, the border between weak localization (WL) and weak anti-localization (WAL) was observed[10]. Following this stimulus, we study the Mn doped Bi_2Te_3 granular thin films grown using a magnetron sputtering system and we study the structural and the magneto-transport properties of this system.

2. Experimental

Granular Bi_2Te_3 doped with Mn thin films were grown by DC magnetron sputtering on Si(111) substrates. The base pressure was about $5x10^{-8}$Torr. During deposition, the sputter Ar pressure was fixed at 3mTorr, without control to the substrate temperature. The Mn concentration has been achieved by repeating ten times an alternating deposition of Bi_2Te_3 with 20nm fixed thickness and Mn(x nm) with x=1, 2.5 and 5nm, from two different targets, at low-deposition rates. EDAX analysis shows that the percentage of Mn concertation inside the Bi_2Te_3 films is 5, 10 and 20% respectively. After the depositions, the films were annealed under the same condition at 300^0C for 30 minutes and after pumping down to $5x10^{-6}$Torr. X-ray diffraction (XRD) measurements were performed at ambient conditions with a Siemens D500 diffractometer, using CuKα radiation. The surface topography was examined using a Field Emission Scanning Electron Microscopy (FESEM), model Jeol JSM-7401F, and Atomic Force Microscopy (AFM) with the NT-MDT NTEGRA model. The film morphology has been inspected by high resolution transmission microscopy (HRTEM), using the Philips CM 20 model. Ohmic contacts were made by room temperature curved silver paste. The magnetotransport properties were measured via a Physical Property Measurement System (PPMS, Quantum design). Magnetoresistance (MR) and Hall resistance (H) measurements were performed in a granular film, with lateral dimensions 4mm by 4mm, using the standard van der Pauw method, in magnetic fields up to 9T, between 3K and 300K.

3. Experimental Results and Discussion

Undoped Bi_2Te_3 (200nm) and doped with Mn, thin films grown on Si(111) substrates and annealed at 300^0C were studied with X-ray diffraction (XRD) measurements. It is worth noting that the films after deposition were amorphous and only peaks from the substrate revealed. Fig. 1 shows that the Bi_2Te_3 film is polycrystalline, and the Bragg-peaks can be allocated to rhombohedral space group where the (00l) indices (l = 3, 6) reveal a predominant texture in Bi_2Te_3. Fig. 2 shows that the intensity of Bragg peaks with (00l) indices (l = 3, 6) tends to decrease with the Mn doping, indicating a small change in the texture of Bi_2T_3 from Mn.

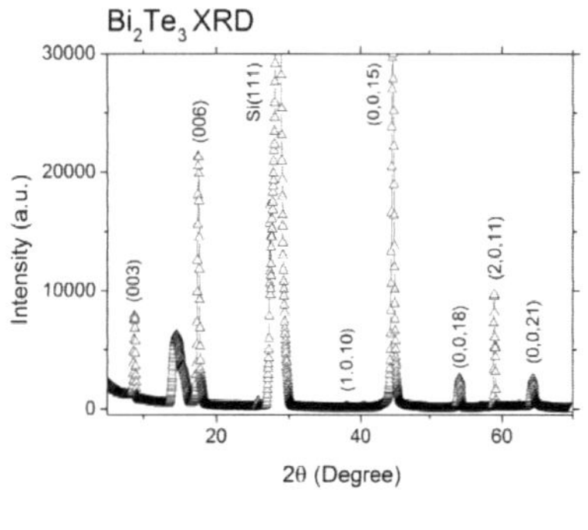

Figure 1: X-ray diffraction θ – 2θ pattern for 200nm thick Bi_2Te_3 single layer

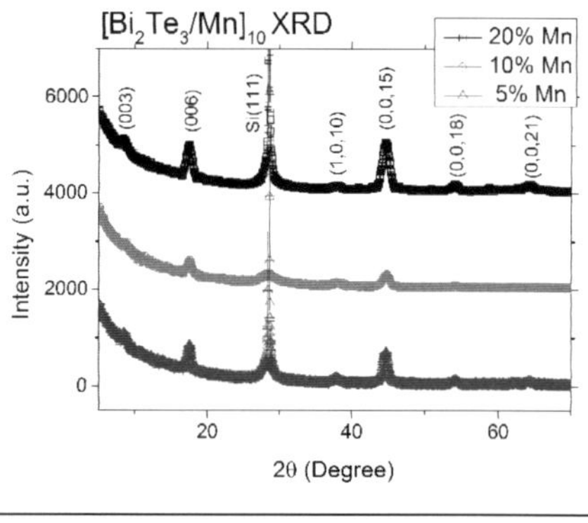

Figure 2: X-ray diffraction θ – 2θ pattern for 200nm thick Bi_2Te_3 films doped with Mn 5-20%.

We have carefully searched for oxides and intermetallic secondary phases such as MnTe and none were observable. A transformation from rhombohedral to hexagonal space group for Bi_2Te_3 has been used in order to estimate the lattice parameters from XRD data:

$$\frac{1}{d^2} = \frac{4}{3}\left(\frac{h^2 + hk + k^2}{a^2}\right) + \frac{l^2}{c^2}$$

The lattice constants of Bi_2Te_3 with Mn 5% were a = 4.381Å, c = 30.300Å determined from the positions of (0 0 6) and (1 0,10) peaks, which were slightly smaller than those of Bi_2Te_3 (a = 4.396Å, c = 30.308Å) due to the smaller Mn atomic radius (1.40Å) than those of Bi (1.60Å). The variations of lattice constants with Mn addition indicate the substitution of Mn in Bi_2Te_3 lattices.

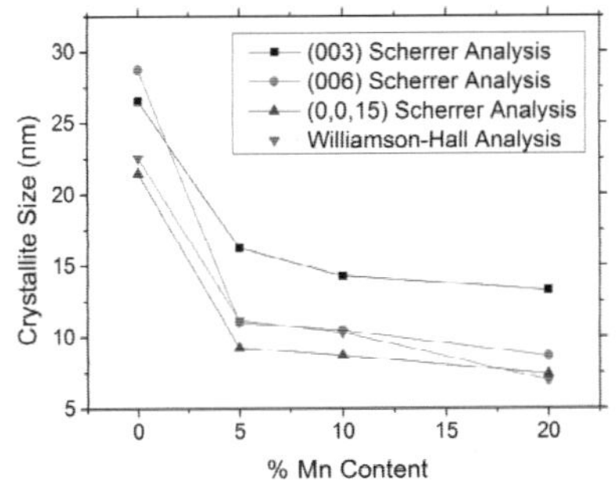

Figure 3: Variation of coherence length with Mn content from 5 to 20%

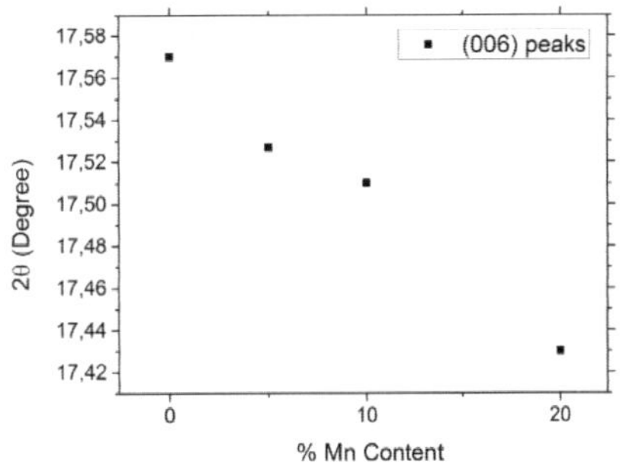

Figure 4: Variation of (006) peaks with Mn content from 5 to 20%

In fig. 3 the coherence length, calculated using the Scherrer equation, using three different peaks and the variation with manganese doping, is presented. The coherence length decreased with Mn percentage. Because the Scherrer formula works mostly in the cubic system, in other crystallographic systems there is about 10% error. In fig. 3 it is also presented the Williamson-Hall analysis, as a more accurate one for the rhombohedral system. In fig. 4 it is presented the shift to lower 2θ angles of the (006) peak, with increasing manganese content.

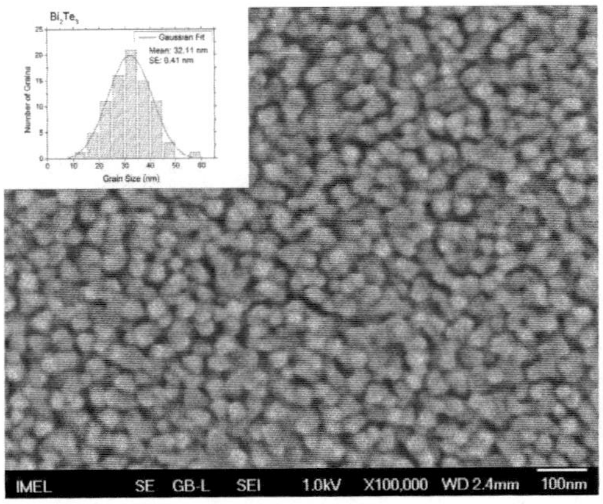

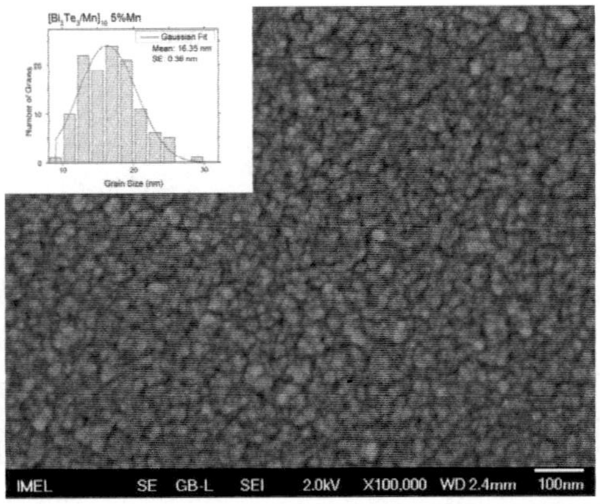

IOP Conf. Series: Journal of Physics: Conf. Series **939** (2017) 012012 doi:10.1088/1742-6596/939/1/012012

Figure 5: FESEM image of 200nm Bi_2Te_3 at 100k magnification (insert figure: distribution of grain size)

Figure 6: FESEM image of 200nm Bi_2Te_3 doped with 5% Mn at 100k magnification (insert figure: distribution of grain size)

Statistical analysis of FESEM images in fig. 5 on a single Bi_2Te_3 200nm layer and in fig. 6 on a 200nm Bi_2Te_3 thin film doped at 5% with Mn, reveal a distribution of grain sizes. The Gaussian fitting line can describe the distribution with mean values of grain sizes equal to D = 32.11nm and D= 16.35nm respectively, with standard deviations of SD=0.41nm and SD=0.36nm, and full-widths at half maximum (FWHM) of 20nm and 10nm respectively. Note that the histograms were obtained from FESEM images with a 100k magnification, and the *ImageJ* software was used for the analysis.

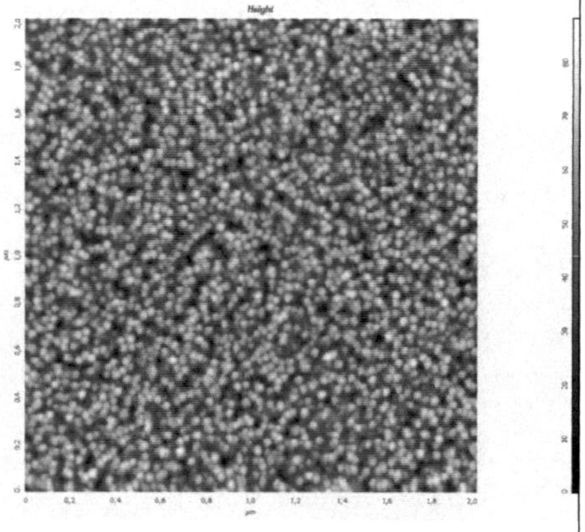

Figure 7: AFM image of 200nm Bi_2Te_3 thin film (2μm x 2μm)

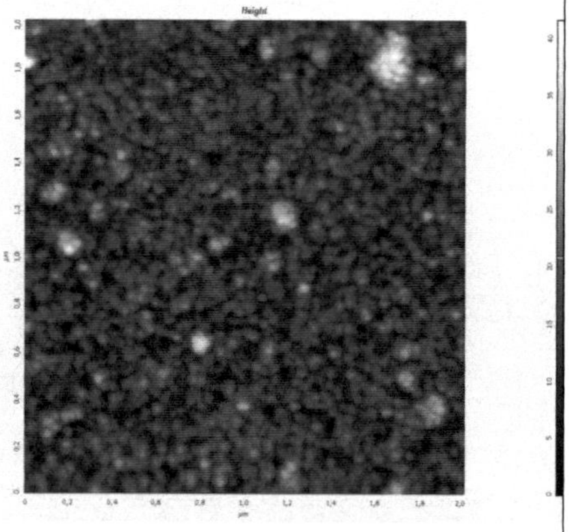

Figure 8: AFM image of 200nm Bi_2Te_3 thin film doped with 5% Mn (2μm x 2μm)

Fig. 7 and 8 show semi-contact mode AFM images from a 200nm Bi_2Te_3 thin film and from a 200nm Bi_2Te_3 thin film doped with 5% Mn, respectively. The root-mean-square (RMS) roughness (standard deviation of the height about the average value) is estimated to be about 12nm for the 200nm Bi_2Te_3 thin film and 5nm for the 200nm Bi_2Te_3 thin film doped with 5% Mn.

IOP Conf. Series: Journal of Physics: Conf. Series **939** (2017) 012012 doi:10.1088/1742-6596/939/1/012012

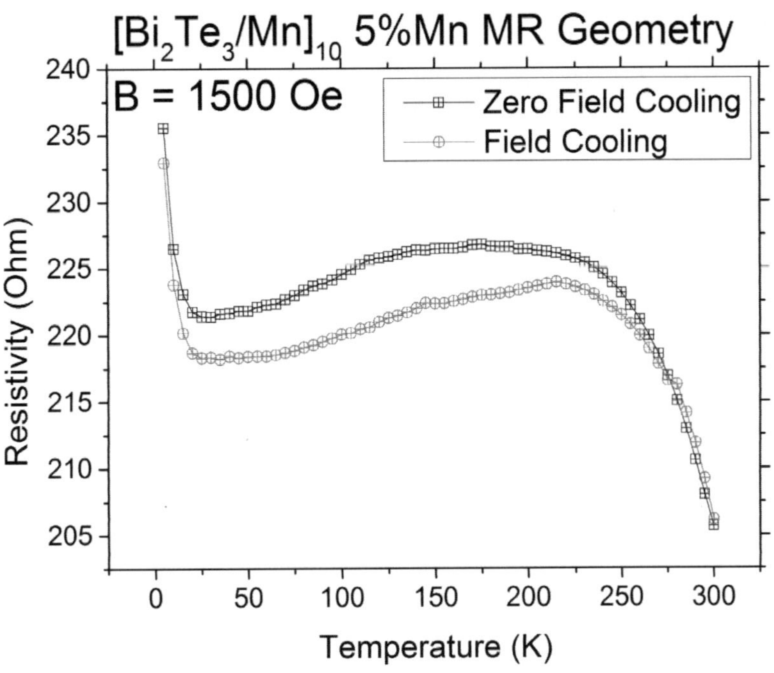

Figure 9: ZfC/FC resistivity temperature dependence

Fig. 9 shows the temperature dependence of resistivity at zero field cooling and field cooling resistivity at B = 0.15T, in MR configuration of voltage leads. The electrical resistivity (R) exhibits a semiconducting behavior below 50K and above 250K, with a plateau between these temperatures.

IOP Conf. Series: Journal of Physics: Conf. Series **939** (2017) 012012 doi:10.1088/1742-6596/939/1/012012

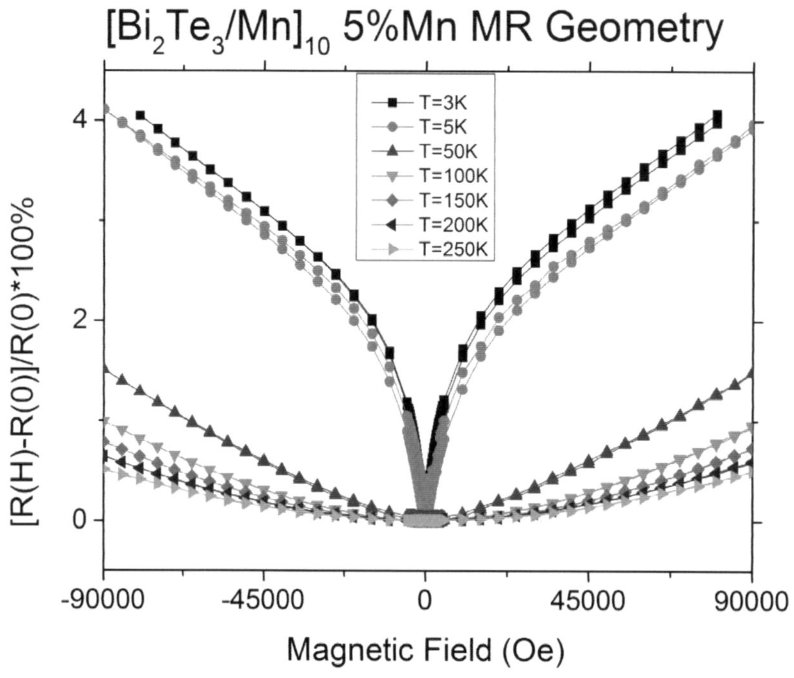

Figure 10: Field dependence of resistivity at selected temperatures, in perpendicular magnetic fields

Fig. 10 shows perpendicular magnetoresistance loops from 200nm thick Bi_2Te_3 thin films doped with 5% Mn, and grown on Si(111) substrates for temperatures from 3 to 250K. The sample was first cooled down to 3K in zero field. The magnetoresistance weakened as the temperature increased. The magnetoresistance at 3 and 5K is most likely the consequence of the weak antilocalization effect of the gapped surface states and, therefore, has a quantum interference origin[11]. As shown in Fig. 10, the perpendicular-field magnetoresistance of Bi_2Te_3 with 5% Mn on the Si(111) substrate exhibited a sharp cusp near zero field. This tip is a characteristic mark of the weak antilocalization effect for the topological surface states due to their nature as massless Dirac fermions.

IOP Conf. Series: Journal of Physics: Conf. Series **939** (2017) 012012 doi:10.1088/1742-6596/939/1/012012

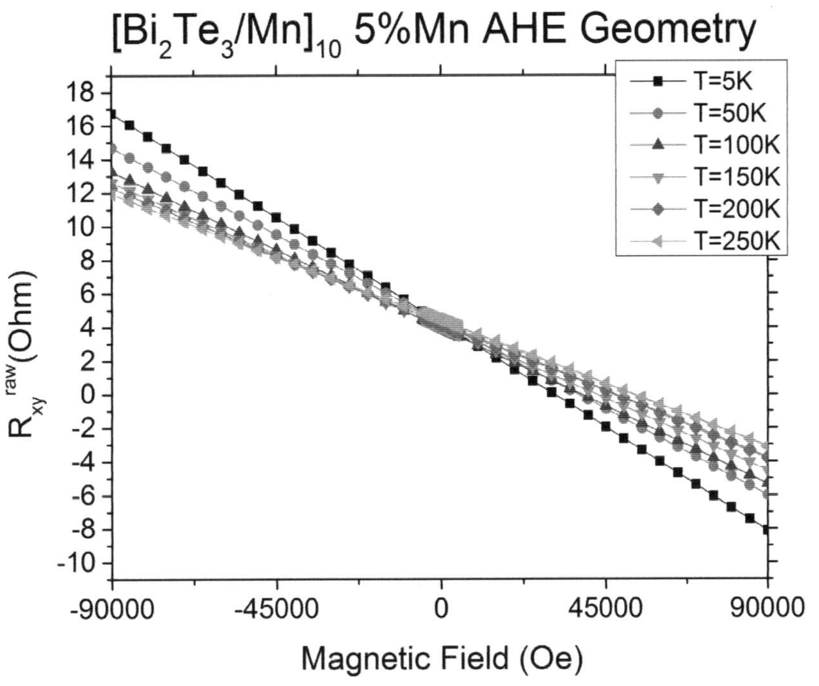

Figure 11: Field dependence of the Hall resistivity at selected temperatures

Fig. 11 presents the field dependence of the Hall resistivity of a Bi_2Te_3 thin film with 5% Mn, between 5 and 250K. The observed Hall resistivity ρ_{xy} is comprised of two contributions:

$$\rho_{xy} = R_0H + R_sM_s$$

were R_0 and R_s are the ordinary and anomalous Hall coefficients respectively, M is the magnetization and H the applied perpendicular magnetic field. The field dependence of ρ_{xy} is made up from a linear term, R_0H and a nonlinear, R_sM_s that reflects the H dependence of M. In the present system and at 5K there is no evidence of the anomalous term and superposed of an n-type H linear term. A detailed study about this will be reported elsewhere.

4. Conclusion

In this work the substitution of Mn in Bi_2Te_3 is demonstrated. From X-ray diffraction measurements, the variation of coherence length with manganese content and the shift of the (006) peak to lower angles with Mn percentage, show a clear doping dependence. The magnetoresistance results of the Mn-doped Bi_2Te_3 establish a tool for magnetic control of the spin dependent transport at the surface of a topological insulator. At 5K the anomalous Hall effect is not observed and this will be studied further with magnetization measurements.

5. References

[1] M. Z. Hasan and C. L. Kane 2010 Rev. Mod. Phys. **82**3045
[2] X. L. Qi and S.-C. Zhang 2011 Rev. Mod. Phys. **83** 1057
[3] L. Fu, C. L. Kane, and E. J. Mele 2008 Phys. Rev. Lett. **98** 106803

[4] G. A. Thomas D. H. Rapkine R. B. Van Dover L. F. Mattheiss W. A. Sunder L. F. Schneemeyer and J. V. Waszczak 1992 Phys. Rev. B **46** 1552
[5] A. R. Mellnik J. S. Lee A. Richardella J. M. Grab P. J. Mintun M. H. Fischer A. Vaezi A Manchon E. A. kim N. Samarth and D. C. Ralph 2014 Nature **511** 449
[6] S. S. Das M. Freedman and C. Nayak 2006 Phys. Today **59** 32
[7] G. E. W. Bauer A. H. MacDonald and S. Maekawa 2010 Solid State Commun. **150** 459
[8] R. Yu W. Zhang H. J. Zhang S. C. Zhang X. Dai and Z. Fang 2010 Science **329** 61
[9] W. K. Tse and A. H. MacDonald 2010 Phys. Rev. Lett. **105** 057401
[10] Minhao Liu Jinsong Zhang Cui-Zu Chan, Zuocheng Zhang Xiao Feng Kang Li Ke He Li-li Wang Xi Chen Xi Dai Zhong Fang Qi-Kun XueXucun M, and Yayu Wang 2012 Phys. Rev. Lett. **108** 036805
[11] Hai-Zhou Lu Junren Shi and Shun-Qing Shen 2011 Phys. Rev. Lett **107**076801

Temperature Dependence of the Seebeck Coefficient in Zinc Oxide Thin Films

Amirreza Noori[1a], Saeed Masoumi[1b] and Najmeh Hashemi[1c]

[1]Electronic Materials Laboratory, Electrical Engineering Department, K.N. Toosi University of Technology, Tehran 16317-14191, Iran.

[a]amirreza.noori@gmail.com, [b]saeed_masoumi86@yahoo.com, [c]najmehhashemi92@gmail.com

Abstract. Thermoelectric devices are reliable tools for converting waste heat into electricity as they last long, produce no noise or vibration, have no moving elements, and their light weight makes them suitable for the outer space usage. Materials with high thermoelectric figure of merit (zT) have the most important role in the fabrication of efficient thermoelectric devices. Metal oxide semiconductors, specially zinc oxide has recently received attention as a material suitable for sensor, optoelectronic and thermoelectric device applications because of their wide direct bandgap, chemical stability, high-energy radiation endurance, transparency and acceptable zT. Understanding the thermoelectric properties of the undoped ZnO thin films can help design better ZnO-based devices. Here, we report the results of our experimental work on the thermoelectric properties of the undoped polycrystalline ZnO thin films. These films are deposited on alumina substrates by thermal evaporation of zinc in vacuum followed by a controlled oxidation process in air carried out at the 350-500 °C temperature range. The experimental setup including gradient heaters, thermometry system and Seebeck voltage measurement equipment for high resistance samples is described. Seebeck voltage and electrical resistivity of the samples are measured at different conditions. The observed temperature dependence of the Seebeck coefficient is discussed.

Keywords: thermoelectric, Seebeck coefficient, polycrystalline semiconductor, temperature dependence, ZnO thin film.

1. Introduction

Thermoelectric (TE) effect directly converts heat to electricity. TE devices have recently attracted considerable attention because of their potential applications in cooling units, power generator, sensors, and waste heat recovery system [1,2]. TE devices are advantages in that they have no mechanical moving parts, no greenhouse gas release, good stability, high reliability and long life. These features make TE devices suitable choices for portable, military and aerospace applications [3-5].

The performance of a TE device depends on the TE material; hence, for decades, researchers have tried to figure out the most efficient TE materials. The efficiency among TE materials is quantitatively compared based on a dimensionless TE figure of merit, defined as $zT = \alpha^2 T/\rho\kappa$, where in α, T, ρ and κ are the Seebeck coefficient of the material, absolute temperature, electrical resistivity and thermal conductivity, respectively [6-9]. An ideal TE material should have large Seebeck coefficient to produce large voltage, low electrical resistivity to reduce Joule heating during operation, and low thermal conductivity to allow establishing temperature differences. Seebeck effect describes the electrical potential difference in terms of the temperature difference across the material. The ratio $\alpha = dV/dT$ is called the Seebeck coefficient. There are two main methods for Seebeck coefficient measurement: differential and integral. In the differential method, a small temperature difference is applied across the sample and the voltage difference across the sample is measured while the absolute temperature of the sample is increasing. In the integral method, the voltage difference across the sample is measured while the temperature difference across the sample is increased [10,11].

Metal oxide semiconductors are promising candidates as TE materials [12-14]. Zinc Oxide is an important II-IV metal oxide semiconductor with wurtzite crystal structure and a good candidate TE material owing to its high chemical stability, high electrical conductivity and acceptable Seebeck coefficient [15]. Oxide semiconductors, because of their remarkable characteristics have various applications such as transparent conductor for solar cells/displays, surface acoustic wave devices, chemical sensors and etc. [16-18]. Here, we report the results of our experimental work on the thermoelectric properties of the undoped polycrystalline ZnO thin films. It is shown that as the temperature increases, the Seebeck coefficient of ZnO increases in absolute value.

2. Experimental

ZnO thin films are deposited on 10 mm × 5 mm alumina substrates by the thermal evaporation of zinc powder in vacuum followed by the controlled oxidation process in air carried out at the 350-500 °C temperature range. Two different types of samples are fabricated, which are produced in the same conditions, but are different in the utilized oxidation temperature. Sample A is oxidized at 450 °C but sample B has experienced 500 °C. Oxidation temperature profiles utilized are shown in figure 1. After oxidation, both samples are transparent indicating the completion of the oxidation process [19].

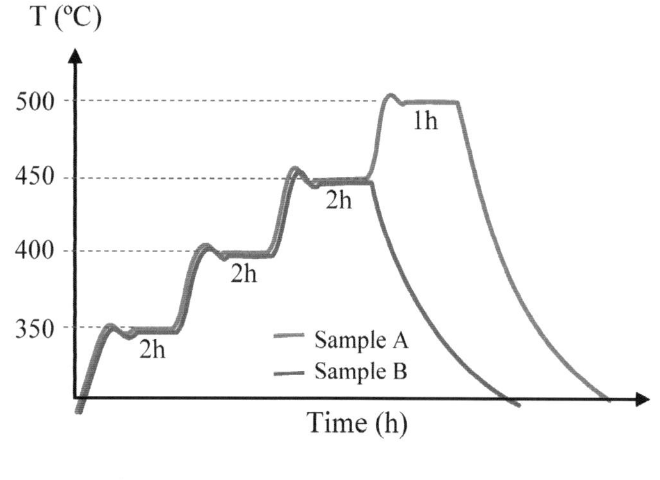

Figure 1. Oxidation temperature profile of the zinc layers deposited on alumina.

A chromium thin film microheater is deposited on the back side of the alumina substrate covering a third of its length. By passing different currents from the heater, temperature differences of different magnitude are created across the sample. All contacts to the microheater and ZnO thin film are made by silver paste printing and silver wire segments (figure 2.a and b) [20,21]. A Ni-Cr wire heater placed on the sample holder under the sample can increase the average sample temperature. This will allow sample resistance measurements at different temperatures. Hence, we are using two different heaters; the first one determines the average temperature of the sample and the second heats up the one third of the sample to create the temperature gradient and generate Seebeck voltage.

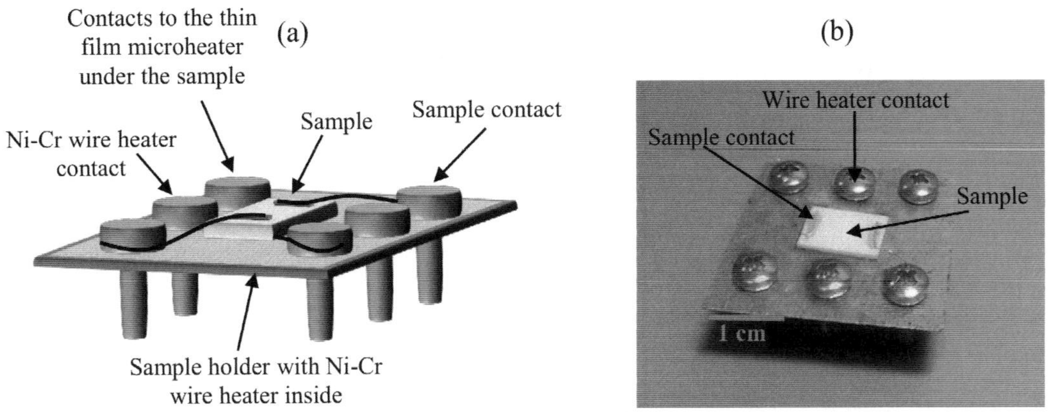

Figure 2. (a) The schematic diagram of the sample positioned on the sample holder. (b) The photograph of the ZnO thin film sample on the sample holder.

The main requirement for an accurate Seebeck coefficient measurement is to make the temperature and voltage readouts simultaneous from the same locations on the sample [22]. To meet this condition, two fine type S thermocouples are placed on ZnO electrical contacts as show in figure 3.b.

The prepared ZnO samples' resistance is in the 10^6 Ω range. Thus, the correct measurement of the Seebeck voltage across our samples requires an ultra-high input impedance voltmeter [23,24]. In the case of a high resistance sample, an alternative method is used in the present work, which is schematically presented in figure 3 [25]. As shown in figure 3.a, a variable power supply and a resistor are connected in series with the sample. Seebeck voltage along the sample is measured by adjusting the voltage source V_s (see figure 3.a) manually till the voltage drop over the resistor R becomes zero. At this point, the power supply voltage is equal to the Seebeck voltage [26].

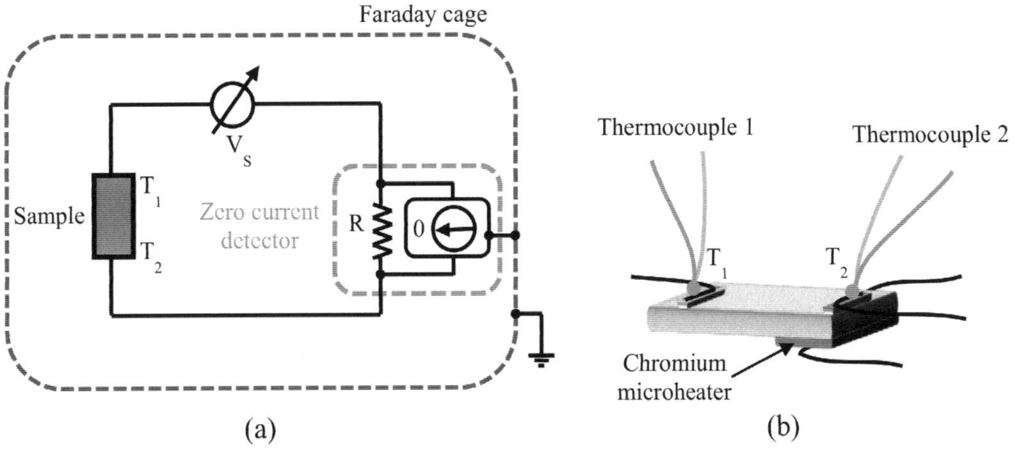

Figure 3. (a) The circuit utilized for the measurement of the Seebeck voltage. (b) The schematic diagram of the sample.

To measure the sample resistance at different temperatures, a dc voltage is applied to Ni-Cr wire heater. The sample temperature becomes stable in a few minutes. The resistance of sample is, then, recorded after 2 minutes. The temperature difference across the sample is require for Seebeck coefficient measurements. This is created by applying a dc voltage to the chromium microheater. Time is allowed for the temperature profile along the sample to become stable. The Seebeck voltage is read a few minutes after this stability.

3. Results and discussion

The variations of the resistance and Seebeck coefficients of the undoped polycrystalline ZnO samples are shown in figure 4.a-c. In figure 4.a the resistance variations in the A and B samples are presented, showing that the span of changes is wider in the A samples. The Arrhenius diagram for resistance is shown in figure 4.b. The conduction activation energy deduced for the sample is 0.50 ± 0.05 eV. As shown in figure 4.b, the activation energy of sample B is marginally lower than that of the sample A. This means that by increasing the oxidation temperature, carriers meet a slightly smaller energy barrier in passing from one grain to another.

Figure 4.c shows the variation of the Seebeck voltage versus the temperature difference across the sample. The slope of the plotted line in figure 4.c determines the Seebeck coefficient. It is clear that the Seebeck coefficient of sample B is larger than that of the sample A. The variations of the Seebeck coefficient of ZnO samples with respect to the average sample temperature is presented in figure 4.d, showing that the Seebeck coefficient in both sample categories increase in absolute value with temperature. That is more profound in the B sample.

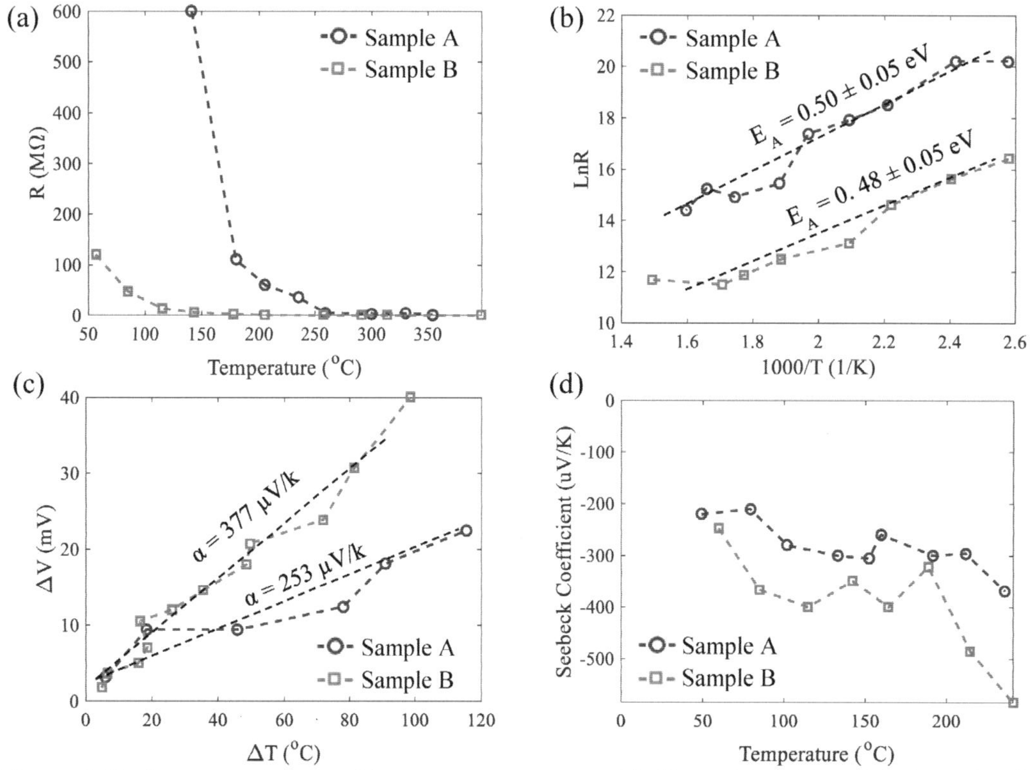

Figure 4. Resistance variations with temperature presented in linear (a) and logarithmic scales (b). (c) Seebeck voltage variations with respect to the established temperature differences across the sample and (d) the variations of the measured Seebeck coefficient with regards to the average sample temperature.

4. Conclusion

The maximum annealing temperature of the ZnO thin films is important in determining the Seebeck effect and zT of the undoped polycrystalline ZnO thin films. In fact, this parameter affects both electrical resistivity and Seebeck coefficient. Samples with higher annealing temperature of 500 °C have lower resistance and higher Seebeck coefficient than those annealed at 450 °C. Therefore, the figure of merit in B sample becomes significantly higher than that in A sample.

References

[1] Elsheikh MH, Shnawah DA, Sabri MFM, Said SBM, Hassan MH, Bashir MBA, Mohamad M 2014 A review on thermoelectric renewable energy: Principle parameters that affect their performance *Renewable and Sustainable Energy Reviews* **30** 337-355

[2] Zhao D, Tan G 2014 A review of thermoelectric cooling: materials, modeling and applications *Applied Thermal Engineering* **66** 15-24

[3] de Rochemont LP 2013 *US Patent App.* 13/602,019

[4] Tritt TM 2011 Thermoelectric phenomena, materials, and applications *Annual review of materials research* **41** 433-448

[5] Yang J, Caillat T 2006 Thermoelectric materials for space and automotive power generation *MRS bulletin* **31** 224-229

[6] Rowe DM. CRC handbook of thermoelectrics: CRC press; 1995

[7] Snyder GJ, Toberer ES 2008 Complex thermoelectric materials *Nature materials* **7**(2)

[8] Vineis CJ, Shakouri A, Majumdar A, Kanatzidis MG 2010 Nanostructured thermoelectrics: big efficiency gains from small features *Advanced Materials* **22**(36)

[9] Dresselhaus MS, Chen G, Tang MY, Yang R, Lee H, Wang D, Ren Z, Fleurial JP, Gogna P 2007 New Directions for Low-Dimensional Thermoelectric Materials *Advanced Materials* **19** 1043-1053

[10] Wood C, Chmielewski A, Zoltan D 1988 Measurement of Seebeck coefficient using a large thermal gradient *Review of scientific instruments* **59** 951-954

[11] Wold PI 1916 The hall effect and allied phenomena in tellurium *Physical Review* **7** 169

[12] Hossein-Babaei F, Lajvardi MM, Boroumand FA 2012 Large area Ag–TiO 2 UV radiation sensor fabricated on a thermally oxidized titanium chip *Sensors and Actuators A: Physical* **173**(1) 116-121

[13] Hossein-Babaei F, Alaei-Sheini N 2016 Electronic Conduction in Ti/Poly-TiO2/Ti Structures *Scientific Reports* **6** 29624

[14] Hossein-Babaei F, Lajvardi MM, Alaei-Sheini N 2015 The energy barrier at noble metal/TiO2 junctions *Applied Physics Letters* **106**(8) 083503-083506

[15] Berger LI. Semiconductor materials: CRC press; 1996

[16] Ko SH, Lee D, Kang HW, Nam KH, Yeo JY, Hong SJ, Grigoropoulos CP, Sung HJ 2011 Nanoforest of hydrothermally grown hierarchical ZnO nanowires for a high efficiency dye-sensitized solar cell *Nano letters* **11** 666-671

[17] Hossein-Babaei F, Paknahad M, Ghafarinia V 2012 A miniature gas analyzer made by integrating a chemoresistor with a microchannel *Lab on a Chip* **12**(10) 1874-1880

[18] Hossein-Babaei F, Amini A 2014 Recognition of complex odors with a single generic tin oxide gas sensor *Sensors and Actuators B: Chemical* **194** 156-163

[19] Chen S, Liu Y, Ma J, Zhao D, Zhi Z, Lu Y, Zhang J, Shen D, Fan X 2002 High-quality ZnO thin films prepared by two-step thermal oxidation of the metallic Zn *Journal of Crystal Growth* **240** 467-472

[20] Hossein-Babaei F, Moghadam S, Masoumi S 2015 Forming ohmic Ag/SnO 2 contacts *Materials Letters* **141** 141-144

[21] Ghalamboran M, Saedi Y 2016 TiO2-TiO2 composite resistive humidity sensor: ethanol crosssensitivity *IOP Conference Series: Materials Science and Engineering* **108** 012039

[22] Martin J, Tritt T, Uher C 2010 High temperature Seebeck coefficient metrology *Journal of Applied Physics* **108** 121101

[23] Srikant V, Clarke DR 1998 On the optical band gap of zinc oxide *Journal of Applied Physics* **83** 5447-5451

[24] Keem J, Honig J 1975 Seebeck measurements and their interpretation in high-resistivity materials—the case of semiconducting V2O3 *Physica Status Solidi (a)* **28** 335-343

[25] Cai H, Cui D, Li Y, Chen X, Zhang L, Sun J 2013 Apparatus for measuring the Seebeck coefficients of highly resistive organic semiconducting materials *Review of Scientific Instruments* **84** 044703

[26] Hossein-Babaei F, Masoumi S 2014 Electrical Resistance and Seebeck Effect in Undoped Polycrystalline Zinc Oxide *Key Engineering Materials* **605** 185-188

Diffusion Mechanisms of Ag atom in ZnO crystal: A First Principles Study

Saeed Masoumi[1a], Amirreza Noori[1b] and Ebrahim Nadimi[2c]

[1]Electronic Materials Laboratory, Electrical Engineering Department, K.N. Toosi University of Technology, Tehran 16317-14191, Iran.
[2]Centre of Micro and Nanoelectronic Modeling, Electrical Engineering Department, K.N. Toosi University of Technology, Tehran 16317-14191, Iran

[a]saeed_masoumi86@yahoo.com, [b]amirreza.noori@gmail.com, [c]nadimi@kntu.ac.ir

Abstract. Zinc oxide (ZnO) is currently under intensive investigation, as a result of its various applications in micro, nano and optoelectronics. However, a stable and reproducible p-type doping of ZnO is still a main challenging issue. Group IB elements such as Au, Cu and Ag, are promising candidates for p-type doping. Particularly, Ag atoms has been shown to be able to easily diffuse through the crystal structure of ZnO and lead to the p-type doping of the host crystal. However, the current understanding of Ag defects and their mobility in the ZnO crystal is still not fully explored. In this work, we report the results of our first-principles calculations based on density functional theory for Ag defects, particularly the interstitial and substitutional defects in ZnO crystal. Defect formation energies are calculated in different charged states as a function of Fermi energy in order to clarify the p-type behaviour of Ag-doped ZnO. We also investigate the diffusion behaviour and migration paths of Ag in ZnO crystal in the framework of density functional theory applying climbing image (CI) nudged elastic band method (NEB).

Keywords: Density functional theory, First-principles, ZnO, P-type, Ag doping, Formation energy, Diffusion barrier, Nudged elastic band.

1. Introduction

Different aspects of oxide semiconductors, owing to their wide and direct energy band gaps and considerable chemical stability at high temperatures and harsh environments are being considered for many electronic applications [1-3]. The main shortcoming of these oxides in this field is the fact that their p-type alternatives have remained elusive [4]. The wurtzite-ZnO, with a wide band gap (3.4 eV) and large exciton binding energy (60 meV), has been considered as an interesting material in wide range of applications in micro, nano and optoelectronics, as well as sensors technology and energy conversion system [5]. ZnO generally shows n-type behavior due to high concentration of oxygen vacancies in fabricated samples. However, in different applications such as light emitting diode (LED) and laser devices having a p-type ZnO is also necessary. Therefore, the fabrication and characterization of p-type ZnO has been the subject of many recent experimental and theoretical works [6].

In spite of many effort to obtain high mobility and high quality low resistance p-type ZnO, they have not been successful except in specific conditions in theoretical and experimental works [7, 8]. The common studied dopants are group-V atom (N, P and As) and group-I atom (Li, Na, K, Au, Cu and Ag) which substitute on the O and Zn sites in ZnO crystal lattice, respectively [9, 10]. However, there are differences and discrepancies among theoretical and experimental results and a general understanding has not yet been achieved. One of the dopant elements which recently has attracted attentions is Ag, which is promising for the achievement of high quality and stable p-type ZnO [11]. The diffusion and stability properties of Ag dopant in ZnO crystal structure, particularly Ag interstitial defect, are very important in order to reach a comprehensive understanding of the p-type behavior of Ag-doped ZnO [12].

In this work, we report the results of our first-principles calculations based on density functional theory for Ag defects, particularly the interstitial defect (Ag_i) in ZnO crystal structure. Four configurations of Ag_i are considered in ZnO crystal structure. The defect formation energy is calculated in different charged states as a function of the Fermi energy in order to determine the most stable site of Ag interstitial defects. We also investigate the diffusion behavior and migration paths of Ag in ZnO crystal in the framework of density functional theory applying climbing image (CI) nudged elastic band method (NEB).

2. Calculation Method

Density functional calculations were carried out using the plane-wave-based Quantum Espresso code [13], based on generalized-gradient approximation (GGA) with the functional of Perdew, Burke and Ernzerhof (PBE) [14]. Norm conserving pseudopotentials were used for all elements considering the $3d^{10}4s^2$, $2s^22p^4$ and $4d^{10}5s^1$ electronic configuration for Zn, O and Ag atoms, respectively. A cutoff energy of 70 Ry was used for real space sampling, whereas the Brillouin zone was sampled with a 6×6×3 k-point grid for ZnO unit cell. It has been shown that a large 3×3×2 supercell with 72 atoms is required to reduce the interaction between defects with their images in the neighbor cells. Thus, the defect calculations were performed in a 3×3×2 supercell with a 2×2×2 k-point sampling. The convergence threshold on all forces is 10^{-6} in relax calculation. The calculated lattice parameters of ZnO are, a= 3.3088 Å, c/a=1.6123 and Zn-O=2.021 Å in good agreement with the experimental values of a=3.25 Å, c/a=1.60 and Zn-O=1.97 Å [15].

The required energy to generate a defect in the perfect lattice is introduced as defect formation energy. The Ag interstitial defects are constructed by adding an Ag atom in the 3×3×2 ZnO supercell. The formation energy was calculated from the following equation:

$$E_{Formation} = E_{Total}^{Defect} - \left[E_{Total}^{Perfect} \pm \sum_i \mu_i\right] + q\left(E_v + E_f\right) \quad (1)$$

Where $E_{Total}^{Perfect}$ and E_{Total}^{Defect} are the total energies of perfect and defected supercells, μ_i is the chemical potential of removed or added elements. q is the fundamental electron charge, E_v is the energy of the valance band edge and E_f is the Fermi energy which could vary between the valence and conduction band edges. The climbing image nudged elastic band method (CI-NEB) [16], as implemented in quantum espresso is used to calculate the energy barrier for Ag_i diffusion path. The images of the CI-NEB were relaxed until the maximum residual force was less than 0.001 eVÅ^{-1}.

3. Results

3.1. Structural configurations of Ag interstitial

The relaxed ZnO supercell structure with 72 atom is shown in figure 1.a.Four different configurations of Ag_i defects are considered to determine the most energetic stable one in ZnO crystal structure. The relaxed supercell structure of the interstitial defects are depicted in figure 1.b-e. These structures show the deformation around the interstitial Ag atom.

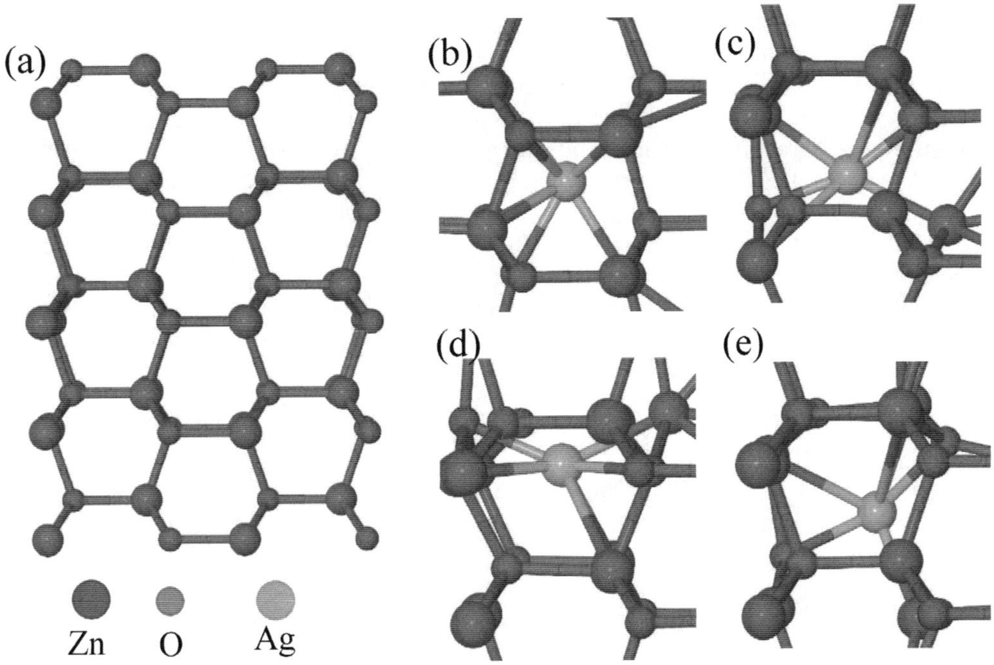

Figure 1. (a) The wurtzite structure of ZnO 3×3×2 supercell, (b) configuration-1 of Ag_i, (c) configuration-2 of Ag_i, (d) configuration-3 of Ag_i, (d) configuration-4 of Ag_i.

3.2. *Ag interstitial formation energy*

The formation energy of defects which are demonstrated in figure 1 are calculated based on equation (1). The formation energies in neutral and charged state are shown in figure 2 as a function of the Fermi energy. Among the different configurations the (d) configuration shows the lowest formation energy and therefore could be considered as the most probable interstitial site for Ag atoms.

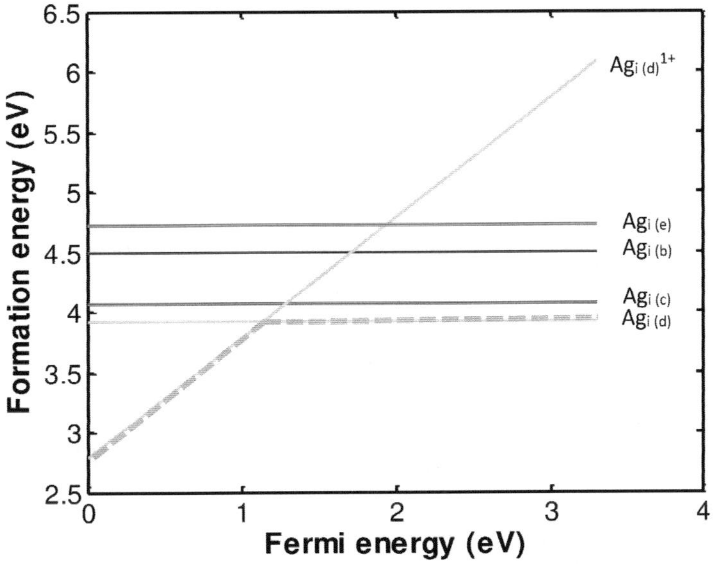

Figure 2. The calculated formation energy of Ag_i for different configurations. Configuration-1 ($Ag_{i\,(b)}$), configuration-2 ($Ag_{i\,(c)}$), configuration-3 ($Ag_{i\,(d)}$), configuration-4 ($Ag_{i\,(e)}$).

IOP Conf. Series: Journal of Physics: Conf. Series **939** (2017) 012014 doi:10.1088/1742-6596/939/1/012014

3.3. Ag interstitial diffusion

Although the Ag ion has been considered as an extremely mobile dopant in metal oxide semiconductor [17, 18] as well as ZnO crystal structure, understanding the diffusion mechanisms of Ag dopant is very important to find out the p-type behavior of ZnO. In this work, we investigate the diffusion behavior of neutral Ag_i atoms through a minimum energy path perpendicular to the c axes of ZnO crystal structure. We calculated the energy barrier, which an interstitial Ag atoms have to overcome in order to move through the crystal structure. The diffusion path of Ag interstitial is illustrated in figure 3. The calculated energy along this diffusion path is shown in figure 4. It is obvious that the highest energy barrier of this diffusion path is 0.36 eV which is in good agreement with previous work [12]. The calculated diffusion barrier shows that this defect can easily diffuse in ZnO lattice. These results indicate that high quality p-type ZnO could be possibly fabricated by implantation or diffusion doping of Ag atoms.

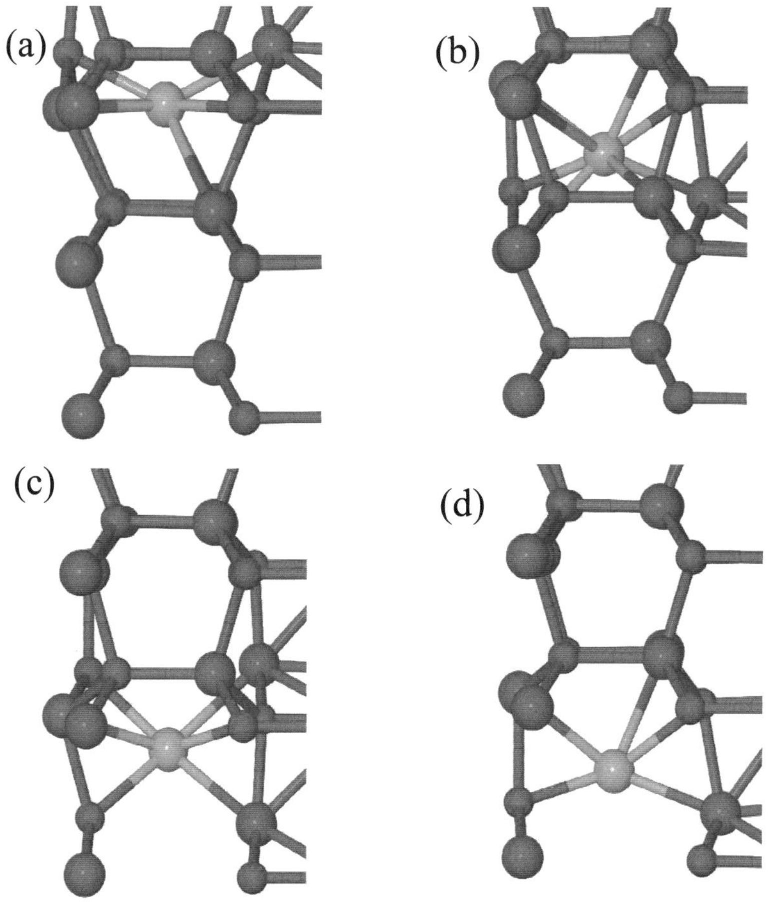

Figure 3. Diffusion path of Ag_i perpendicular to the c axes of ZnO crystal structure in four Image. (a) Image 1, (b) Image 2, (c) Image 3, (d) Image 4.

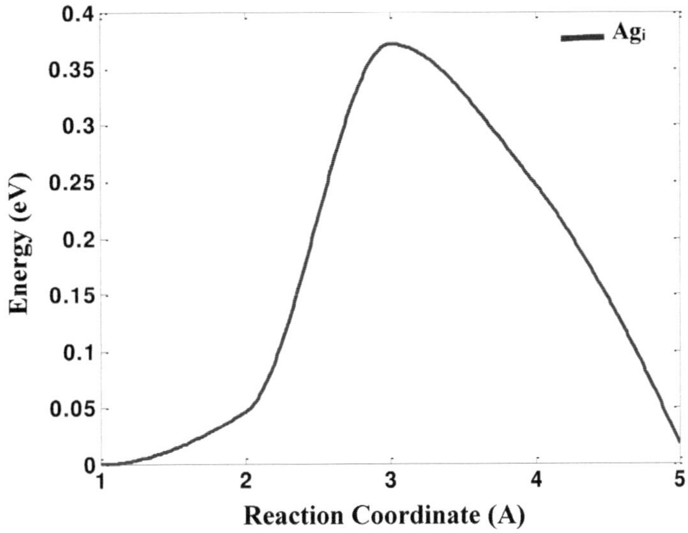

Figure 4. Calculated energy along the migration paths perpendicular to the c axis in the same octahedral interstitial cage for Ag.

4. Conclusion

We report the results of our first-principles calculations based on density functional theory for Ag defect, particularly the interstitial defect (Ag_i) in ZnO crystal structure. Four configurations of Ag_i are considered in ZnO crystal structure and defects formation energies are calculated in neutral and charged states as a function of the Fermi energy in order to determine the most stable site of Ag interstitial defects. The results show that the (d) configuration with a formation energy of 2.4 eV is the most stable one. The calculated diffusion barrier for this stable Ag_i defect perpendicular to the c axis of ZnO crystal is 0.36 eV in good agreement with the previous work [12]. This low energy barrier could lead to an easy diffusion of Ag dopant in ZnO lattice, which in turn could pave the way for fabricating p-type ZnO.

References

[1] Hossein-Babaei F, Alaei-Sheini N and Lajvardi M M 2015 Oxygen adsorption at noble metal/TiO2 junctions *Appl. Phys. Lett* **106** 083503.
[2] Hossein-Babaei F, Lajvardi M M and Boroumand F A 2012 Large area Ag–TiO 2 UV radiation sensor fabricated on a thermally oxidized titanium chip *Sens. Actuators, A* **173** 116.
[3] Hossein-Babaei F and Alaei-Sheini N 2016 Electronic Conduction in Ti/Poly-TiO2/Ti Structures *Sci. Rep.* **6** 29624.
[4] Hossein-Babaei F, Keshmiri M, Kakavand M and Troczynski T 2005 A resistive gas sensor based on undoped p-type anatase *Sens. Actuators, B* **110** 28.
[5] Hossein-Babaei F and Masoumi S 2014 Electrical Resistance and Seebeck Effect in Undoped Polycrystalline Zinc Oxide *Key Eng. Mater.* **605** 185.
[6] Fana J C, Sreekantha K M, Xied Z, , Change S L, , Raoa K V 2013 p-Type ZnO materials: Theory, growth, properties and devices *Prog. Mater. Sci.* **58** 874.
[7] Look D C, Reynolds D C, Litton C W, Jones R L, Eason D B and Cantwell G 2002 Characterization of homoepitaxial p-type ZnO grown by molecular beam epitaxy *Appl. Phys. Lett* **81** 1830.
[8] Liu L, Xu J, Wang D, Jiang M, Wang S, Li B and Zhang Z 2012 p-Type Conductivity in N-Doped ZnO: The Role of the NZn-VO Complex *Phys. Rev. Lett.* **108** 215501.

IOP Conf. Series: Journal of Physics: Conf. Series **939** (2017) 012014 doi:10.1088/1742-6596/939/1/012014

[9] Wardle M G, Goss J P and Briddon P R 2005 Theory of Li in ZnO: A limitation for Li-based p-type doping *Phys. Rev. B* **71** 155205.
[10] Catellani A, Ruini A, Cicero G and Calzolari A 2013 First principles description of the electronic properties of doped ZnO *Phys. Status Solidi B* **250** 2106.
[11] Li Y, Zhao X and Fan W 2010 Structural, Electronic, and Optical Properties of Ag-Doped ZnO Nanowires: First Principles Study *J. Phys. Chem. C* **115** 3552.
[12] Huang G, Wang C, and Wang J 2009 First-principles study of diffusion of Li, Na, K and Ag in ZnO *J. Phys.: Condens. Matter* **21** 345802.
[13] Giannozzi P, Baroni S, Bonini N, Calandra M, Car R, Cavazzoni C, Ceresoli D, Chiarotti G L, Cococcioni M, Dabo I, Corso A D, Gironcoli S D, Fabris S, Fratesi G, Gebauer R, Gerstmann U, Gougoussis C, Kokalj A, Lazzeri M, Martin-Samos L, Marzari N, Mauri F, Mazzarello R, Paolini S, Pasquarello A, Paulatto L, Sbraccia C, Scandolo S, Sclauzero G, Seitsonen A P, Smogunov A, Umari P and Wentzcovitch R M 2009 QUANTUM ESPRESSO: a modular and open-source software project for quantum simulations of materials *J. Phys.: Condens. Matter* **21** 395502.
[14] Perdew J P, Burke K and Ernzerhof M 1997 Generalized Gradient Approximation Made Simple *Phys. Rev. Lett.* **77** 3865.
[15] Sowa H, Ahsbahs H. 2006 High-pressure X-ray investigation of zincite ZnO single crystals using diamond anvils with an improved shape *J. Appl. Crystallogr.* **39** 169.
[16] Henkelman G, Uberuaga B P and Jónsson H 2000 A climbing image nudged elastic band method for finding saddle points and minimum energy paths *J. Chem. Phys.* **113** 9901.
[17] Hossein-Babaei F, Moghadam S and Masoumi S 2015 Forming ohmic Ag/SnO 2 contacts *Mater. Lett.* **141** 141.
[18] Hossein-Babaei F, 2008 Porous silver-TiO 2 Schottky-type chemical sensor fabricated on thermally oxidised titanium *Electron. Lett.* **44** 161.

IOP Conf. Series: Journal of Physics: Conf. Series **939** (2017) 012015 doi:10.1088/1742-6596/939/1/012015

Electronic properties of hybrid monolayer-multilayer MoS_2 nanostructured materials

Vladan Mlinar

Research Institute for Advanced Materials Design, Providence, RI 02906, USA

E-mail: vladan.mlinar@riamd.org

Abstract. The remarkable, layer-dependent properties of molybdenum disulphide (MoS_2), such as an appropriately small and sizable bandgap or interplay between spin and the valley degrees of freedom, make it an attractive candidate for photodetectors, electrominescent devices, valleytronic devices, etc. Using nanostructuring to manipulate the size in lateral direction or number of layers of MoS_2, we are opening a new playground for exploring and tuning properties of such systems. Here, we theoretically study the electronic properties of nanostructured MoS_2 systems consisting of monolayer and multilayer MoS_2 regions. In our analysis we consider hybrid systems in which monolayer region is surrounded by multilayer region and vice versa. We show how energy spectra and localization of carriers are influenced by the size and shape of the regions in lateral direction, number of MoS_2 layers in the multilayer region, and the edge structure. Finally, we demonstrate how to control localization of carriers in these hybrid systems, which could make them appealing candidates for optoelectronic devices. Our findings are extracted from a tight-binding model that includes non-orthogonal sp^3d^5 orbitals, nearest-neighbor hopping matrix elements, and spin-orbit coupling.

1. Introduction

Molybdenum disulphide (MoS_2), a transition metal dichalcogenide semiconductor, is a layered material in which each layer consists of hexagonal plane of Mo atoms sandwiched between two hexagonal planes of S atoms [1, 2, 3]. Layers are bounded by weak van der Waals type interactions, enabling a relatively straightforward exfoliation into two-dimensional (2D) monolayers. Consequently, MoS_2 exhibits remarkable, layer-dependent features, which makes it a promising candidate for a range of applications [2, 3, 4, 5, 6].

For example, *in the monolayer limit,* MoS_2 has a 1.8 eV direct band gap, located at the corners of the Brillouin zone, strong spin-orbit splitting, which is caused by the high mass of Mo and S, lack of inversion symmetry and confinement of carriers in a 2D layer, and also exhibit inequivalent valleys in the electronic structure. This offers prospects for optoelectronic-device applications [3], spintronic devices [4], or valleytronics [4, 3, 2].

Bilayer and multilayer MoS_2, have indirect band gaps. Interestingly, in the case of bilayer, the band gap can be tuned by an electric field applied perpendicular to the bilayer [7]. Multilayer MoS_2 have found their application in field-effect transistors [8, 5], where high mobilities and robust current saturation in a large voltage window were reported [8].

Furthermore, there have been experimental and theoretical studies on MoS_2 nanostructured materials. Examples include, mono- and multi-layer triangular nanoplatelets with side length up to 3.4 nm [9, 10], nano-flakes [5], and mono- and multi-layer nanowire arrays [11]. Actually,

IOP Conf. Series: Journal of Physics: Conf. Series **939** (2017) 012015 doi:10.1088/1742-6596/939/1/012015

nanostructuring opens the whole new playground to engineer MoS_2-based systems with different size and shape in lateral direction and varying number of layers within the structure.

Here, we theoretically study the electronic properties of nanostructured MoS_2 systems consisting of monolayer and multilayer MoS_2 regions. In our analysis we consider hybrid systems in which monolayer region is surrounded by multilayer region and vice versa. Theoretical predictions of electronic properties in those systems can provide guidelines for further experimental studies and potential practical optoelectronic applications [2, 6], including their targeted design [12]. Given an immense progress in fabrication techniques of reduced dimensionality MoS_2 nanostructures, it not unreasonable to expect experimental realization and characterization of basically any arbitrary-shaped nanostructured MoS_2 materials.

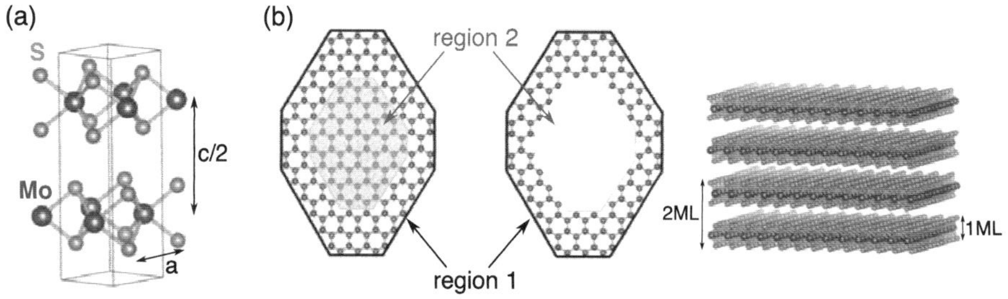

Figure 1. (a) MoS_2 structure, a and c are lattice parameters. Black lines depict the unit cell of bulk MoS_2; (b) model of hybrid monolayer-multilayer MoS_2 nanostructured materials. The structures are characterized by two regions, 1ML and a multilayer region, specified by their size and shape and number of layer in the case of multilayer region.

2. Method

Model of the atomic structure of pristine MoS_2 is shown in Figure 1(a). The bulk compound has 2H-MoS_2 structure with two MoS_2 layers per unit cell. Fig. 1(b) depicts model hybrid monolayer-multilayer MoS_2 nanostructured materials analyzed in this work.

To resolve electronic structure of these hybrid systems, we employ an, in-house developed, tight-binding (TB) model that includes *non-orthogonal* sp^3d^5 orbitals and spin-orbit coupling [13, 14, 15, 16]. For intra-layer S-S, Mo-S, and Mo-Mo interactions we consider nearest-neighbor and for inter-layer S-S interactions the second nearest neighbor hopping and overlap matrix elements. The spin-orbit interaction is assumed to arise only from the coupling of intra-atomic states with nonzero angular momentum [17, 18]. For our calculations we use a set of 96 TB parameters proposed by Zahid *et al.* [19, 20], which were fitted to the band structure of monolayer (1ML), bilayer (2ML), and bulk MoS_2 obtained from density functional theory. Our calculated band structures for 1ML and 2ML pristine MoS_2 are given in the Appendix.

3. Results and Discussion

Figure 2 shows calculated electronic properties of periodic arrays of MoS_2 nanostructured materials and holes. Fig. 2(a) shows the calculated total electronic density of states (DOS) for pristine monolayer MoS_2 [top panel] and how it changes when we introduce periodic array of 2ML disks [middle panel], and periodic array of holes [bottom panel]. Calculated electronic structure for 1ML of MoS_2 with a periodic array of 2ML disks and holes is given in Fig. 2(b), left and right panels, respectively.

We see that presence of periodic arrays of either disks or holes changes states in the band gap, making these hybrid systems metallic-like. Also, the periodic array of holes has more pronounced effect on DOS than the periodic array of 2ML disks.

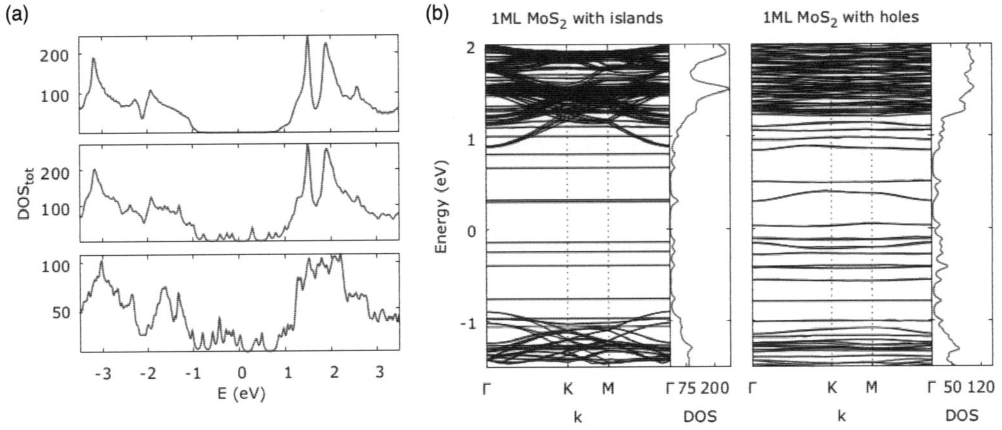

Figure 2. Calculated DOS (a) and electronic structure (b) of periodic arrays of MoS_2 nanostructured materials and holes (see text). Note that the energy range shown in (b) is smaller than in (a). In our calculation we used 6×6 supercell size, where 5×5 part is mapped to geometry shown in Fig. 1.

Figure 3 shows the calculated DOS of hybrid monolayer-multilayer MoS_2 nanostructured isolated disks [shape shown in Fig. 1]. If the region 2 is identical to the region 1 [see Fig. 1(b)], disk becomes uniform 1ML or 2ML disk, with calculated DOS shown in Fig.3(a). Analysis of

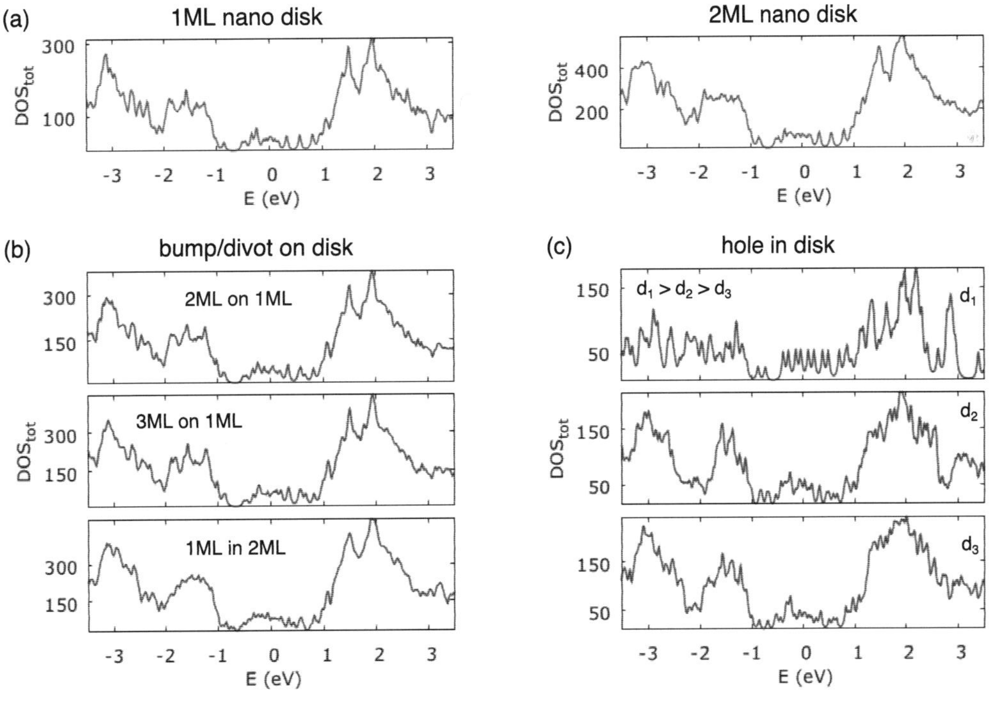

Figure 3. (a) DOS of isolated MoS_2 1ML (left) and 2ML thick nano disk; (b) Influence of bump and divot in the region 2 of the disk [Fig. 1(b)] on DOS of MoS_2 disk (see text); (c) Calculated DOS of nano disk, where the region 2 is hole, d_1, d_2, and d_3 are different sizes of the hole.

partial DOS (not shown here) reveals that edge structure of the disk is responsible for those DOS variations. These findings are also in the agreement with the conclusions of Ref. [10].

Next, if we introduce 2ML and 3ML bump on a 1ML disk, or 1ML divot on 2ML disk, we obtain DOS as shown in Fig. 3(b). Compared to pure nano disks, we notice small differences. Thus, for the analyzed systems, the electronic structure is modified by the confinement of the nano disk, but not significantly influenced by the variation of the thickness in the region 2 of the structure.

On the other hand, DOS is significantly altered by adopting the region 2 to be a hole, Fig. 1(b). We vary the size of the hole, changing it from d_1, order of a, via $d_2 \sim 4a$, to d_3, which is chosen to be as large as possible not to break to break the ring of MoS_2 atoms. We see that the presence of the hole significantly changes DOS. Simply, this is a direct consequence of making the edge states much more dominant compared to the case of the disk.

In the case of hybrid monolayer-multilayer MoS_2 nanostructured isolated disks considered here [Fig.3(b)], variation of number of layers in the region 2 does not significantly influence DOS. However, presence of the hole and variation of its size [Fig. 1(c)], is obviously reflected in DOS. Experimentally, it could be possible to trace these differences in DOS, e.g., via dark/bright pattern contrast in STM images.

4. Conclusions

We theoretically studied the electronic properties of periodic and isolated nanostructured MoS_2 systems consisting of monolayer, multilayer MoS_2 regions, and holes. We found that by introducing a periodic array of disks or holes we change the character of the system to metallic-like. We showed that presence of a hole and its size in the nano disk is directly reflected in DOS, whereas DOS of the hybrid monolayer-multilayer MoS_2 nanostructured isolated disks is not significantly influenced by the variation of number of layers. Theoretical predictions of electronic properties in those systems coupled with the immense progress in fabrication techniques of reduced dimensionality MoS_2 nanostructures could open pathways for discovery and optimization of MoS_2-based optoelectronic devices.

Acknowledgments

We thank F. Zahid for the discussion regarding the set of tight-binding parameters for MoS_2 [20].

Appendix

Fig. A1(a) depicts the hexagonal Brillouin zone with the labeled high-symmetry points. Fig. A1(b) shows calculated band structure of single monolayer (1ML) and bilayer (2ML) MoS_2. 1ML MoS_2 is a direct band gap material, with valence band maximum (VBM) and conduction band minimum (CBM) located at high-symmetry point K and 2ML MoS_2 has an indirect band gap, with VBM at Γ and CBM at K.

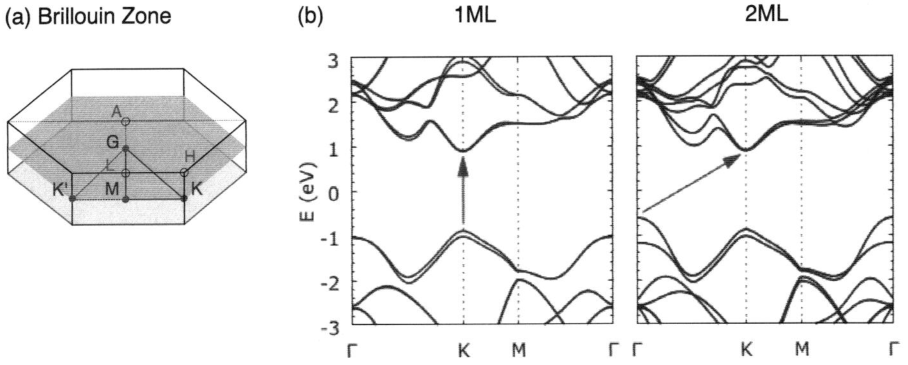

Figure A1. (a) The hexagonal Brillouin zone with labeled high-symmetry points. Red lines indicate paths along which the band structure are calculated(path Γ-K-M-Γ); (b) Band structure of monolayer (1ML) and bilayer (2ML) as extracted from our tight-binding calculations. The red arrows point out the lowest energy transitions.

References

[1] Mak K F, Lee C, j Hone, Shan J and Heinz T F 2010 *Phys. Rev. Lett.* **105** 136805
[2] Wang Q H, Kalantar-Zadeh K, Kis A, Coleman J N and Strano M S 2012 *Nat. Nanotechnol.* **7** 699–712
[3] Cao T, Wang G, Han W, Ye H, Zhu C, Shi J, Niu Q, Tan P, Wang E, Liu B and Feng J 2012 *Nat. Commun.* **3** 887:1–887:5
[4] Feng W, Yao Y, Zhu W, Zhou J, Yao W and Xiao D 2012 *Phys. Rev. B* **86** 165108:1 – 165108:6
[5] Bao W, Cai X, Kim D, Sridhara K and Fuhrer M S 2013 *Appl. Phys. Lett.* **102** 042104:1 – 042104:4
[6] Mlinar V 2013 *Nanotechnology* **24** 042001
[7] Ramasubramaniam A, DHaveh and Towe E 2011 *Phys. Rev. B* **84** 205325:1 – 205325:10
[8] Kim S, Konar A, Hwang W S, Lee J H, Lee J, Yang J, Jung C, Kim H, Yoo J B, Choi J Y, Jin Y W, Lee S Y, Jena D, Choi W and Kim K 2012 *Nat. Commun.* **3** 1011:1 – 1011:7
[9] Helveg S, Lauritsen J V, Laegsgaard E, Stensgaard I, Nørskov J K, Clausen B S, Topsøe H and Besenbacher F 2000 *Phys. Rev. Lett.* **84** 951:1 – 951:4
[10] Li T and Galli G 2007 *J. Chem. Phys. C* **111** 16192–16196
[11] Chaudhari A, Ghoshal T, Shaw M T, O'Connell J, Kelly R A, Glynn C, O'Dwyer C, Holmes J D and Morris M A 2016 *Adv. Mater. Interfaces* **3** 1500596:1 – 1500596:9
[12] Mlinar V 2015 *Ann. Phys. (Berlin)* **527** 187–204
[13] Slater J C and Koster G F 1954 *Phys. Rev.* **94** 1498–1524
[14] Papaconstantopoulos D A and Mehl M J 2003 *J. Phys.: Condens. Matter* **15** R413–R440
[15] Hegde G, Povolotskyi M, Kubis T, Boykin T and Klimeck G 2014 *J. Appl. Phys.* **115** 123703
[16] Mlinar V 2012 *J. Mater. Chem.* **22** 1724–1732
[17] Chadi D J 1977 *Phys. Rev. B* **16** 790
[18] Roldán R, López-Sancho M P, Guinea F, Cappelluti E, Silva-Guillén J A and Ordejón P 2014 *2D Materials* **1** 034003:1–034003:21
[19] Zahid F Private communication
[20] Zahid F, Liu L, Zhu Y, Wang J and Guo H 2013 *AIP Adv.* **3** 052111

IOP Conf. Series: Journal of Physics: Conf. Series **939** (2017) 012016 doi:10.1088/1742-6596/939/1/012016

The electrophoretic deposition of ZnO on highly oriented pyrolytic graphite

Milad Ghalamboran[1(c)], Mojtaba Jahangiri[2], and Ehsan Yousefiazari[3]

Electronic Materials Laboratory, Electrical Engineering Department, K. N. Toosi University of Technology, Tehran 16317-14191, Iran.

[1(c)]miladghalamboran@gmail.com [2]mojtaba.jahangiri@gmail.com
[3]ehsanyousefiazari@gmail.com

Abstract. Intensive research has been conducted on ZnO thin and thick films in recent years. Such layers, used in different electronic devices, are deposited utilizing various methods, but electrophoretic deposition (EPD) has been chosen because of the advantages like low energy consumption, economical superiority, ecofriendliness, controllability, and high deposition rate. Here, we report electrophoretically depositing ZnO layers onto highly oriented pyrolytic graphite. Well-dispersed and stable ZnO suspensions are used for the deposition of continuous and even layers of ZnO on the substrate. ZnO powder is dispersed in acetone. The electric field applied is in the 250 V/cm to 2000 V/cm range. The morphology of the deposits are studied by SEM at the different stages of the deposition process.

1. Introduction

Electrophoretic deposition (EPD) is one of the most attractive methods employed to form thin or thick films of ceramic materials on conductive substrates [1-3]. Compared to other deposition techniques, EPD has advantages of lower energy consumption, economical superiority, higher growth rate, and ecofriendliness. The method is applicable for production of both dense and porous coatings on a variety of substrates with different shapes, sizes, and surface properties [4, 5]. In addition, nanostructured materials such as carbon nanotubes [6], graphene nanosheets [7], and metal oxide (MO) nanorods [8] have been successfully developed by EPD processing.

MOs are the building blocks for fabrication of chemical sensors, effectively functioning as substrates, heating elements, and sensing pellets in their structure [9-15]. Furthermore, MOs have shown promising efficacy in energy harvesting field and are ideal candidates for utilization in numerous optoelectronic devices including radiation detectors, solar cells, and display panels [16-20]. Zinc oxide, having wide applications in electronic and optoelectronic devices, is of particular interest [21].

Graphene, a 2-dimensional network of carbon hexagons, is viewed as the way to atomic-scale electronics. Excellent properties such as visible transparency, high thermal and electrical conductivity, mechanical strength, and chemical stability have given rise to intense research in the field of graphene electronics [22, 23]. Highly oriented pyrolytic graphite (HOPG) is a layered semimetal, consisting of many graphene sheets stacked on each other. In this work, HOPG is utilized to assess the functionality of the EPD process for forming oxide layers on graphene sheets. ZnO is electrophoretically deposited on the conductive HOPG substrates and the deposits are characterized by scanning electron microscopy (SEM) images.

IOP Conf. Series: Journal of Physics: Conf. Series **939** (2017) 012016 doi:10.1088/1742-6596/939/1/012016

2. Experimental

Analytical grade ZnO powder is used as purchased (Merck Millipore, 108846). Absolute acetone is used for the EPD process. In order to attain 5×10^{-4} wt% suspensions, 1 mg of ZnO powder is dispersed in 250 mL of acetone. The container of the suspension is placed in an ultrasonic bath for 30 minutes. The electrophoresis cell includes a 50 mL borosilicate glass beaker and the aluminum electrodes. Electrodes are tightly covered by aluminium foils, and an HOPG flake (3×3 mm^2) is placed on the cathode electrode. The apparent surface areas of both cathode and the anode are 2 cm^2. The distance between cathode and anode is 4 mm. The schematic diagram of the experimental setup is shown in figure 1. The voltage source is a regulated DC power supply, and current variation is recorded by monitoring the voltage drop on a contact resistor.

3. Results

The temporal variation of the EPD cell current is presented in figure 2. These curves can be used for the calculation of cell conductivity and determining contamination levels. Current variations are recorded immediately after applying the EPD voltage between the cathode and anode. According to this curve, the current decreases with the duration of deposition. The measurements and the tests are carried out at room temperature (25°C).

Figure 3 shows the SEM image of the deposits on the HOPG substrate. Images (a) and (b) are related to the deposition at 250 V/cm and 2000 V/cm, respectively. According to the previous reports [24], the electrophoretic deposition rate increases with the applied voltage up to a certain level, it decreases with voltage afterwards. Figure 4 shows the relationship between the weight of ZnO deposit on the cathode and the established electric field. ZnO particles formed a porous surface layer at 250 V/cm. the deposited layer becomes denser by increasing the field.

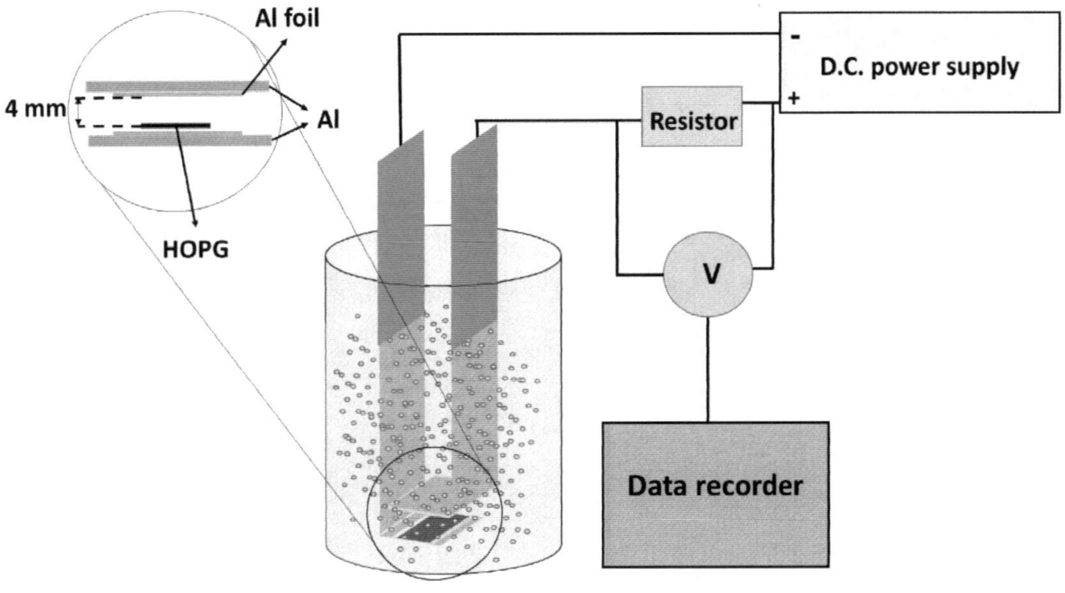

Figure 1. Schematic diagram of the experimental setup used for the EPD of ZnO.

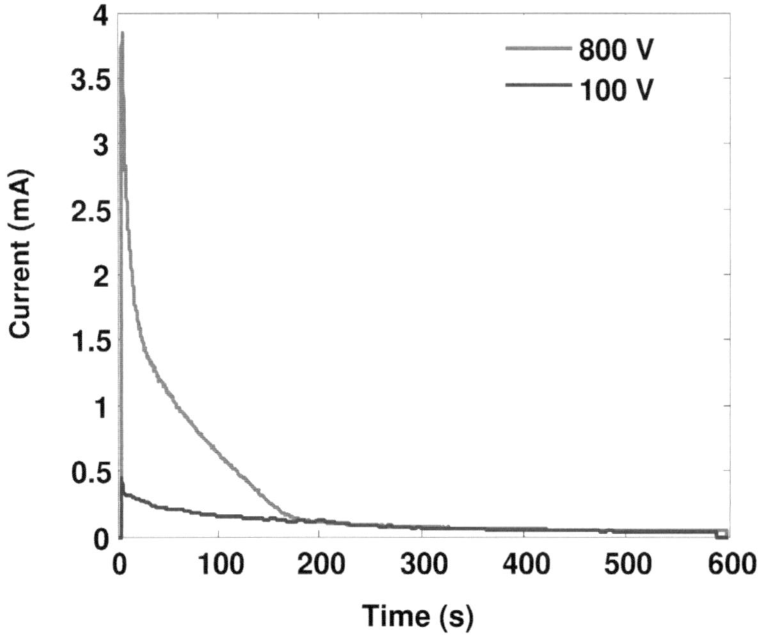

Figure 2. The variation of the cell current with the time of deposition at the stated applied voltages.

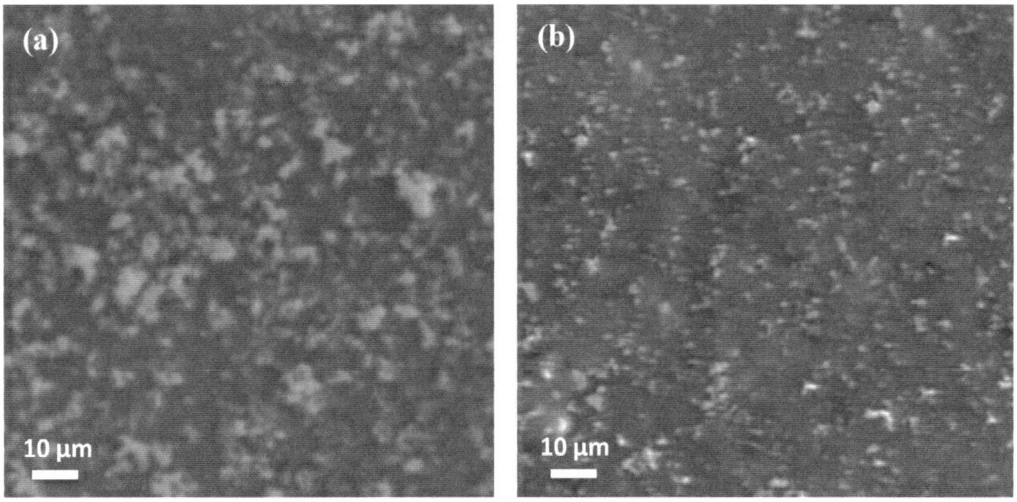

Figure 3. The SEM micrographs of ZnO on multilayer graphene formed 10 minutes after the applicate of the cell voltage, which is 100V in (a) and 800V in (b).

IOP Conf. Series: Journal of Physics: Conf. Series **939** (2017) 012016 doi:10.1088/1742-6596/939/1/012016

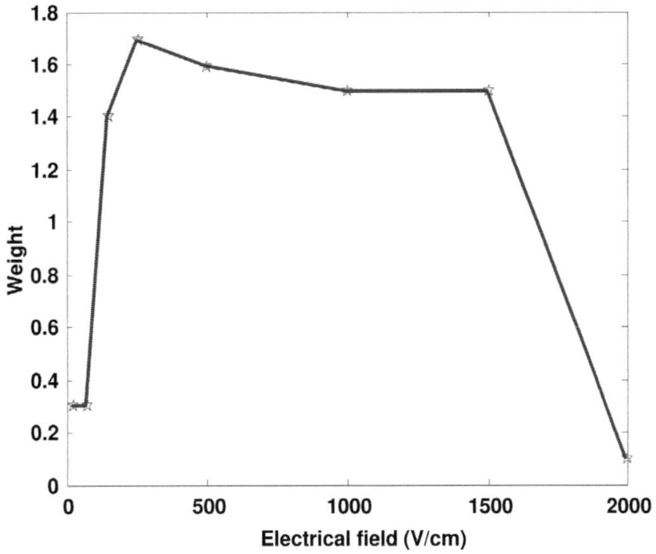

Figure 4. The weight of the ZnO deposit on the cathode, with an area of 2 cm^2, at various applied electrical fields.

4. Conclusions

The results of our preliminary experimental work on the electrophoretic deposition of ZnO on multilayer graphene is reported. It was shown that the deposition rate increases with the field applied up to a certain level. Afterwards, the trend is different and deposition rate decreases with the field. The reason for this behaviour is presently under investigation.

References

[1] Hossein-Babaei F and Raissidehkordi B 2000 Electrophoretic deposition of MgO thick films from an acetone suspension *J. Eur. Ceram. Soc.* **20** 2165-8.

[2] Hossein-Babaei F and Taghibakhsh F 2000 Electrophoretically deposited zinc oxide thick film gas sensor *Electron. Lett.* **36** 1815-6.

[3] Sarkar P and Nicholson P S 1996 Electrophoretic deposition (EPD): mechanisms, kinetics, and application to ceramics *J. Am. Ceram. Soc.* **79** 1987-2002.

[4] Besra L and Liu M 2007. A review on fundamentals and applications of electrophoretic deposition (EPD) *Prog. Mater. Sci.* **52** 1-61.

[5] Corni I, Ryan M P and Boccaccini A R 2008 Electrophoretic deposition: from traditional ceramics to nanotechnology *J. Eur. Ceram. Soc.* **28** 1353-67.

[6] Chen L, He H, Yu H, Cao Y, Lei D, Menggen Q, Wu C and Hu L 2014 Electron field emission characteristics of graphene/carbon nanotubes hybrid field emitter *J. Alloys Compd.* **610** 659-64.

[7] Chavez-Valdez A Shaffer M S P and Boccaccini A R 2012 Applications of graphene electrophoretic deposition. A review *J. Phys. Chem. B* **117** 1502-15.

[8] Limmer S J and Cao G 2003 Sol–gel electrophoretic deposition for the growth of oxide nanorods *Adv. Mater.* **15** 427-31.

[9] Hossein-Babaei F and Orvatinia M 2003 Thickness dependence of sensitivity in thin film tin oxide gas sensors deposited by vapor pyrolysis *Int. J. Eng., Trans. B* **16** 33-40.

[10] Bai Z, Wang A and Xie C 2006 Laser grooving of Al_2O_3 plate by a pulsed Nd:YAG laser: characteristics and application to the manufacture of gas sensors array heater. *Mater. Sci. Eng. A* **435** 418-24.

[11] Hossein-Babaei F, Paknahad M and Ghafarinia V 2012 A miniature gas analyzer made by

integrating a chemoresistor with a microchannel *Lab Chip* **12** 1874-80.
[12] Hosseini-Golgoo S M and Hossein-Babaei F 2011 Assessing the diagnostic information in the response patterns of a temperature-modulated tin oxide gas sensor *Meas. Sci. Technol.* **22** 035201.
[13] Ghalamboran M and Saedi Y 2016 TiO_2-TiO_2 composite resistive humidity sensor: ethanol crosssensitivity *IOP Conf. Ser.: Mater. Sci. Eng.* **108** 012039.
[14] Hossein-Babaei F and Rahbarpour S 2014 Alteration of pore size distribution by sol-gel impregnation for dynamic range and sensitivity adjustment in Kelvin condensation-based humidity sensors *Sens. Actuators, B* **191** 572-8.
[15] Gharesi M and Ansari M 2016 Tin oxide microheater for chemical sensors *IOP Conf. Ser.: Mater. Sci. Eng.* **108** 012018.
[16] Ohta H, Kim S, Mune Y, Mizoguchi T, Nomura K, Ohta S, Nomura T, Nakanishi Y, Ikuhara Y, Hirano M and Hosono H 2007 Giant thermoelectric Seebeck coefficient of a two-dimensional electron gas in $SrTiO_3$ *Nat. Mater.* **6** 129-34.
[17] Lajvardi M M and Jahangiri M 2016 Ni/TiO_2 ultraviolet detector *IOP Conf. Ser.: Mater. Sci. Eng.* **108** 012031.
[18] You J, Meng L, Song T B, Guo T F, Yang Y M, Chang W H, Hong Z, Chen H, Zhou H, Chen Q and Liu Y 2016 Improved air stability of perovskite solar cells via solution-processed metal oxide transport layers *Nat. Nanotechnol.* **11** 75-81.
[19] Hossein-Babaei F, Lajvardi M M and Alaei-Sheini N 2015 The energy barrier at noble metal/TiO_2 junctions *Appl. Phys. Lett.* **106** 083503.
[20] Park S, Kim K H, Jo J W, Sung S, Kim K T, Lee W J, Kim J, Kim H J, Yi G R, Kim Y H and Yoon M H 2015 In-depth studies on rapid photochemical activation of various sol–gel metal oxide films for flexible transparent electronics *Adv. Funct. Mater.* **25** 2807-15.
[21] Krzywiecki M, Grządziel L, Sarfraz A, Iqbal D, Szwajca A and Erbe A 2015 Zinc oxide as a defect-dominated material in thin films for photovoltaic applications–experimental determination of defect levels, quantification of composition, and construction of band diagram *Phys. Chem. Chem. Phys.* **17** 10004-13.
[22] Ma Y and Chen Y 2015 Three-dimensional graphene networks: synthesis, properties and applications *Natl. Sci. Rev.* **2** 40-53.
[23] Liu R, You X C, Fu X W, Lin F, Meng J, Yu D P and Liao Z M 2015 Gate modulation of graphene-ZnO nanowire Schottky diode *Sci. Rep.* **5** 10125.
[24] Gutierrez C P, Mosley J R and Wallace T C 1962 Electrophoretic deposition: a versatile coating method *J. Electrochem. Soc.* **109** 923-7.

Concentric-electrode organic electrochemical transistors: case study for selective hydrazine sensing

S Pecqueur*, S Lenfant, D Guérin, F Alibart and D Vuillaume
Institut d'Electronique, Micro-électronique et Nanotechnologie, CNRS, CS 60069, avenue Poincaré, 59652 Cedex, Villeneuve d'Ascq, France

* E-mail ; sebastien.pecqueur@*iemn.univ-lille1*.fr

Abstract. We report on hydrazine-sensing organic electrochemical transistors (OECTs) with a design consisting in concentric annular electrodes. The design engineering of these OECTs was motivated by the great potential of using OECT sensing arrays in fields such as bioelectronics. In this work, PEDOT:PSS-based OECTs have been studied as aqueous sensors, specifically sensitive to the lethal hydrazine molecule. These amperometric sensors have many relevant features for the development of hydrazine sensors, such as a sensitivity down to 10^{-5} M of hydrazine in water, an order of magnitude higher selectivity for hydrazine than for 9 other water soluble common analytes, the capability to recover entirely its base signal after water flushing and a very low voltage operation. The specificity for hydrazine to be sensed by our OECTs is caused by its catalytic oxidation at the gate electrode and enables increasing the output current modulation of the devices. This has permitted the device-geometry study of the whole series of 80 micrometric OECT devices with sub-20-nm PEDOT:PSS layers, channel lengths down to 1 μm and a specific device geometry of coplanar and concentric electrodes. The numerous geometries unravel new aspects of the OECT mechanisms governing the electrochemical sensing behaviours of the device, more particularly the effect of the contacts which are inherent at the micro-scale. By lowering the device cross-talking, micrometric gate-integrated radial OECTs shall contribute to the diminishing of the readout invasiveness and therefore promotes further the development of OECT biosensors.

1. Introduction

Polymeric-semiconductor devices have already shown great promises to substitute many technologies behind consumer products such as electronics **[1]**, photovoltaics **[2]**, lightings **[3]** and displays **[4]**. Recently, efforts have been devoted to employ the outstanding features of these plastic electronics technologies to create smarter sensors, such as electrolyte-gated organic field-effect transistors **[5]** or organic photodetectors **[6]**. More specifically, the rich chemistry of molecular and polymeric semiconductors offers the possibility to build biocompatible sensors. Indeed, organic electrochemical transistors are a type of electrolyte-gated transistors in which the constituting semiconducting ionomer blend has a good affinity with cations. As such, this platform is extensively studied as a bio-electronical interface to sense cellular action potentials **[7-9]**. Nowadays, one of the key challenges for this technological tool is to interpret complex readouts of dense networks. In order to control a whole array, one should first engineer the OECT units and second understand the mechanism of these units responsible for their sensitivity.

Efforts have been devoted to theoretically study on capacitive sensors having a radial symmetry **[10]** but these geometries are not extensively applied for practical applications. Here, we report results on OECTs with a circular geometry offering a higher sensing isotropy and a large geometry variation to interpret the device functioning. Also, the whole device comprises its individual gate electrode, which is locally patterned on the substrate in an overall diameter down to 100 μm. Despites the high current densities obtained at very low operated voltage, the dimension restrictions strongly limit the active areas of the sensor which hinders its gate-controlled response. Therefore, we activated the large-gate output-current modulation by the introduction of hydrazine. The OECTs show to be highly sensitive and selective to hydrazine. Hydrazine is a 0.1-ppm-level genotoxic for which the human carcinogenicity has not been evidenced but high caution is still recommended on inhalation-induced lung-cancer risks **[11]**. Thus, the ability to realise this kind of sensor is then of the clearest interest. Moreover, these OECTs have shown various performances depending on their geometry and we have explored in this study the hydrazine-enabled gate OECT on a population of 80 devices with different geometries to better understand their functioning.

2. Experimental procedure

The devices have been fabricated by e-beam lithography (see figure 1 a to d), adapted from a previously-reported process **[12]**. Platinum electrodes have been patterned by e-beam lithography and lift-off on top of a 200 nm thermally-grown SiO_2 / n-type Si substrate, using a resist of poly(methyl methacrylate). 10 nm of titanium were evaporated as an adhesion layer prior the 70 nm of platinum. Platinum has been chosen as a contact metal for both source/drain and gate because of its chemical inertia and its high work-function forming ohmic contacts with organic hole transport materials. Then, SiO_2 was functionalized with (3-glycidyloxypropyl)trimethoxysilane (22 mM-concentrated in a 40 mL 200:1-toluene/acetic-acid solution) as an adhesion promoter, prior the processing of a 2-μm-thick poly(chloro-*para*-xylylene) (or parylene C) capping layer by chemical vapor deposition using the Gorham process (Dichloro[2,2]-*para*-cyclophane was used as a precursor in a C20S deposition chamber from Comelec). Then, the device apertures were patterned by e-beam lithography with a sacrificial 3.7-μm-thick resist (a modified UV210 from Rohm and Haas Electronic Materials LLC) before dry-etching the surface with an oxygen plasma. Then, a surfactant (2 % v/v Micro-90 in deionized water) was dispensed on top of the substrate by spin-coating, before depositing a subsequent 2-μm-thick parylene C layer (under similar conditions as previously). Another e-beam lithography and dry-etching step, with similar conditions as previously, was carried out in order to etch the channel cavities and mask the gate apertures. The channel cavities were filled by a filtered and spincoated PEDOT:PSS solution (76 % v/v Clevios PH1000 Heraeus, 19 % v/v ethylene glycol, 4 % v/v dodecylbenzenesulfonic acid, 1 % v/v 3- glycidyloxypropyl)trimethoxysilane) and the underlying parylene C layer carefully stripped out with some adhesive tape, before an hour long hard baking at 140°C.

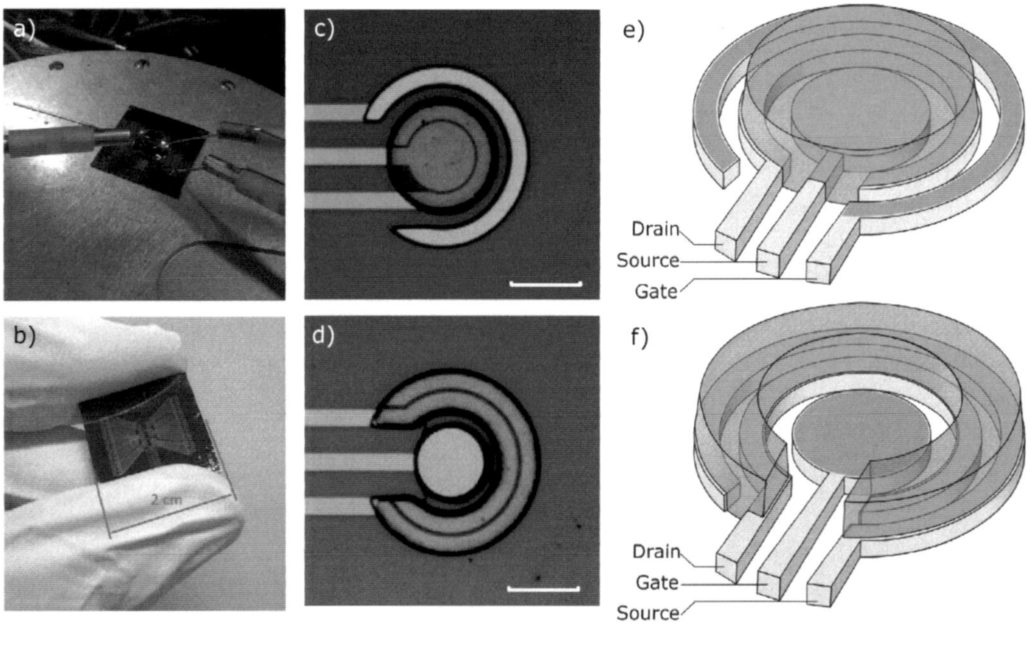

Figure 1. a) and b) Electrical setup including the 4-cm^2 OECT-patterned silicon substrate with the electrolyte drop dispensed on top of it. Source-centred (c and e) and the gate-centred (d and f) OECTs: Device microscope pictures (c and d, scale bar: 40 μm) and depiction of their architecture (e and f, the representation of the passivating parylene C has been omitted for sake of clarity).

The electrical characterization was performed on an Agilent 4155 parameter analyser, OECTs grounded at the gate. Before measurements and to ensure the highest reproducibility **[13]**, the PEDOT:PSS patches were stressed 3 times with a grounded platinum wire gate at -700 mV in a 1 M KCl aqueous solution. Although the potentials are not referenced, we set the highest polarisation to 700 mV in absolute value, Pt gate grounded, in order to lower the risk of water electrolysis (no significant steady-state current density at the gate electrode was observed in the study).

3. Discussion on the OECT as a hydrazine sensor

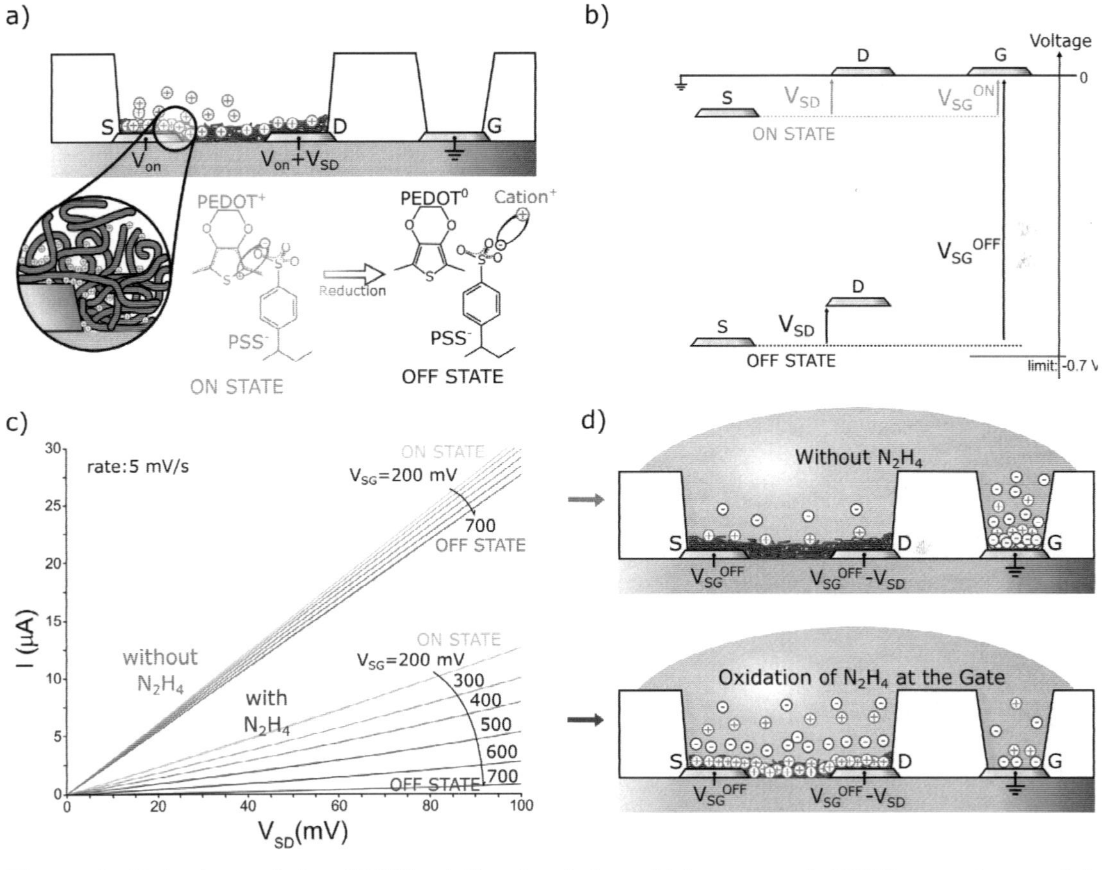

Figure 2. Behaviour of the OECT: a) Sketch of the widely accepted OECT operation model, implicating the participation of cations in the PEDOT dedoping mechanism upon channel/gate polarisation b) Diagram of the applied voltage biases, necessary in our experiments to read the OECT ON and OFF states c) Typical current-voltage characteristics of an OECT without and with hydrazine (conditions: 1 M KCl, 0.1 M N_2H_4, source-centred, L = 5 μm and R_{int} = 20 μm) d) Schematic of two cases of different ionic distributions on an OECT under operation. At the top, the case for which the voltage drops mainly at the gate/electrolyte interface. At the bottom, the case for which the voltage drops mainly at the PEDOT/electrolyte interface.

Central symmetry OECTs have been fabricated on either a source-centred[1] or gate-centred configuration (see figure 1 e and f). The geometry of the devices was modulated by the channel length L (20, 10, 5, 2 and 1 μm) and the radius of the central electrode R_{int} (50, 40, 30 and 20 μm). Practically, the round-shaped PEDOT:PSS patches on the source-centred OECTs were patterned easier than the "C"-shaped ones on the gate-centred OECTs, which required a preferential direction for the parylene C peeling as well as a fine tuning of the baking and peeling conditions to have an optimised yield.

The patches of PEDOT:PSS overlap the whole top area of the source and drain electrodes on a width of at least 10 μm in order to ensure the maximal injection of current in the channel. Based on a well-accepted mechanism **[14]**, the channel conductance can be modulated upon source-and-drain/gate polarisation, such as the PEDOT semiconductor is dedoped from the PSS by reduction, supported by cations which penetrate the ionomer and chelate the sulfonates (see figure 2 a and b). Both the presence of labile cations and the device polarisation are essential but not sufficient in order to dedope of the PEDOT.

Indeed, first tests on our OECTs in hydrazine-free aqueous electrolytes of KCl showed weak conductance modulations in the studied voltage windows, even for concentrations up to 10 M of KCl (see figure 2 c). As published, the OECT dedoping efficiency depends on multiple parameters other than ionic concentration and applied voltages, such as the ratio of the channel area (A_{ch}) over the area of the gate (A_g) **[15,16]**, the nature of the gate metal **[17]** and also the volume of PEDOT:PSS to dedope **[18,19]**. In order to achieve a significant dedoping of the whole PEDOT:PSS material with a small gate, the volume of PEDOT:PSS was kept as low as possible. Layers thinner than 20 nm of PEDOT:PSS were realised by spinning the formulation at velocities higher than 4000 rpm, followed by a subsequent spinning of deionised (DI) water to thinner it further before substrate baking. Despite the small thickness of polymer, rather good performances were obtained, reaching currents up to 60 μA. Also, the lack of hysteresis and the linearity of the curves indicate respectively a rapid dedoping at a scan rate of 5 mV/s and the currents to be ohmic (see figure 2 c).

Despite the low volume of PEDOT:PSS, the usage of Pt as gate metal and the small A_{ch}/A_g poorly modulated the output-current density. It has been proposed that the electrochemical inertia of the gate metal in addition of its small size is at the origin of the lack of field-effect, causing the voltage to drop mainly at the platinum gate interface, screening the channel from polarisation **[17]** (see figure 2 d, upper skech). Also, in order to counterbalance this effect for the same A_{ch}/A_g, strategies such as using a fast redox-system gate like Ag/AgCl **[17]** or introducing a redox active analyte such as H_2O_2 in the electrolyte **[15]** have been used. In the case of H_2O_2 introduction, it has been reported that the field-effect on the OECT depends on the concentration of H_2O_2 introduced in the electrolyte, such as the OECT can be used as a H_2O_2 sensor down to 10^{-5} M (level of the mg/L) **[15]**.

[1]The term "source-centred" refers strictly to a device geometry and not the polarisation of the source with respect to the drain. This information is specified by the term "direct polarisation", opposed to "reversed polarisation".

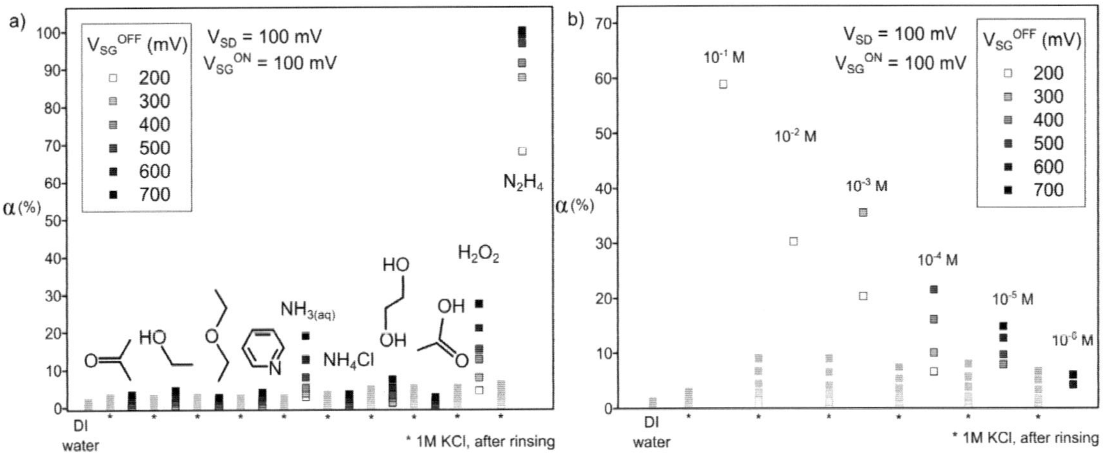

Figure 3. Selectivity and sensitivity of an OECT (conditions: source-centred, L = 5 μm and R_{int} = 20 μm) to hydrazine at different source-gate potentials. a) Study of different 0.1 M concentrated analytes, diluted in 0.1 M KCl solutions. b) Study of N_2H_4, diluted in 1 M KCl solutions. Data on each graph are displayed from left to right in the chronological order.

In this study, we demonstrate the OECT to be far more selective to hydrazine. As a strong reductant, its electro-catalysed oxidation at the Pt gate supports the reduction of the PEDOT at the channel upon polarisation, creating N_2 and $N_2H_5^+$ as byproducts **[20]**. Considering the oxidation products which do not deposit on the gate, the chemical potential at the Pt-gate shall remain stable over time and the mirrored potential shall not depend on the sensor usage.

The electrical measurements of a source-centred OECT were performed with N_2H_4 at different electrical and chemical conditions, after the device being immersed overnight in deionised water (see figures 3).

If $\alpha(V_{SG}^{OFF},V_{SG}^{ON},V_{SD})$ is the amplitude of current modulation between the off-state and the on-state, defined such as:

$$\alpha = \frac{I(V_{SG}{}^{ON};V_{SD})-I(V_{SG}{}^{OFF};V_{SD})}{I(V_{SG}{}^{ON};V_{SD})} = 1 - \frac{I_{OFF}}{I_{ON}} \quad (1)$$

and fixing the values V_{SD} and V_{SG}^{ON} at 100 mV, the sensor shows current density modulations $\alpha(V_{SG}^{OFF})$ below 3% with deionised water and 1 M concentrated KCl (see figure 3). This demonstrates that in the absence of an electroactive analyte, the output is barely sensitive to the concentration of KCl. Successively, different analytes have been introduced and tested on the OECT, followed by a rinsing and test on a 1 M KCl concentrated blank (see figure 3.a). After each analyte test, the sensor recovers down to a α = 5±2 at V_{SG}^{OFF} = 700 mV, indicating the OECT to not be damaged by any analyte. The sensor displays a α for acetone, ethanol, diethylether, pyridine, ammonium chloride and acetic acid very near the one of the blanks, showing the insensitivity of the OECT for 0.1 M of these analytes in water. A slight modulation with 0.1 M of ethylene glycol was observed but without significant change in the PEDOT:PSS current density (since this analyte has the property to boost the conductivity of PEDOT:PSS **[21]**, the lack of sensitivity to this molecule is of the highest importance). Moderated modulations with 0.1 M of aqua ammonia or hydrogen peroxide were observed. Confirming the applicability of H_2O_2 sensing with an OECT, it is worth noticing that other metabolites such as ammonia can also be detected at the same level. While 0.1 M of ammonia or hydrogen peroxide modulate with α = 25±5% at V_{SG}^{OFF} = 700 mV, hydrazine promotes α = 65±5% at as few as V_{SG}^{OFF} = 200 mV at the same concentration. The channel is completely closed with I_{OFF} = 3±2 nA when V_{SG}^{OFF} = 700 mV, resulting to α ≈ 100%.

IOP Conf. Series: Journal of Physics: Conf. Series **939** (2017) 012017 doi:10.1088/1742-6596/939/1/012017

Dry hydrazine has the property to reduce p-doped polythiophenes such as P3HT **[22]**. An eventual thermodynamically favoured reduction of the PEDOT by N_2H_4 could explain the observed decrease of I_{off} with the concentration of N_2H_4 (see figures 2 a). Nevertheless, this single property does not sufficiently justify the V_{SG}^{OFF} dependency of the PEDOT conductivity, since the electrode polarisation under device operation promotes the electrolyte oxidation at the gate electrode and not at the PEDOT channel. The OECT sensor presents also an acute sensitivity to hydrazine (see figure 3.b). Successive dilutions of fresh hydrazine aqueous KCl solutions shows the sensor to be sensitive to at least 10^{-5} M of N_2H_4 at the studied voltages: The sensor responds positively for an applied V_{SG} from 200 mV to 700 mV when exposed to an electrolyte of N_2H_4 of concentrations from 0.1 M to 10^{-5} M. The response α remains at the base level when immersed to a 10^{-6} M concentrated N_2H_4 electrolyte. We note that the sensor regeneration to $\alpha = 9\pm1$ strongly limits the device sensitivity at 10^{-6} M. Another operation protocol on the device rinsing step or on the electrical inputs could be thought to improve the sensor regeneration and therefore to increase the sensitivity range of the sensor, but it has not been explored in this work.

4. Discussion on the OECT structure-property relationship

Studies on millimetre-scale OECTs demonstrated that the sensitivity α of the devices to ions was highly dependent to gate electrode area **[15]**. Therefore, a parameter $\gamma=A_{ch}/A_g$ was defined as a preferred geometry parameter for the sensitivity of OECTs to ions at a given concentration and voltage **[15]**. Although this study enabled to correlate the device geometry to a dedoping strictly occurring in the channel, most practical micrometric bottom-contacts transistors require a large area for the source-and-drain contacts to be exposed to the electrolyte **[7]**. This implies that the contact area has to be taken into account as counter-electrode for the gate, and therefore, impacts on α by both dedoping of the channel and the contacts, affecting both the channel conductivity and the Schottky barriers at the contacts. Since the conductivity and the Schottky barrier have a different dependency on the hole carrier density in the PEDOT **[23]**, we believe that α of the OECTs can be severely impacted by parameters which are not exclusively channel-related, such as the contacts geometry.

To characterise it, tests of 10^{-2} M N_2H_4 in 10^{-1} M KCl_{aq} were realised for all the device geometries, with V_{SG}^{ON} = 700 mV and $V_{SG}^{OFF} = V_{SD}$ = 100 mV (see figure 4). All the devices show a sensitivity α from 30 to 85% for both the source-centred geometries and the gate-centred ones (see figure 4 a). When referenced to 1 M KCl_{aq}, one sees the base modulation α without hydrazine to be located between 0 and 20% for both the source-centred and the gate-centred OECTs.

When plotting the sensitivity α against the geometric γ parameter of each device (see figure 4 b), one can distinguish a different amplitude for α of each OECT between operation with hydrazine and without hydrazine, showing all the devices to be operational and sensitive to it. Additionally, α of the OECT with reversed SD-polarisation are similar to the one of the direct SD-polarisation, indicating the insensitivity of α to the surface area of the hole injecting electrode, even for the largest R_{int} source-centred devices. For the source-centred, a trend for α to decrease linearly with γ can be seen for a given channel length, and the slope of this trend seems to be channel length dependent.

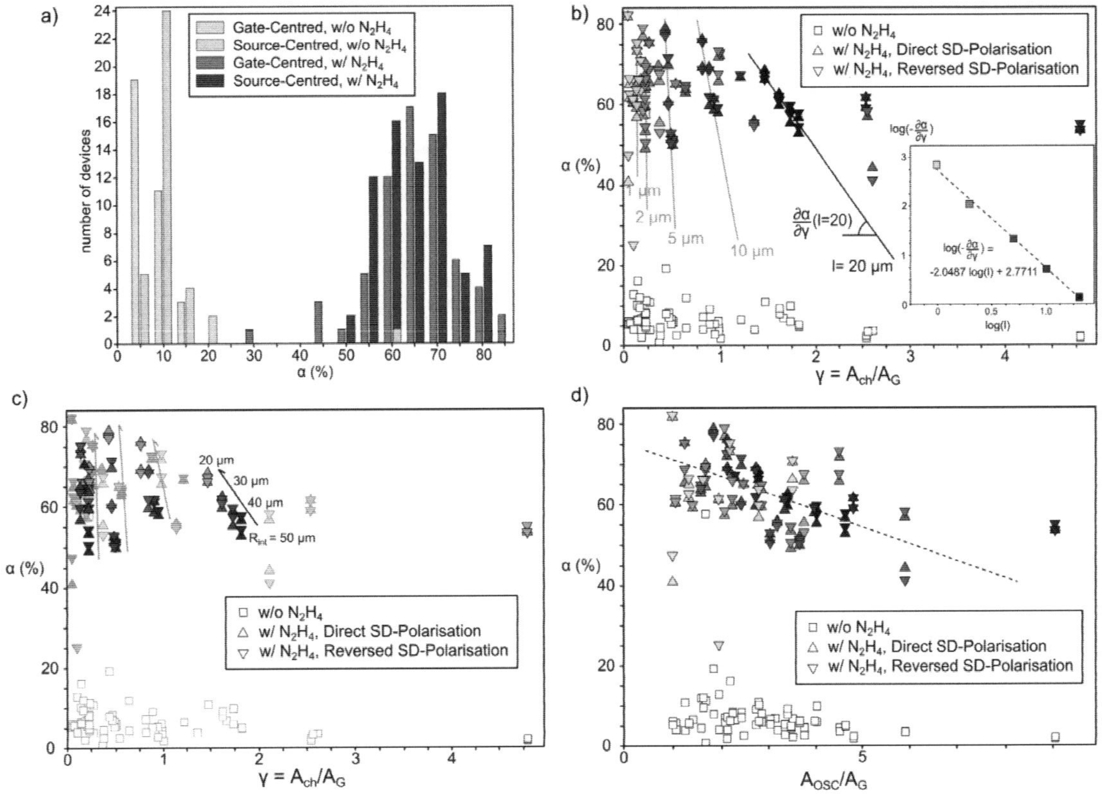

Figure 4. Study of the geometry-dependent sensitivity of OECTs. a) Statistical study on the sensitivity α for all geometries of OECTs, without and with N_2H_4. b) Plot of α(γ) showing the channel length dependency of the derivative for the source-centred OECTs (the inset plot shows the quadratic dependency). c) Plot of α(γ) showing the dependency of α with the radius of the central electrode for the source-centred OECTs. d) Plot of $\alpha(A_{osc}/A_G)$ showing a better correlation when considering a contact-area depending parameter (the dashed line is a guide for the eye). For all graphs, red accounts for gate-centred OECTs and blue for source-centred ones. In b) and d), the colour gradients are such as the lighter, the shorter channel and for c) the lighter, the smaller central electrode.

When plotting the logarithm of the slope for each |α|(γ) curves and for the different source-centred channel lengths, a slope equals to -2 can be seen (see the inset graph of figure 4 b), indicating the source-centred OECTs to have a sensitivity particularly dependent on the channel length, such as the variation of α with γ increases in absolute value with $1/L^2$. But the fact that the 5 series of different channel length with different γ have comparable values of α under the same electrochemical conditions indicates that α depends on a geometric constraints other than $\gamma = A_{ch}/A_G$. When looking at the radius of the central electrode for each channel-length series of source-centred OECTs (see figure 4 c), one sees a monotonic trend for the sensitivity α to increase when the radius of the central electrode decreases. This indicates α to be source-and-drain-contact-area dependent and therefore the overlapping of the PEDOT:PSS with the contacts shall be considered for the design of electrochemical OECT sensors.

As an example, the figure 4 d shows that when we include the contact area in the discriminating geometric parameter, we see the series for both gate-centred and source-centred OECTs have the same sensitivity trends, with an overall trend for α to decrease with the area ratio PEDOT:PSS/gate (A_{osc}/A_G).

In the end, α for the gate-centred series remain more dispersed around the trend than the source-centred ones, when plotted either against γ or A_{osc}/A_G. This dispersion of α values for the gate-centred devices do not allow us to interpolate obvious trends with the channel lengths or the contact area as for the source-centred devices. Since the dimensions of both geometries are very comparable the one with the other (and so their γ and A_{osc}/A_G values), we believe the "C"-shaped PEDOT:PSS patches on the gate-centred devices are more subjects to variabilities than the round ones on the source-centred.

Regarding the trend for α to increase with the decrease of the contact area for source-centred OECTs, we would also recommend to diminish it to further increase the OECT sensitivity.

5. Conclusion

We report both a technological and a scientific result in the field of PEDOT:PSS-based organic electrochemical transistors for sensing. First, we demonstrate that this platform can selectively sense hydrazine, and as such, can be a promising structure to sense toxics in liquids. Second, a structure-property relationship showed the OECT to be highly contact sensitive. Future investigations on models taking into account the contact area of the devices in their functioning could help us to further understand their mechanism. This whole study was performed on micrometric concentric electrode OECTs: a set of engineered devices which might promote the development of higher definition electrochemical transistors.

6. Acknowledgments

The authors wish to thank the European Commission under the H2020-FETOPEN-2014-2015-RIA program, RECORD-IT project GA 664786, and the French National Nanofabrication Network RENATECH for financial supports. We thank also the IEMN cleanroom staff of their advices and support.

7. References

[1] Dodabalapur A 2006 Organic and polymer transitors for electronics *Mater. Today* **9** 24–30
[2] Yu J, Zheng Y and Huang J 2014 Towards high performance organic photovoltaic cells: a review of recent development in organic photovoltaics *Polymers* **6** 2473–509
[3] Gong X, Wang S, Moses D, Bazan G C and Heeger A J 2005 Multilayer polymer light-emitting diodes: white-light emission with high efficiency *Adv. Mater.* **17** 2053-8
[4] Liang J, Li L, Niu X, Yu Z and Pei Q 2013 Elastomeric polymer light-emitting devices and displays *Nat. Photon.* **7** 817-24
[5] Kergoat L, Herlogsson L, Braga D, Piro B, Pham M-C, Crispin X, Berggren M and Horowitz G 2010 A water-gated organic field-effect transistor *Adv. Mater.* **22** 2565-9
[6] Arca F, Tede S F, Sramek M, Rauh J, Lugli P and Hayden O 2013 Interface trap states in organic photodiodes *Sci. Rep.* **3** 1324
[7] Khodagholy D et al. 2013 In vivo recordings of brain activity using organic transistors *Nat. Commun.* **4** 1575
[8] Rivnay J, Owens R M and Malliaras G G 2014 The rise of organic bioelectronics *Chem. Mater.* **26** 679-85
[9] Williamson A et al. 2015 Localized neuron stimulation with organic electrochemical transistors on delaminating depth probes *Adv. Mater.* **27** 4405-10
[10] Cheng H, Zhang Y, Huang X, Rogers A R and Huang Y 2013 Analysis of a concentric coplanar capacitor for epidermal hydration sensing *Sens. Actuators* A **203** 149–53
[11] European Commision 2010/08 Recommendation for the scientific committee on occupational exposure limits for hydrazine *SCOEL/SUM/164*
[12] Sessolo M, Khodagholy D, Rivnay J, Maddalena F, Gleyzes M, Steidl E, Buisson B and Malliaras G G. 2013 Easy-to-fabricate conducting polymer microelectrode array *Adv. Mater.* **25** 2135-9
[13] Nguyen V C and Lee P S 2016 Resistive switching memory phenomena in PEDOT PSS: Coexistence of switchable diode effect and write once read many memory *Sci. Rep.* **6** 19594
[14] Bernards D A and Malliaras G G 2007 Steady-state and transient behaviour of organic electrochemical transistors *Adv. Funct. Mater.* **17** 3538–44

IOP Conf. Series: Journal of Physics: Conf. Series **939** (2017) 012017 doi:10.1088/1742-6596/939/1/012017

[15] Cicoira F, Sessolo M, Yaghmazadeh O, DeFranco J A, Yang S Y and Malliaras G G 2010 Influence of device geometry on sensor characteristics of planar organic electrochemical transistors *Adv. Mater.* **22** 1012–16

[16] Yaghmazadeh O, Cicoira F, Bernards D A, Yang S Y, Bonnassieux Y and Malliaras G G 2011 Optimization of organic electrochemical transistors for sensor applications *J. Polym. Sci., Part* B **49** 34–9

[17] Tarabella G, Santato C, Yang S Y, Iannotta S, Malliaras G. G. and Cicoira F. 2010 Effect of the gate electrode on the response of organic electrochemical transistors *Appl. Phys. Lett.* **97** 123304

[18] Rivnay J et al. 2015 High-performance transistors for bioelectronics through tuning of channel thickness *Sci. Adv.* **1** e1400251

[19] Hütter P C, Fian A, Gatterer K and Stadlober B 2016 Efficiency of the Switching Process in Organic Electrochemical Transistors *ACS. Appl. Mater. Interfaces* (DOI: 10.1021/acsami.6b02698)

[20] Aldous L and Compton R G 2011 The mechanism of hydrazine electro-oxidation revealed by platinum microelectrodes: role of residual oxides *Phys. Chem. Chem. Phys.* **13** 5279-87

[21] Zhang S, Kumar P, Nouas A S, Fontaine L, Tang H and Cicoira F 2015 Solvent-induced changes in PEDOT:PSS films for organic electrochemical transistors *APL Mater.* **3** 014911

[22] Ellis D L, Zakin M R, Bernstein L S and Rubner M F. 1996 Chemical sensors for hydrazine and monomethylhydrazine vapor *Anal. Chem.* **98** 817–22

[23] Sze S M 1969 *Physics of Semiconductor Devices* (New York: Wiley–Interscience)

IOP Conf. Series: Journal of Physics: Conf. Series **939** (2017) 012018 doi:10.1088/1742-6596/939/1/012018

HOPG/ZnO/HOPG pressure sensor

Mojtaba Jahangiri[1, a], Ehsan Yousefiazari[2, b], Milad Ghalamboran[3, c]

[1,2,3]Electronic Materials Laboratory, Electrical Engineering Department, K. N. Toosi University of Technology, Tehran, Iran.

[a]mojtaba.jahangiri@gmail.com; [b]ehsanyousefiazari@gmail.com; [c]miladghalamboran@gmail.com;

Abstract. Pressure sensor is one of the most commonly used sensors in the research laboratories and industries. These are generally categorized in three different classes of absolute pressure sensors, gauge pressure sensors, and differential pressure sensors. In this paper, we fabricate and assess the pressure sensitivity of the current vs. voltage diagrams in a graphite/ZnO/graphite structure. Zinc oxide layers are deposited on highly oriented pyrolytic graphite (HOPG) substrates by sputtering a zinc target under oxygen plasma. The top electrode is also a slice of HOPG which is placed on the ZnO layer and connected to the outside electronic circuits. By recording the I-V characteristics of the device under different forces applied to the top HOPG electrode, the pressure sensitivity is demonstrated; at the optimum biasing voltage, the device current changes 10 times upon changing the pressure level on the top electrode by 20 times. Repeatability and reproducibility of the observed effect is studied on the same and different samples. All the materials used for the fabrication of this pressure sensor are biocompatible, the fabricated device is anticipated to find potential applications in biomedical engineering.

1. Introduction

Nowadays various sensors are in demand for different scientific and industrial applications [1-6]. New materials and novel studying are required for sensors performing in harsh environments with specified features [6-8]. Sensors are expected to have substantially smaller size and lower weight, leading to higher sensitivity, better responsivity and exceptional stability. Pressure sensors are required to detect deformations and structural changes occurring in the surrounding infrastructure, as well as the biological activities in living organisms [1, 9, 10]. An efficient pressure sensor or strain detector is made utilizing a material that can exhibit larger structural changes in response to a small applied pressure. One of the most compelling materials is this respect is two-dimensional carbon. Graphene has attracted enormous attention in recent years due to its wonderful mechanical and electrical properties. These properties make graphene attractive for pressure sensor research as well [11]. Graphene two-dimensional structure has an advantage in scalable device fabrication via top-down approaches, which is compatible with the existing semiconductor fabrication technology [12]. Compared with other pressure sensors, Owing to the single atom layer structure, graphene was predicted to be a suitable platform for pressure sensor fabrication graphene-based nano diaphragm pressure detectors are examples [13, 14]. Furthermore, it is important to understand the interaction between graphene and other building blocks in such sensors. In a pressure sensor designed for biomedical applications, biocompatibility of all components of the sensor is of prime importance. Magnetron sputtering of zinc oxide films yields a polycrystalline films which has many advantages in comparison with metallic and other semiconductor materials such as low cost, low toxicity, facile deposition, biocompatibility and biodegradability [15, 16]. These properties make it attractive for application in bioelectronics and biosensing. A biomedical compatible pressure sensor can be used for human motion detector among many other applications [17-19]. In this paper, a pressure sensor with highly oriented pyrolytic graphite (HOPG)/ zinc oxide structure is fabricated.

2. Experimental

The schematic structure of the proposed pressure sensor is shown in figure 1; it includes two HOPG electrodes placed at sides of a ZnO layer. The electrodes are connected to the outside electronic circuit. When a pressure is applied on top electrode, the I-V characteristics of the device varies according to the force applied. This sensor can be used for human motion detection by assembling them on clothings. figure 1 illustrates the three-dimensional schematic structure of graphene-based pressure sensor fabricated in this study. The device is fabricated on an HOPG substrate. DC magnetron sputtering has been used to deposit 2μm ZnO on a HOPG substrate. The way to deposit a compound material (in our case oxide) is called reactive sputtering a metallic target is placed in an atmosphere of argon and oxygen with 50~100 mTorr pressure [20]. The oxide layer thickness depends on the deposition time. In this work, the sputtering process lasts 7 hours. 400V DC is applied between target and substrate, and the current is around 300mA. For DC reactive sputtering, higher oxygen partial pressures are necessary compared to RF sputtering.

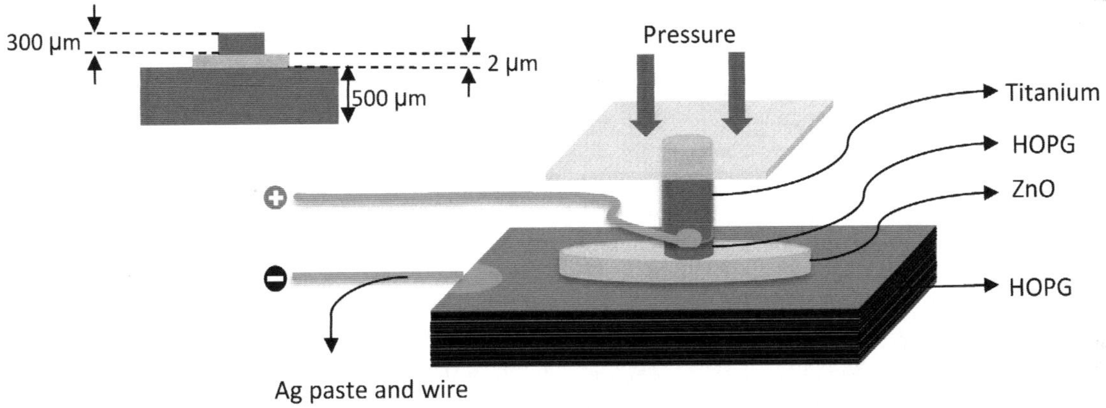

Figure 1. Schematic diagram of HOPG/ZnO/HOPG pressure detector.

The SEM, XRD and EDAX images of the structure confirm the dominance of the (100) and (101) ZnO facets on the substrate. The SEM image is given in figure 2 [21]. The device deposited layer is annealed at 300°C in air for an hour. The top electrode is an spring loaded HOPG layer with 2mm radius attached to a titanium rod which applies a constant pressure to the top electrode. Ag paste and Ag wires are used to connect the electrodes to the outside circuit.

IOP Conf. Series: Journal of Physics: Conf. Series **939** (2017) 012018 doi:10.1088/1742-6596/939/1/012018

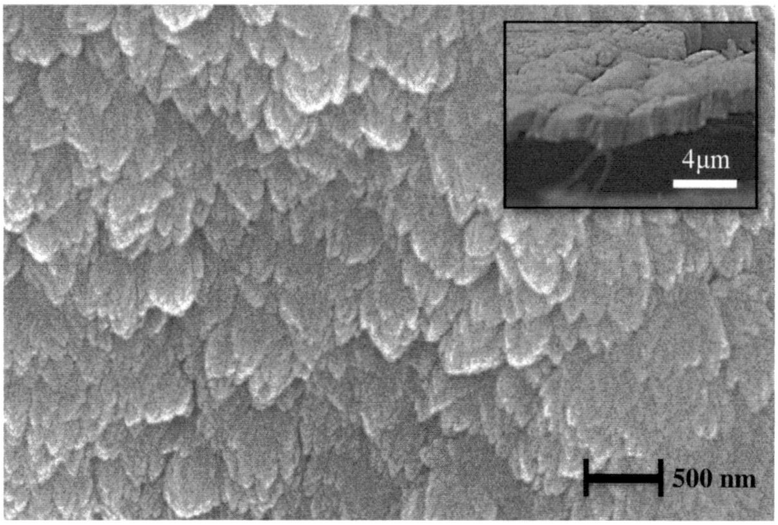

Figure 2. The SEM micrograph of the DC sputtered zinc oxide layer on HOPG substrate. The cross-sectional view is given as an inset.

For pressure measurement, triangular voltage waveform with 2V amplitude and at 10Hz frequency is applied to the device. The current passing through the device is continuously monitored. At constant biasing voltage the current changes with the exerted pressure. The device is calibrated by recording the current change caused by known forces.

3. Results and discussion

The fabricated sensor is connected to I-V data logger with an accuracy of 10^{-8}A. I-V diagrams are plotted at different pressure levels. The pressure is calculated by dividing the exerted force to the area of the top electrode. The results are shown in figure 3. The current increases with the pressure; as an example, in the fabricated prototype the no load current amplitude (the base line) is 10^{-7}A, while, by applying 5kPa pressure, it increases to 0.5μA. The I-V diagram shown in figure 3 indicate current amplitude changes from 0.5μA to 50μA for 5kPa and 100kPa respectively.

IOP Conf. Series: Journal of Physics: Conf. Series **939** (2017) 012018 doi:10.1088/1742-6596/939/1/012018

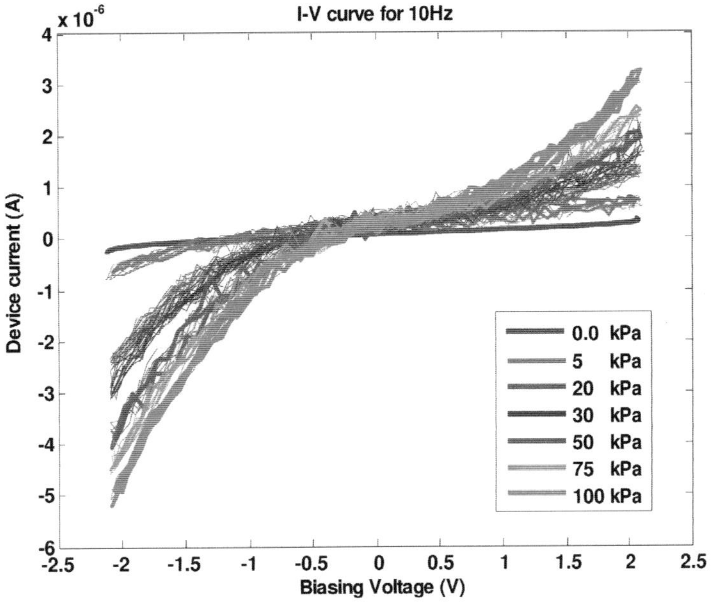

Figure 3. The IV diagrams of device recorded under the stated pressure levels.

Figure 4 presents the device performance at higher pressures. Figure 3 and figure 4 together demonstrate the wide dynamic range of our pressure sensor.

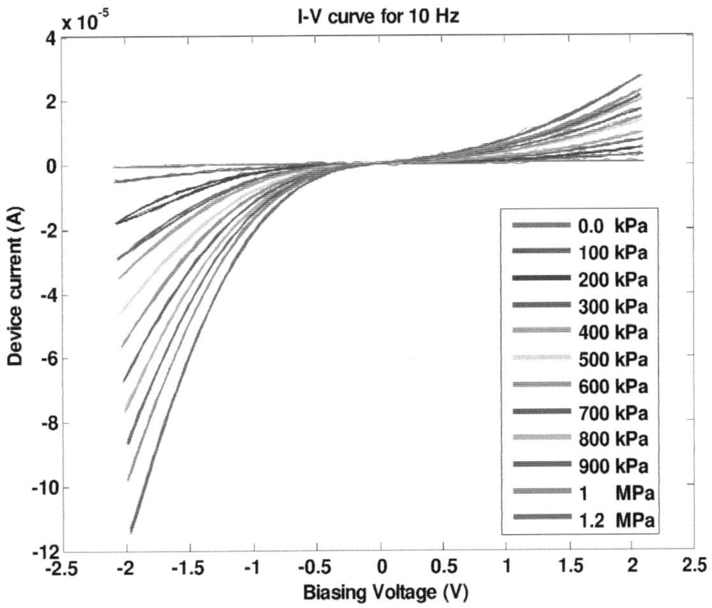

Figure 4. The IV diagrams of device recorded under the state pressures.

The sensing mechanism is related to the contact area between the ZnO layer and the top HOPG electrode. The deposited ZnO has nano-needle microstructure. By applying pressure, needles penetrate into the HOPG sheets and increase the effective contact area, which leads to current increase. Another possible mechanism is the change in the oxygen vacancy distribution. In the ZnO layer cross-section which can alter the device current [22-24].

4. Conclusion

In summary, the fabrication and the result of the preliminary test on a pressure sensor is reported. The device structure consists of a HOPG/ZnO/HOPG diode whose I-V characteristics change upon force exertion. The dynamic range of the device was shown to be wide; operation in the range of 5 kPa – 1.2 MPa was demonstrated.

References

[1] H. Tian, Y. Shu, Y.-L. Cui, W.-T. Mi, Y. Yang, D. Xie, *et al.* 2014 *Nanoscale* "Scalable fabrication of high-performance and flexible graphene strain sensors," **6** 699

[2] F. Hossein-Babaei, M. Paknahad, and V. Ghafarinia 2012 *Lab on a Chip* "A miniature gas analyzer made by integrating a chemoresistor with a microchannel,"**12** 1874

[3] Z. Jing, Z. Guang-Yu, and S. Dong-Xia 2013 *Chin. Phys. B* "Review of graphene-based strain sensors," **22** 057701

[4] M. M. Lajvardi and M. Jahangiri 2016 *IOP Conf. Ser.: Mater. Sci. Eng.* **108** 012031

[5] F. Hossein-Babaei and S. Rahbarpour 2014 *Sens. Actuators, B: Chemical* "Alteration of pore size distribution by sol–gel impregnation for dynamic range and sensitivity adjustment in Kelvin condensation-based humidity sensors," **191** 572

[6] F. Hossein-Babaei, M. M. Lajvardi, and F. A. Boroumand 2012 *Sens. Actuators, A: Physical* "Large area Ag–TiO 2 UV radiation sensor fabricated on a thermally oxidized titanium chip," **173** 116

[7] M. Ghalamboran and Y. Saedi, 2016 *IOP Conf. Ser.: Mater. Sci. Eng.* "TiO2-TiO2 composite resistive humidity sensor: ethanol crosssensitivity," **108** 012039

[8] N. A. Sheini and M. Rohani, 2016 *IOP Conf. Ser.: Mater. Sci. Eng.* "Ag-doped titanium dioxide gas sensor," **108** 012033

[9] V. Sorkin and Y. W. Zhang 2011 *J. molecular modeling* "Graphene-based pressure nano-sensors," **17** 2825

[10] S.-H. Bae, Y. Lee, B. K. Sharma, H.-J. Lee, J.-H. Kim, and J.-H. Ahn 2013 *Carbon* "Graphene-based transparent strain sensor," **51** 236

[11] X. Li, R. Zhang, W. Yu, K. Wang, J. Wei, D. Wu, *et al.* 2012 *Sci. Rep.* "Stretchable and highly sensitive graphene-on-polymer strain sensors," **2** 870

[12] A. Di Bartolomeo, 2016 *Phys. Rep.* "Graphene Schottky diodes: An experimental review of the rectifying graphene/semiconductor heterojunction," **606** 1

[13] J. Ma, W. Jin, H. L. Ho, and J. Y. Dai, 2012 *Opt. Lett.* "High-sensitivity fiber-tip pressure sensor with graphene diaphragm," **37** 2493

[14] H. Tian, Y. Shu, X.-F. Wang, M. A. Mohammad, Z. Bie, Q.-Y. Xie, *et al.* 2015 *Sci. Rep.* "A graphene-based resistive pressure sensor with record-high sensitivity in a wide pressure range," **5**

[15] R. Yatskiv, J. Grym, K. Zdansky, and K. Piksova, 2012 *Carbon* "Semimetal graphite/ZnO Schottky diodes and their use for hydrogen sensing," **50** 3928

[16] A. Kołodziejczak-Radzimska and T. Jesionowski 2014 *Mater.* "Zinc oxide—from synthesis to application: a review," **7** 2833

[17] T. Yamada, Y. Hayamizu, Y. Yamamoto, Y. Yomogida, A. Izadi-Najafabadi, D. N. Futaba, *et al.* 2011 *Nat. Nanotechnol.* "A stretchable carbon nanotube strain sensor for human-motion detection," **6** 296

[18] D. J. Lipomi, M. Vosgueritchian, B. C. Tee, S. L. Hellstrom, J. A. Lee, C. H. Fox, *et al.* 2011 *Nat. Nanotechnol.* "Skin-like pressure and strain sensors based on transparent elastic films of carbon nanotubes," **6** 788

[19] Ö. Ümit, D. Hofstetter, and H. Morkoc 2010 *Proc. IEEE* "ZnO devices and applications: a review of current status and future prospects," **98** 1255

[20] K. Ellmer, 2000 *J. Phys. D: Appl. Phys.* "Magnetron sputtering of transparent conductive zinc oxide: relation between the sputtering parameters and the electronic properties," **33** R17

[21] Ü. Özgür, Y. I. Alivov, C. Liu, A. Teke, M. Reshchikov, S. Doğan, *et al.* 2005 *J. Appl. Phys.* "A comprehensive review of ZnO materials and devices," **98** 041301

[22] F. Hossein-Babaei and N. Alaei-Sheini 2016 *Sci. Rep.* "Electronic Conduction in Ti/Poly-TiO2/Ti Structures," **6**

[23] F. Hossein-Babaei, M. M. Lajvardi, and N. Alaei-Sheini 2015 *Appl. Phys. Lett.* "The energy barrier at noble metal/TiO2 junctions," **106** 083503

[24] F. Hossein-Babaei, N. Alaei-Sheini, and M. M. Lajvardi 2016 *IOP Conf. Ser.: Mater. Sci. Eng.* "Oxygen adsorption at noble metal/TiO2 junctions," **108** 012030

IOP Conf. Series: Journal of Physics: Conf. Series **939** (2017) 012019 doi:10.1088/1742-6596/939/1/012019

First-principles simulation on thermoelectric properties in Bi-Sb System

M S El-Asfoury[1], K Nakamura[1,2] and A Abdel-Moneim[1]

[1]Department of Materials Science and Engineering,
Egypt-Japan University of Science and Technology, New Borg El-Arab,
Alexandria 21934, Egypt
[2]Center for the Promotion of Interdisciplinary Education and Research,
Kyoto University, Kyoto 615-8540, Japan

E-mail: koichi@cpier.kyoto-u.ac.jp

Abstract. Thermoelectric properties of bismuth-antimony (Bi-Sb) alloy system were simulated on the basis of first-principles calculation, to discuss the potential for thermoelectric devices. Atomistic model structures of Bi-Sb alloy system were devised in the forms of single-crystal bulk and one-dimensional nanowire under the periodic boundary condition. The cell parameters of the bulk model were simulated with respect to temperature by the quasi-harmonic approximation through phonon calculation, and dependences of the Seebeck coefficient on composition, surface condition, and temperature have been demonstrated by using our original methodology in terms of the electronic state of Bi-Sb alloy system. For the single-crystal bulk Bi-Sb models, a meaningful effect of the composition on the Seebeck coefficient has not been observed, whereas a clear difference in phonon dispersion was confirmed between pure Bi and Sb-substituted Bi, leading to the significant difference in thermal conductivity. We clarified that the surface condition is a key point to control the Seebeck coefficient for the nanowire form.

1. Introduction

Application of thermoelectric devices is spreading over a variety of special environments and conditions in power generation and refrigeration. For the low-temperature environment, bismuth-antimony (Bi-Sb) alloy is one of the most promising thermoelectric materials with highly efficient energy conversion in the temperature range less than 200 K [1-7]. Generally, the thermoelectric performance of materials is evaluated by the dimensionless figure of merit, $ZT = (S^2\sigma/\kappa)T$, where S is the Seebeck coefficient, σ is the electric conductivity, and κ is the thermal conductivity [8]. Even at low temperature T, the dimensionless ZT for some appropriate Bi-Sb alloy systems comes up to an efficient value; for example, $ZT = 0.6$ has been reported for the single-crystal $Bi_{1-x}Sb_x$ alloy with the compositions of $Bi_{0.91}Sb_{0.09}$ and $Bi_{0.85}Sb_{0.15}$ [1]. In addition, the development of Bi-Sb alloy-based composite materials have been an important target to add a merit of mechanical point to the unique thermoelectric performance at low temperature. We have succeeded in fabricating a novel composite material based on $Bi_{0.85}Sb_{0.15}$ and graphene with good thermoelectric performance [2].

To enhance energy conversion efficiency, it is well known that the miniaturization of thermoelectric materials to low dimension at nanoscale is effective technique [3,9-13]. Rabina et al. has simulated ZT of $Bi_{1-x}Sb_x$ nanowire numerically based on the conventional procedures [3]. We have already developed the *ab initio* procedure without any experimental information to evaluate the Seebeck coefficient on the basis of first-principle electronic band structure [9,10,14]. In this study, thermoelectric properties of single-crystal bulk $Bi_{1-x}Sb_x$ alloy and $Bi_{1-x}Sb_x$ nanowire with small x were simulated by using our original methodology, to discuss the potential for thermoelectric devices.

2. Simulation method

2.1. First-principles calculation

The electronic band structures of the periodic boundary models for bulk and nanowire were calculated using the VASP program package [15,16] based on the density functional theory (DFT) [17]. For the DFT exchange-correlation interaction, the generalized-gradient approximation method was used with the Perdew-Wang (PW91) functional [18,19]. The three-dimensional supercell approximation technique was adopted with ultrasoft pseudopotentials [20]. The cutoff energy for wave functions of electrons with plane-wave expansion was set at 40 Ry (544 eV) for bulk models and 30 Ry (408 eV) for nanowire models.

To investigate the free energy for the geometry optimization of bulk models under the condition of finite temperature, the Helmholtz free energy $F(V,T)$ with phonon free energy at temperature T was derived from the quasi-harmonic approximation (QHA) [21,22];

$$F(V,T) = E(V,0) + F_{ph}(V,T), \tag{1}$$

where $E(V,0)$ is a total energy of the model structure with the geometry V at 0 K, and

$$F_{\mathrm{ph}}(V,T) = k_{\mathrm{B}}T\sum_{q}\sum_{j}\ln\left[2\sinh\frac{\hbar\omega_j(\vec{q})}{2k_{\mathrm{B}}T}\right]. \tag{2}$$

In (2), k_B is the Boltzmann constant, $\hbar$ is equal to Planck's constant divided by 2π, and $\omega_j(\vec{q})$ is a frequency of jth phonon band at $\vec{q}$ point. The first-principles phonon frequencies were calculated by diagonalization of the dynamical matrix with numerical force constants from the $2 \times 2 \times 2$ supercell. The results due to the QHA have been known as to be in good agreement with experimental behaviors for thermal expansion and contraction [23-28].

2.2. Bulk and nanowire models for Bi-Sb systems

Bi and Sb belong to the same group in the periodic table, Group 15, and it is well known that the most stable crystal structure of Bi and Sb is a trigonal structure [4]. We have devised the single-crystal bulk Bi model with the primitive unit cell including two Bi atoms (Bi_2) and its extended-cell model including eight Bi atoms (Bi_8). By replacing some Bi atom(s) of the Bi_8 model with Sb, the Bi_7Sb_1 model and three kinds of Bi_6Sb_2 models were prepared as the compositions of x = 0.125 and x = 0.25 for $Bi_{1-x}Sb_x$ alloy, respectively, as shown in figure 1.

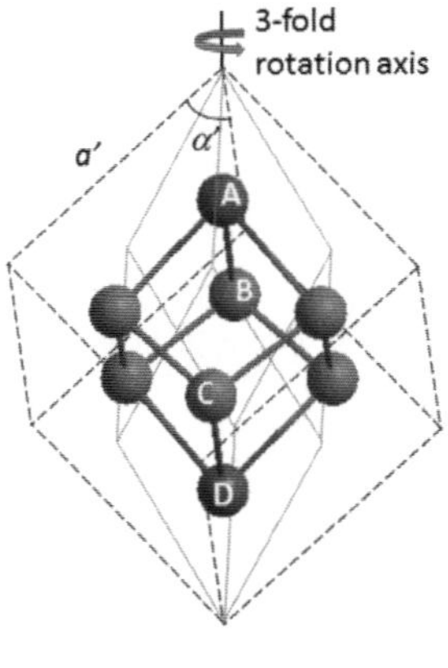

Figure 1. The 8-atom extended cell of bulk Bi-Sb system: Bi_8 model where all atoms are Bi, Bi_7Sb_1 model where atom A is Sb, Bi_6Sb_2-a model where A and B are Sb, Bi_6Sb_2-b model where A and C are Sb, and Bi_6Sb_2-c model where A and D are Sb. Dashed lines are the unit cell of the 8-atom extended cell, and blue solid lines are the primitive unit cell in the case of pure bulk Bi.

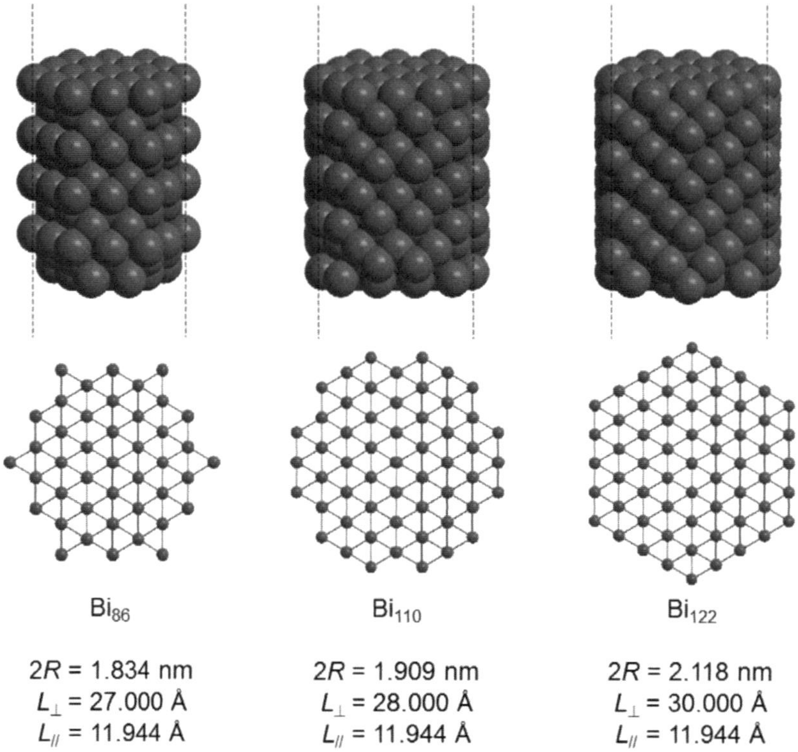

Figure 2. Side and top views of Bi nanowire models. Side views consist of 2 unit cells.

Bi nanowire models were devised by cutting out a fragment with one-dimensional periodic boundary condition from the optimized trigonal bulk Bi. In this paper, the direction of the fragment with the one-dimensional periodic boundary is set to the three-fold rotation axis of trigonal bulk Bi and can be defined as the longitudinal direction of Bi nanowire, as shown in figure 2. Wire diameters of nanowire models, $2R$, were set to about 2 nm, and a periodic boundary condition along the transverse or perpendicular direction to the nanowire in the three-dimensional hexagonal supercell was given by inserting sufficient space between nanowires with a large cell parameter $L_{\perp}$, as listed in figure 2. Similarly as bulk models, $Bi_{1-x}Sb_x$ nanowire models can be introduced by replacing some Bi atoms with Sb, as explained in detail later. On the condition that sufficient space in the $Bi_{1-x}Sb_x$ nanowire model disturbs the interaction between nanowires, the electronic band structure of the $Bi_{1-x}Sb_x$ nanowire model can be reduced to be one dimensional, which is dependent on only one reciprocal coordinate, k_z.

2.3. *Evaluation of Seebeck coefficient*

The Seebeeck coefficient S is given by the ratio of electromotive force $\vec{E}$ to temperature gradient $\vec{\nabla}_{\vec{r}}T$ incorporated implicitly in the Boltzmann transport equation [29,30]. In the stationary state, velocity distribution function of carriers f in semiconductor should be derived from the Boltzmann transport equation by use of relaxation time τ as

$$f(\varepsilon)-f_0(\varepsilon)=\tau(\varepsilon)\vec{v}(\varepsilon)\left(df_0/d\varepsilon\right)\cdot\left\{e\vec{E}+[(\varepsilon-\varepsilon_{\mathrm{F}})/T]\vec{\nabla}_{\vec{r}}T\right\}, \tag{3}$$

where e is the elementary charge, $\vec{v}$ is carrier velocity, f_0 is the Fermi-Dirac distribution, and ε_F is the Fermi energy. Then, S can be evaluated from the information from band structure of the material in terms of integrals including carrier DOS, $g(\varepsilon)$, with respect to energy ε by

$$S = -(1/eT)\left\{\left[\int g(\varepsilon)\tau(\varepsilon)\varepsilon v^2(\varepsilon)(df_0/d\varepsilon)d\varepsilon \Big/ \int g(\varepsilon)\tau(\varepsilon)v^2(\varepsilon)(df_0/d\varepsilon)d\varepsilon\right] - \varepsilon_{\mathrm{F}}\right\}, \qquad (4)$$

where interval of the integrals corresponds to the region of carrier occupation [11]. All physical properties in the above equation except for the relaxation time can be easily determined from the first-principles electronic state [9,10,14]. In this paper, $v^2(\varepsilon)$ was evaluated from

$$\vec{v}_j(\varepsilon_{j,\vec{k}}) = (1/\hbar)\left(\partial\varepsilon_{j,\vec{k}} / \partial\vec{k}\right), \qquad (5)$$

where $\vec{v}_j(\varepsilon_{j,\vec{k}})$ is the carrier velocity for jth subband at $\vec{k}$ point with band energy $\varepsilon_{j,\vec{k}}$. At the case that the relaxation time is constant regardless of ε, the effect of τ on S is canceled.

Bi and Sb are typical semimetals. In practical n- or p-doped semimetals, the total number of carriers per unit cell, $\delta = nV$, must be a few orders less than 1, where n is the carrier concentration and V is the volume of structure in the unit cell. Under the condition that a small amount of carrier occupation does not cause significant change in the band structure, δ can be given by an appropriate shift of ε_{F} as

$$\delta = \pm\left[\int 2g(\varepsilon)f_0(\varepsilon)d\varepsilon - N\right] = \pm\left[2\sum_{\mathrm{All}\,j}\sum_{\vec{k}} w_{\vec{k}} f_0(\varepsilon_{j,\vec{k}}) - N\right], \qquad (6)$$

where N is the total number of valence electrons in the system, $w_{\vec{k}}$ is the k-point weight for $\vec{k}$, and plus-minus sign means plus for n-doped state or minus for p-doped state [14]. In this study, the Fermi energy ε_{F} was determined according to (6) for each condition, and then the value of S was evaluated by using (4) under the condition that the relaxation time is constant. Practically, (4) can be rewritten as

$$S = -(1/eT)\left\{\left[\sum_j\sum_{\vec{k}} w_{\vec{k}} v^2(\varepsilon_{j,\vec{k}})\varepsilon_{j,\vec{k}} \frac{df_0(\varepsilon_{j,\vec{k}})}{d\varepsilon} \Big/ \sum_j\sum_{\vec{k}} w_{\vec{k}} v^2(\varepsilon_{j,\vec{k}}) \frac{df_0(\varepsilon_{j,\vec{k}})}{d\varepsilon}\right] - \varepsilon_{\mathrm{F}}\right\}. \qquad (7)$$

The sampling for $w_{\vec{k}}$ is performed using the three-dimensional Gauss-Legendre quadrature [31] with $29 \times 29 \times 29$ $\vec{k}$ points for bulk $Bi_{1-x}Sb_x$ models and the Newton-Cotes formula [32] with 21 k_z points for $Bi_{1-x}Sb_x$ nanowire models.

3. Results and discussion

3.1. Optimized structure of $Bi_{1-x}Sb_x$ models

In the regular first-principles calculation, lattice constants of the optimized structure of Bi_2 model were obtained as a = 4.767 Å and α = 57.7°, under the Monkhorst-Pack 6 × 6 × 6 $\vec{k}$ point set. These values are close to the measured values, a = 4.746 Å and α = 57.2° [4], so we judged that our calculation level is reasonable for the simulation in this paper. Each Bi atom is bonding to six neighboring Bi atoms, and its six Bi–Bi bonds are classified into three short bonds (3.138 Å) and three long bonds (3.502 Å). By using the QHA, the lattice constant a at 100 K was obtained as 4.772 Å. The calculated linear coefficient of thermal expansion, 10.5×10^{-6} K^{-1}, has the same order as the measured value at room temperature, 13.4×10^{-6} K^{-1} [33].

For the extended-cell models of single-crystal bulk $Bi_{1-x}Sb_x$, the lattice constants of each optimized structure are tabulated in table 1. The result of optimized structure of Bi_8 model is not completely identical with that of Bi_2 model owing to difference of the calculation level such as the fast-Fourier-transform grids, but the lattice constants are also close to the measured values for bulk Bi. In the Sb-

IOP Conf. Series: Journal of Physics: Conf. Series **939** (2017) 012019 doi:10.1088/1742-6596/939/1/012019

substituted models, six Sb–Bi and Sb–Sb bonds for each Sb atom are also classified three short bonds and three long bonds as shown in table 2, and significant change of internal coordinates for Sb atoms

Table 1. Cell parameters of the 8-atom extended-cell models.

Model	a' (Å)	α' (deg)
Bi_8	6.614 (a = 4.758 Å)	88.0 (α = 57.7°)
Bi_7Sb_1	6.590	87.8
Bi_6Sb_2-a	6.548; 6.588	87.6; 88.1
Bi_6Sb_2-b	6.555; 6.556	87.9; 87.9
Bi_6Sb_2-c	6.565	87.7

Table 2. Bond lengths of the 8-atom extended-cell models (Å).

Model	Short Bi–Bi	Long Bi–Bi	Short Sb–Bi	Long Sb–Bi	Sb–Sb
Bi_8	3.154	3.475			
Bi_7Sb_1	3.144–3.152	3.456–3.459	3.057	3.553	2.923
Bi_6Sb_2-a	3.129–3.150	3.420–3.470	3.070	3.504–3.682	
Bi_6Sb_2-b	3.145–3.151	3.424–3.425	3.059–3.066	3.510–3.514	
Bi_6Sb_2-c	3.147	3.441	3.051	3.533	

owing to its smaller atomic radius makes the short Sb–Bi bonds shorter and the long Sb–Bi bonds longer than the short and long Bi–Bi bonds, respectively, in bulk Bi. Therefore, it is clearly found that the long Sb–Bi bonds in bulk $Bi_{1-x}Sb_x$ are effected by much weaker interaction than the long Bi–Bi bonds in bulk Bi, and the weaker interaction would affect the phonon structure in $Bi_{1-x}Sb_x$ alloy with small x.

Figure 3 shows the phonon dispersion diagrams for the single-crystal Bi_8 and Bi_7Sb_1 models. The difference in frequencies between Bi_8 and Bi_7Sb_1 is mainly due to mass of Sb atom, but some effects by the peculiar Sb–Bi interaction should be involved in the phonon structure. The clear difference in phonon dispersion leads to the significant difference in thermal conductivity [11,29].

3.2. Seebeck coefficient of single-crystal bulk $Bi_{1-x}Sb_x$ models

The electronic band diagrams near the Fermi energy for the extended-cell models of single-crystal bulk $Bi_{1-x}Sb_x$ are shown in figure 4. For all models, carriers are mainly located in the vicinity of Z point in the Brilloin zone of the extended unit cell. The band diagrams closely resembles one another, owing to similarity of electronic state in Bi and Sb as the members of Group 15.

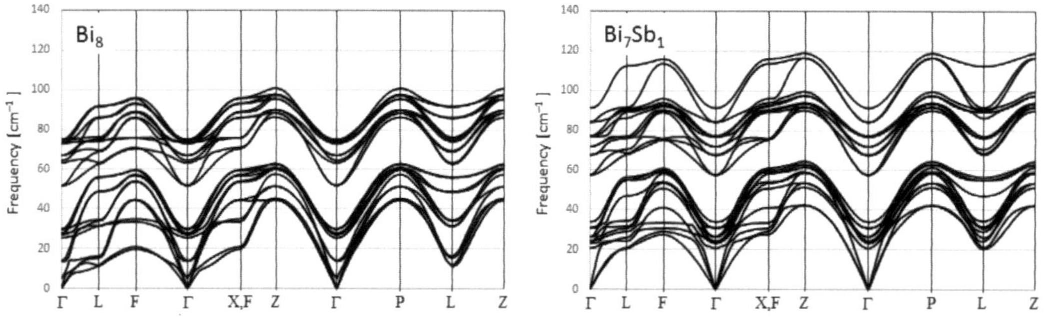

Figure 3. Phonon dispersion diagrams of the Bi_8 and Bi_7Sb_1 models. Paths in the reciprocal lattice space are referred to figure 4.

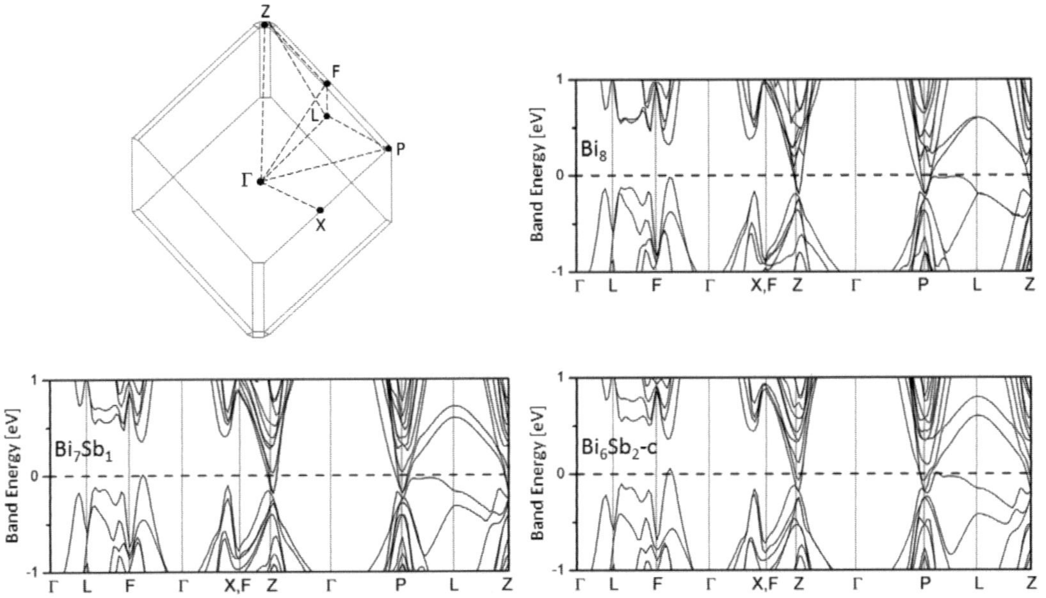

Figure 4. Brillouin zone of the 8-atom extended cell and electronic band diagrams of the Bi_8, Bi_7Sb_1, and Bi_6Sb_2-c models. Zero energy is defined at the Fermi level.

Temperature and doped-state dependences of the calculated Seebeck coefficients for the extended-cell models of bulk $Bi_{1-x}Sb_x$ are shown in figure 5. The effect of volume expansion owing to temperature increase is quite small, so the geometries of model structures listed in tables 1 and 2 were used for the Seebeck coefficient simulation. A meaningful effect by the composition change on the Seebeck coefficient has not been observed in our simulation. As common characteristics in the Bi_8 and Sb-substituted models from 100 to 200 K, absolute value of the Seebeck coefficient for the intrinsic semi-metal state is small, and the Seebeck coefficients for n- and p-doped states respectively show negative and positive values, similarly as the cases for semiconducting materials [9,10,29]. Practically, the calculated Seebeck coefficients are quantitatively very sensitive to the k-point sampling, and the quantity of them is smaller than that of measurement values for bulk $Bi_{1-x}Sb_x$, – 110 ± 10 μV K^{-1}, with

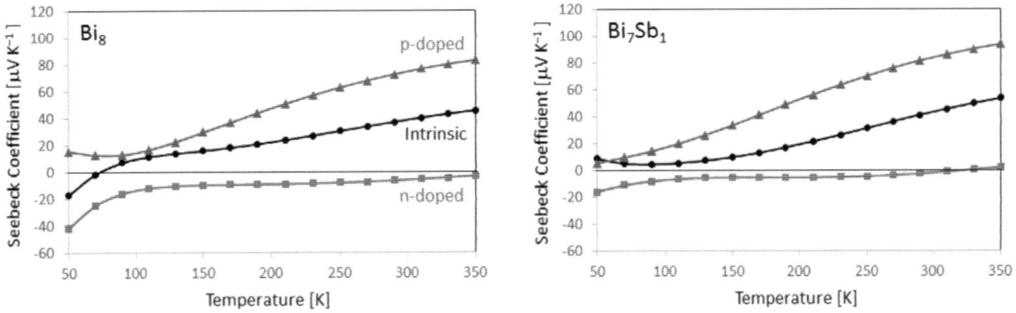

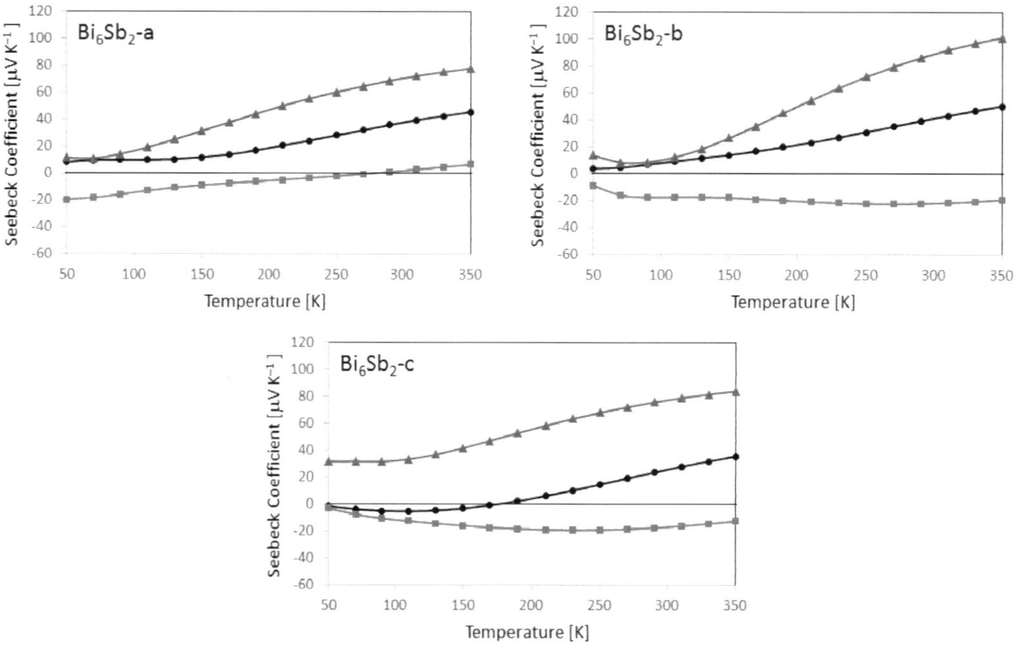

Figure 5. Temperature and doped-state dependences of the Seebeck coefficients for the Bi_8, Bi_7Sb_1, Bi_6Sb_2-a, Bi_6Sb_2-b, and Bi_6Sb_2-c models. Circle (black), square (red), and triangle (blue) plots respectively correspond to the intrinsic state, n- and p-doped state with 10^{20} cm^{-3} carrier concentration.

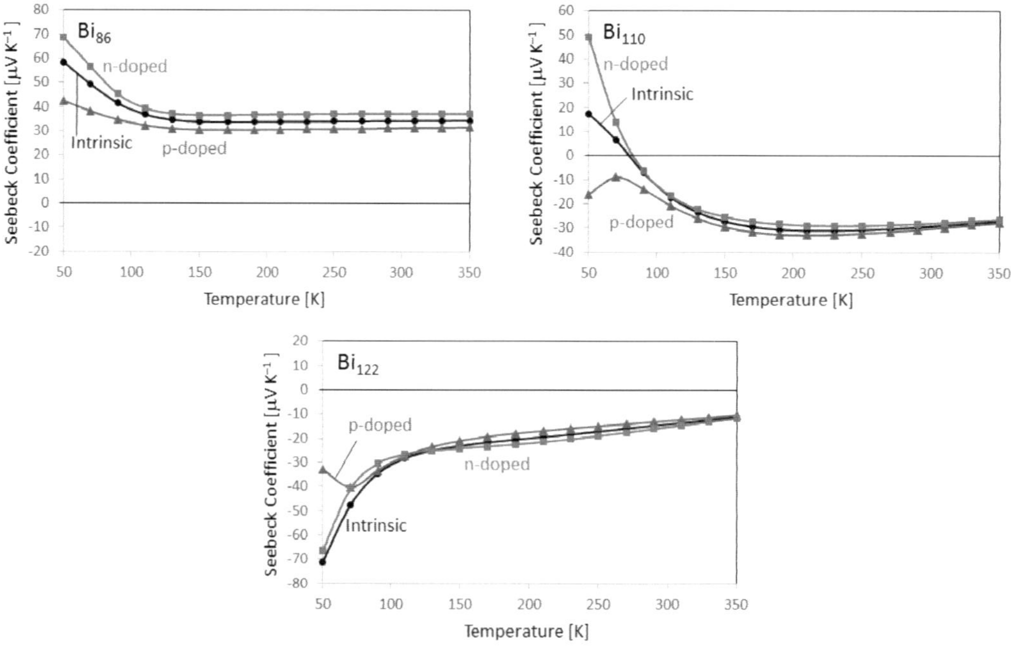

Figure 6. Temperature and doped-state dependences of the Seebeck coefficients for the Bi_{86}, Bi_{110}, and Bi_{122} nanowire models. Circle (black), square (red), and triangle (blue) plots respectively correspond to the intrinsic state, n- and p-doped state with 10^{20} cm^{-3} carrier concentration.

x = 0.05 – 0.12 [5,6]. The reason of underestimate in simulation would be due to the single-crystal geometries of the Bi_8 and Sb-substituted models and inevitable drops in electronic band energies for subbands upper than the Fermi level in DFT calculation.

3.3. Seebeck coefficient of $Bi_{1-x}Sb_x$ nanowire models

Figure 6 shows temperature and doped-state dependences of the calculated Seebeck coefficients for the Bi nanowire models. In our simulation, the nanowire model with uneven surface such as Bi_{86} has positive Seebeck coefficients. On the contrary, as the surface condition gets smooth like the Bi_{122} nanowire model, the Seebeck coefficients tend to show negative values. Accordingly, the thermoelectric properties should be strongly affected by the surface condition of Bi. For the application to thermoelectric devices, the surface condition of device materials should be taken into careful consideration.

$Bi_{1-x}Sb_x$ alloys have negative Seebeck coefficient in the measurement [5,6] as mentioned above, so we tried to simulate the Seebeck coefficient in $Bi_{1-x}Sb_x$ nanowires, based on the Bi_{122} model. We have introduced four models of $Bi_{0.85}Sb_{0.15}$ nanowire in the case of x = 0.15 by replacing eighteen Bi atoms with Sb, as shown in figure 7(a), with keeping the three-fold rotation symmetry. The calculated Seebeck coefficients for the $Bi_{104}Sb_{18}$ nanowire models are shown in figure 7(b) and (c). Similarly as the simulation in the bulk models, a meaningful effect of the atom replacement on the Seebeck coefficient has not been observed. Among these models, the $Bi_{104}Sb_{18}$-D model, where some surface Bi atoms have been replaced with Sb, has a relatively large quantity of negative Seebeck coefficient in the range from 100 to 200 K.

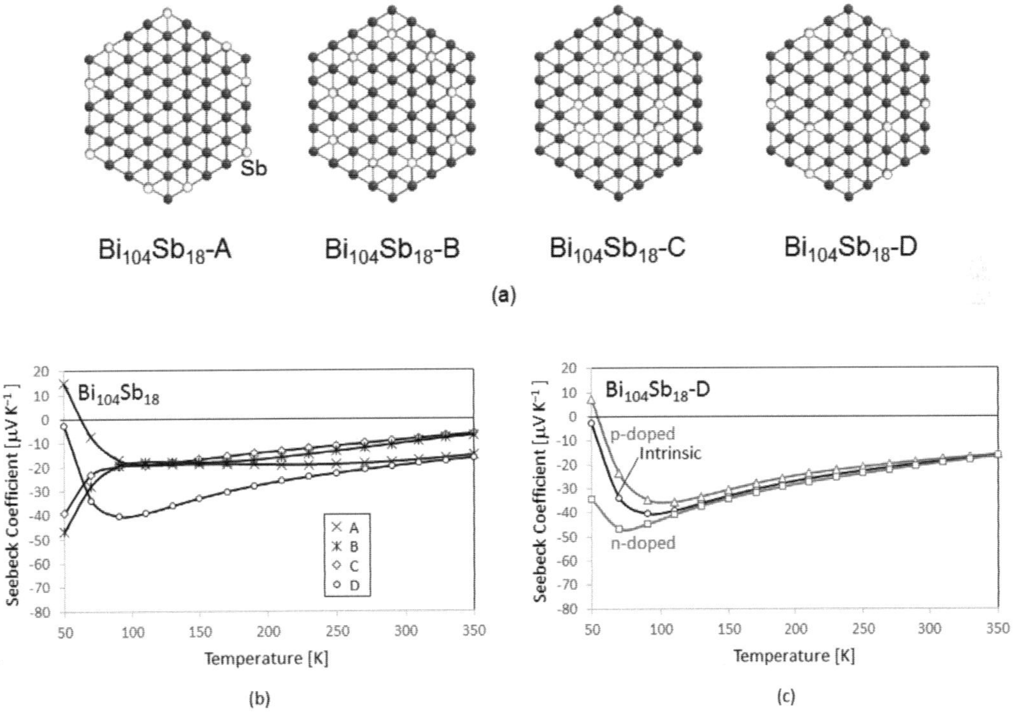

Figure 7. The Seebeck coefficients for the $Bi_{104}Sb_{18}$ nanowire models: (a) top views of the unit cell of $Bi_{104}Sb_{18}$-A, $Bi_{104}Sb_{18}$-B, $Bi_{104}Sb_{18}$-C, and $Bi_{104}Sb_{18}$-D nanowire models, where 9 yellow atoms are Sb, and other 9 Sb atoms are located point-symmetrically on the back side; (b) temperature dependence of the Seebeck coefficients for them; (c) temperature and doped-state dependences of the Seebeck coefficients for the $Bi_{104}Sb_{18}$-D nanowire model.

4. Summary

We simulated thermoelectric properties of Bi-Sb alloy system on the basis of first-principles calculation. Atomistic model structures of $Bi_{1-x}Sb_x$ alloy were devised in the forms of single-crystal bulk (x = 0, 0.125, and 0.25) and one-dimensional nanowire (x = 0 and 0.15) under the periodic boundary condition. The cell parameters of the bulk model were simulated with respect to temperature by the QHA [21,22] through phonon calculation, and dependences of the Seebeck coefficient on composition, surface condition, and temperature have been demonstrated by using our original methodology [9,10,14] in terms of the electronic state of $Bi_{1-x}Sb_x$ alloy models. For the single-crystal bulk $Bi_{1-x}Sb_x$ models, a meaningful effect of the composition on the Seebeck coefficient has not been observed, whereas a clear difference in phonon dispersion was confirmed between pure Bi and Sb-substituted Bi, leading to the significant difference in thermal conductivity [11,29]. Furthermore, we clarified that the surface condition is a key point to control the Seebeck coefficient for the nanowire form. For the application to thermoelectric devices, the surface condition of device materials should be taken into careful consideration.

Acknowledgment

This work was partially supported by JSPS KAKENHI Grant Number 26420665, for which K. N. expresses his gratitude.

References

[1] Zemskov V S, Belaya A D, Beluy U S and Kozhemyakin G N 2000 *J. Cryst. Growth* **212** 161
[2] El-Asfoury M S, Nasr M N A, Nakamura K and Abdel-Moneim A 2016 *Jpn. J. Appl. Phys.* **55** 045802
[3] Rabina O, Lin Y-M and Dresselhaus M S 2001 *Appl. Phys. Lett.* **79** 81
[4] Lenoir B, Scherrer H and Caillat T 2001 *Semicond. Semimet.* **69** 101
[5] Smith G E and Wolfe R 1962 *J. Appl. Phys.* **33** 841
[6] Goldsmid H J 1970 *Phys. Stat. Sol. A* **1** 7
[7] Teramoto T, Komine T, Yamamoto S, Kuraishi M, Sugita R, Hasagawa Y and Nakamura H 2008 *J Appl. Phys.* **104** 053714
[8] Goldsmid H J 1964 *Thermoelectric Refrigeration* (New York: Plenum)
[9] Nakamura K 2016 *Jpn. J. Appl. Phys.* **55** 06GJ07
[10] Nakamura K 2017 *Key Eng. Mater.*, in press
[11] Pichanusakorn P and Bandaru B 2010 *Mater. Sci. Eng. R* **67** 19
[12] Hicks L D and Dresselhaus M S 1993 *Phys. Rev. B* **47** 12727
[13] Hicks L D and Dresselhaus M S 1993 *Phys. Rev. B* **47** 16631
[14] Nakamura K 2016 *IOP Conf. Ser.: Mater. Sci. Eng.* **108** 012040
[15] Kresse G and Hafner J 1993 *Phys. Rev. B* **47** 558(R)
[16] Kresse G and Furthmüller J 1996 *Phys. Rev. B* **54** 11169
[17] Hohenberg P and Kohn W 1964 *Phys. Rev.* **136** B864
[18] Perdew J P and Wang Y 1992 *Phys. Rev. B* **45** 13244
[19] Perdew J P, Chevary J A, Vosko S H, Jackson K A, Pederson M R, Singh D J and Fiolhais C 1992 *Phys. Rev. B* **46** 6671
[20] Vanderbilt D 1990 *Phys. Rev. B* **41** 7892
[21] Allen R E and de Wette F W 1969 *Phys. Rev.* **179** 873
[22] Maradudin A A, Montroll E W, Weiss G H and Ipatva I P 1971 *Theory of Lattice Dynamics in the Harmonic Approximation, 2nd ed.* (New York: Academic)
[23] Xie J, de Gironcoli S, Baroni S and Scheffler M 1999 *Phys. Rev. B* **59** 965
[24] Kuwabara A, Tohei T, Yamamoto T and Tanaka I 2005 *Phys. Rev. B* **71** 064301
[25] Frankcombe T J and Kroes G-J 2006 *Phys. Rev. B* **73** 174302
[26] Minamoto S, Kato M, Konashi K and Kawazoe Y 2009 *J. Nucl. Mater.* **385** 18
[27] Nishitani S R, Takeda R, Ishii H, Yamamoto Y and Kaneko T 2009 *J. Jpn. Inst. Metals* **73** 566

[28] Xiang S, Xi F, Bi Y, Xu J, Geng H, Cai L, Jing F and Liu J 2010 *Phys. Rev. B* **81** 014301
[29] Ioffe A F 1957 *Semiconductor Thermoelements and Thermoelectric Cooling* (London: Infosearch)
[30] Lundstrom M 2000 *Fundamentals of Carrier Transport* (Cambridge: Cambridge Press)
[31] Hildebrand F B 1956 *Introduction to Numerical Analysis* (New York: McGraw-Hill) p 323
[32] Hildebrand F B 1956 *Introduction to Numerical Analysis* (New York: McGraw-Hill) p 160
[33] Lide D R 2003 *AIP Physics Desk Reference* 3rd edition, ed E R Cohen, D R Lide and G L Trigg (New York: Springer-Verlag) chapter 27 p 826

IOP Conf. Series: Journal of Physics: Conf. Series **939** (2017) 012020 doi:10.1088/1742-6596/939/1/012020

Self-powered heat-resistant polymeric 1D nanowires and 3D micro/nanowire assemblies in a pressure-crystallized size-distributed graphene oxide/poly (vinylidene fluoride) composite

Pengfei Tian [1], Jun Lyu [1, a, b], Rui Huang [2], Chaoliang Zhang [3]

[1] Key Laboratory of Advanced Technologies of Materials, Ministry of Education, School of Materials Science and Engineering, Southwest Jiaotong University, Chengdu 610031, Sichuan, China
[2] College of Polymer Materials Science and Engineering, Sichuan University, Chengdu 610065, Sichuan, China
[3] State Key Laboratory of Oral Diseases, West China Hospital of Stomatology, Sichuan University, Chengdu 610041, China

[a] Corresponding author's e-mail address: junluprc@hotmail.com

[b] Other used names Jun Lu and Jun Lv

Abstract. Piezoelectric one- (1D) and three-dimensional (3D) hybrid micro/nanostructured materials have received intense research interest because of their ability in capturing trace amounts of energy and transforming it into electrical energy. In this work, a size-distributed graphene oxide (GO) was utilized for the concurrent growth of both the 1D nanowires and 3D micro/nanowire architectures of poly (vinylidene fluoride) (PVDF) with piezoelectricity. The in situ formation of the polymeric micro/nanostructures, with crystalline beta phase, was achieved by the high-pressure crystallization of a well dispersed GO/PVDF composite, fabricated by an environmentally friendly physical approach. Particularly, by controlling the crystallization conditions of the binary composite at high pressure, the melting point of the polymeric micro/nanowires, which further constructed the 3D micro/nanoarchitectures, was nearly 30°C higher than that of the original PVDF. The large scale simultaneous formation of the 1D and 3D micro/nanostructures was attributed to a size-dependent catalysis of the GOs in the pressure-treated composite system. The as-fabricated heat-resistant hybrid micro/nanoarchitectures, consisting of GOs and piezoelectric PVDF micro/nanowires, may permit niche applications in self-powered micro/nanodevices for energy scavenging from their working environments.

1. Introduction

Effective mechanical-to-electrical conversion is of great importance in powering small electronic devices and systems using the vibrational energy available in their working environments [1-3]. Of the current developed energy scavenging techniques, piezoelectric transduction, based on self-powered 1D micro/nanowires or their 3D assemblies, has received the largest attention, and already shown exciting application potentials in electronics, optoelectronics, sensors and biological science [1, 2]. Compared with their inorganic counterparts, such as ZnO, PZT and $BaTiO_3$ [3], the lightweight and conformable micro/nano-structured piezoelectric polymeric materials have more advantages in the realization of all-organic-based self-powered electronic devices [4-6].

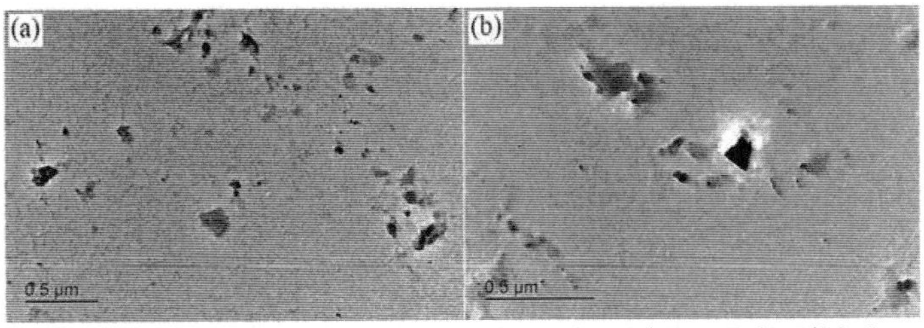

Figure 1. Typical TEM micrographs of the as fabricated GO/PVDF (0.5/99.5, wt/wt) composite.

PVDF is one of the limited known polymers with large piezoelectric coefficients, exhibiting a pronounced polymorphism that transforms between several crystal forms under specified conditions, and successful development of piezoelectric devices with the polymer depends on the effective fabrication of polar crystalline structures, such as β and γ [4-6]. Recently, nanostructured carbon allotropes, such as fullerenes [7, 8], carbon nanotubes [9, 10], graphene and its nanoplatelet [11], and graphene quantum dots [12], were utilized to induce the high-pressure self-assembly of PVDF molecules into polymeric micro/nano-architectures with piezoelectric crystalline β phase. In this study, GO, a 2D carbon-based nanomaterial, with carboxylic acid moieties on the basal planes and edges of its sheets [13], was used in achieving control over the size, morphology and molecular ordering of PVDF micro/nanostructures. By the crystallization of a well dispersed binary GO/PVDF composite at high pressure, a large scale formation of the self-powered GO-based hybrid 3D micro/nanowire assemblies was realized with remarkably improved heat resistance.

2. Experimental

GO (>99% wt/wt purity), with 0.55-1.2nm thickness, was supplied by Chengdu Organic Chemicals Co., Ltd, Chinese Academy of Science, and used as received. PVDF powder, commercial-grade Solef 6010, was supplied by Solvay Co., Ltd, Belgium. The corresponding weight-average molecular weight, Mw, was 322000g/mol. GO and PVDF were pre-mixed at 23000rpm and room temperature in a commercial Joyoung JYL-C012 blender. This was followed by melt compounding at 60rpm, 200°C for 15min using a ZJL-300 torque rheometer.

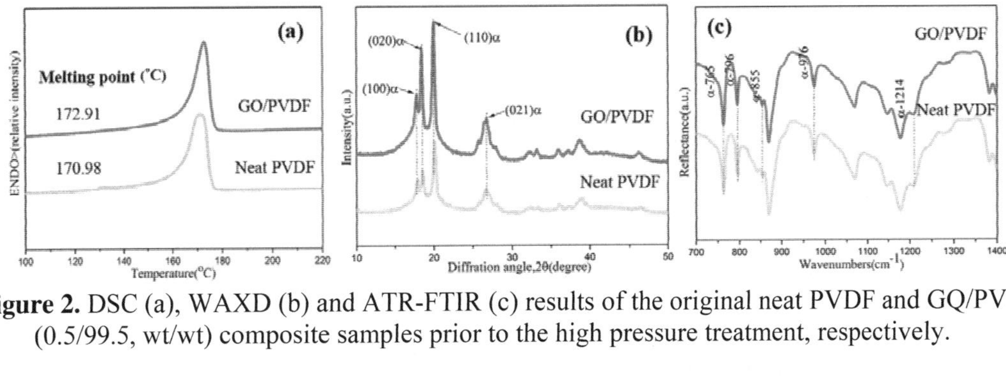

Figure 2. DSC (a), WAXD (b) and ATR-FTIR (c) results of the original neat PVDF and GQ/PVDF (0.5/99.5, wt/wt) composite samples prior to the high pressure treatment, respectively.

High-pressure experiments for the GO/PVDF composite were carried out with a self-made piston-cylinder apparatus [7]. The following procedure for crystallization was used. After loading the samples, the temperature was increased to 200°C to allow them to be melted. Then a low pressure (150MPa) was applied, and the temperature was raised to a predetermined level. After equilibrium was established, the pressure was further raised to the predetermined value. These samples were kept under these conditions for a scheduled time, and then quenched down to ambient condition.

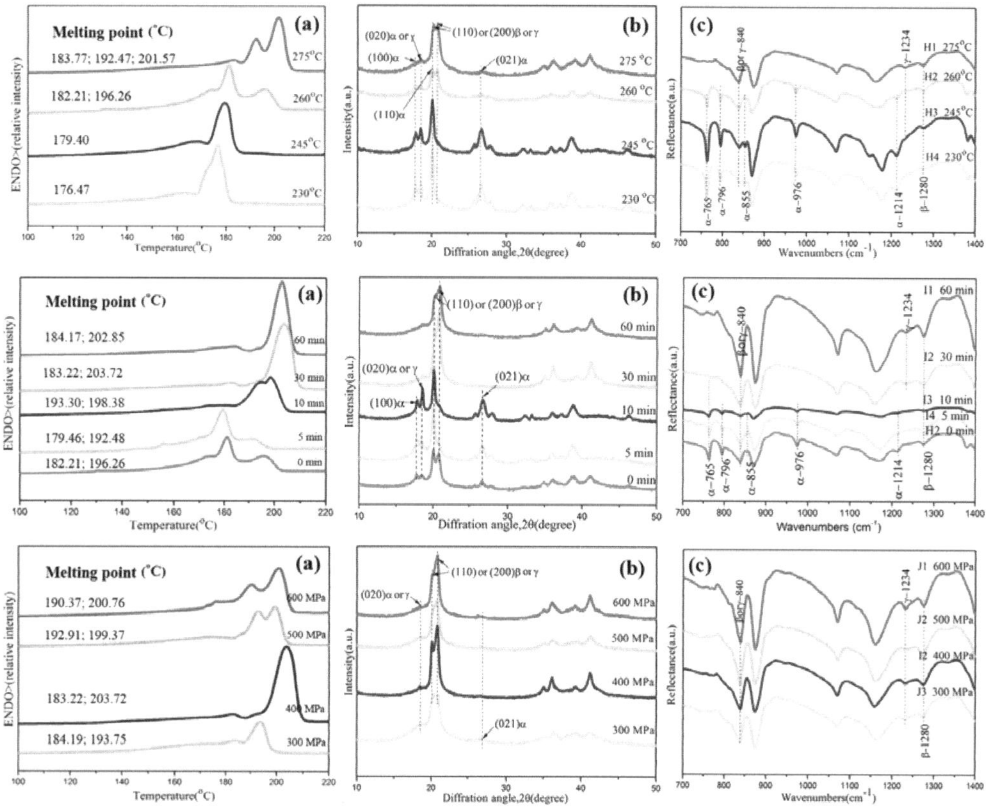

Figure 3. DSC (a), WAXD (b) and ATR-FTIR (c) results of the GO/PVDF (0.5/99.5, wt/wt) composite samples crystallized at 400MPa and different maximum quenching temperature (top), 400MPa, 260°C for different time (middle), and different pressure, 260°C for 30min (bottom).

Transmission electron microscopy (TEM) detections were performed with a JEOL JEM-2100F apparatus. Differential scanning calorimetry (DSC) measurements were conducted at atmospheric pressure by using a TA-Q20 instrument. Wide-angle X-ray diffraction (WAXD) results were obtained at room temperature with a PANalytical X'pert PRO diffractometer. Attenuated total reflectance Fourier transform infrared spectroscopy (ATR-FTIR) data were obtained using a Nicolet 5700 spectrometer. After the WAXD and ATR-FTIR characterizations, the sample surfaces were etched, and then coated with gold for scanning electron microscopy (SEM) observations using a JSM-6330F apparatus.

3. Results and discussion

Figure 1 shows the typical TEM micrographs of the as fabricated size-distributed GO/PVDF (0.5/99.5, wt/wt) composite, just before the applied treatment by pressure. As can be seen, an overall good dispersion of the GOs, with lateral dimensions from a few nanometers to several hundreds of nanometers, was achieved in PVDF matrix by the used easy physical and mechanical route.

A comparison experiment was conducted for the original pristine PVDF and the GO/PVDF composite samples, prior to the high pressure crystallization, and the results are shown in Figure 2. Although a slight increase of the melting point and melting enthalpy was observed by the introduction of GOs, no crystalline β form, essential in the creation of piezoelectricity of the polymer, was crystallized in the GO/PVDF composite. The characteristic bands of IR and WAXD were almost the same, for the PVDF and its GO composite, which should be attributed to the reflections of monoclinic α phase.

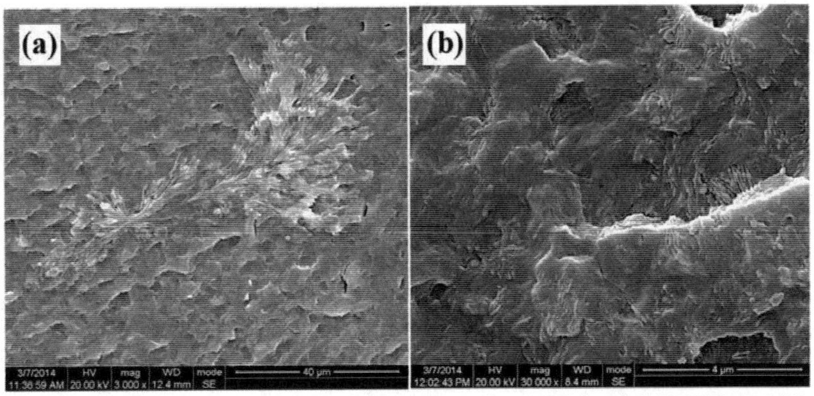

Figure 4. SEM micrographs of the etched fracture surfaces of the GO/PVDF (0.5/99.5, wt/wt) composite samples, crystallized at 400MPa and different maximum quenching temperature: (a) 230°C; (b) 275°C.

Figure 3 gives out the DSC, WAXD and IR results of the GO/PVDF (0.5/99.5, wt/wt) composite samples after the high pressure treatment by varying temperature, pressure and crystallization time. With the increase of crystallization temperature, the melting point at high-temperature side increased, as revealed by the DSC data (Figure 3a, top). Also, the WAXD (Figure 3b, top) and IR (Figure 3c, top) reflections suggested that the content of α and β phase in GO/PVDF decreased and increased, respectively, and γ phase began to appear when the maximum quenching temperature reached 275°C. Multiple melting behaviors were observed for the composite samples crystallized at the same pressure and temperature for an increased time (Figure 3a, middle). Particularly, the main-peak melting point of the GO/PVDF sample crystallized at 400MPa, 260°C for 30min attained 203.72°C, which is around 30°C higher than that of the original PVDF. No α phase was detected for the composite samples crystallized for more than 30min, and their crystalline structures were totally with β and γ phase lamellae (Figure 3b and c, middle). The inappropriate increase of pressure hindered the melting point from further increasing (Figure 3a, bottom). However, α phase crystals were completely removed from the GO/PVDF simply by the pressure increase (Figure 3b and c, bottom). We noted that GO/PVDF composite samples with melting points above 200°C, corresponding to the formation of polar extended-chain crystals of the polymer, could be obtained through an appropriate combination of crystallization conditions.

The above DSC, WAXD and IR data were further confirmed by the morphological observation of the high pressure crystallized GO/PVDF samples. For example, featured α form spherulites covered the etched surface of the sample crystallized at 400MPa and 230°C, and no trace of γ phase crystallite was observed (Figure 4a). Moreover, the number of fiber-like structures of β crystallites increased with the increase of crystallization temperature. Especially, β-form fibrils with striated appearance, the most characteristic feature for polymer extended-chain crystals, were clearly observed in the sample crystallized at 275°C (Figure 4b).

Piezoelectric β form crystalline 1D nanowires of PVDF were formed in the high pressure crystallized size-distributed GQ/PVDF (0.5/99.5, wt/wt) composite, and the typical morphology of such nanowires is represented in Figure 5. The nanowires, embedded in the polymer matrix, were observed only by the etching of the amorphous part of the fracture surface, and they totally belong to a crystalline entity. Furthermore, the diameter, length and morphology of the PVDF nanowires were found to be very sensitive to the crystallization conditions. Sometimes the abnormal growth of the nanowires was observed, and they lost their size uniformity. Nevertheless, the polygonal shaped nanowires, with uniform size and morphology, could be crystallized by an appropriate pressure treatment.

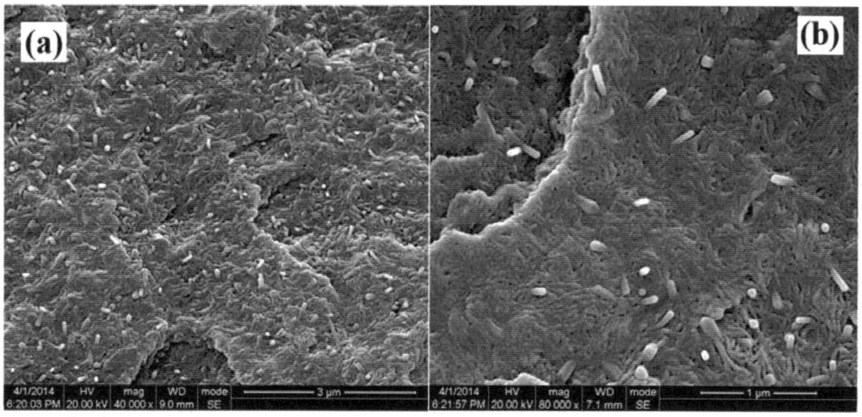

Figure 5. Typical SEM micrographs of the 1D polymeric nanowires in the GO/PVDF (0.5/99.5, wt/wt) composite samples, crystallized at different pressure, 260°C for 30min: (a) 500MPa; (b) 300MPa.

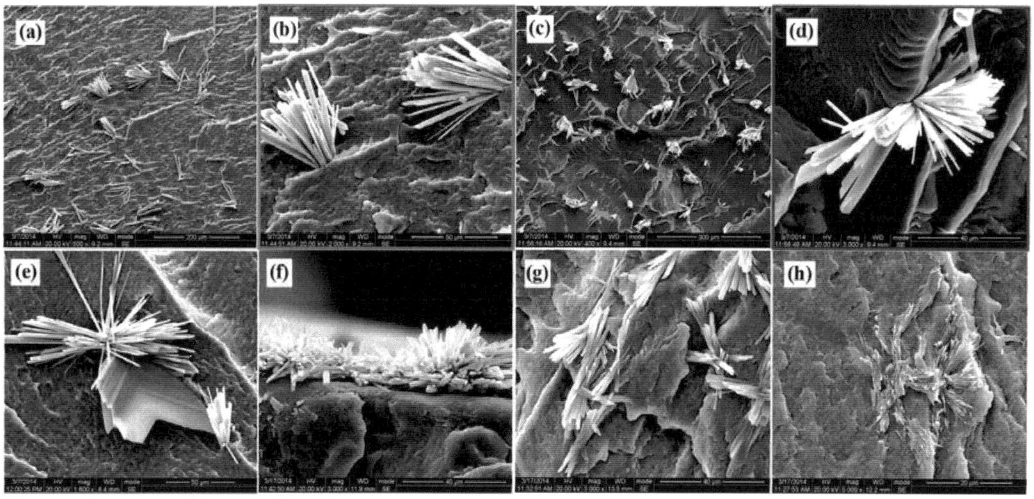

Figure 6. SEM micrographs of the 3D polymeric micro/nanowire architectures in the GO/PVDF (0.5/99.5, wt/wt) composite samples, crystallized at 400MPa and 245 °C (a and b), 260°C (c and d) and 275°C (e) maximum quenching temperature, and 400MPa, 260°C for 10min (f), 30min (g) and 60min (h), respectively.

Figure 6 shows the representative SEM micrographs of the piezoelectric 1D micro/nanowire-assembled 3D architectures in the high-pressure crystallized GO/PVDF composite. The 3D micro/nanowire assemblies, with crystalline β form, were obtained with diversified morphologies and substructures by the variation of crystallization parameters. The DSC, WAXD and IR suggested that the melting point of such micro/nanowire architectures, as well as their molecular ordering, could be increased by the appropriate increase of crystallization temperature and time at the same pressure. Consistently, the substructures of the 3D assemblies were transformed from folded-chain lamellae into extended-chain lamellae. This finally resulted in the growth of the unique heat-resistant 3D micro/nanowire assemblies, as shown in Figure 6e, g and h.

The concurrent growth of the piezoelectric 1D and 3D polymeric micro/nanostructures, with super heat resistance, should be attributed to the self-assembly of PVDF macromolecular chains, induced by

IOP Conf. Series: Journal of Physics: Conf. Series **939** (2017) 012020 doi:10.1088/1742-6596/939/1/012020

a size-dependent catalysis of the GOs in the pressure-treated composite system [12, 14]. The GOs with smaller lateral dimension catalyzed the assembly of the polymer into micro/nanowires, and these with larger size induced the simultaneous formation of numerous 1D micro/nanowires that finally evolved into the 3D assemblies. Moreover, the existence of carboxylic acid groups in GOs may enhance the interactions of GOs with PVDF molecules in the growth of the GO-based hybrid micro/nanostructures with piezoelectricity.

4. Conclusion

In summary, concurrent in situ growth of 1D nanowires and 3D micro/nanowire assemblies, with crystalline β form, was achieved by the high-pressure crystallization of a well dispersed size-distributed GO/PVDF composite. The size, morphology and substructures of the micro/nano-architectures were controllable by the variation of crystallization conditions, which finally resulted in a large scale formation of piezoelectric polymeric 1D and 3D micro/nanostructures with remarkably improved heat resistance. The simultaneous formation of the 1D and 3D micro/nanostructures was attributed to a size-dependent catalysis effect of the GOs, enhanced further by the multiple interactions between GO and PVDF. The as fabricated GO-based hybrid micro/nanoarchitectures, with excellent thermal stability of their piezoelectric activity, may diversify niche applications in self-powered micro/nanodevices for energy harvesting, especially for these working in higher temperature environments.

Acknowledgements
This project was supported by the National Natural Science Foundation of China (51373139).

References

[1] Wang Z L and Song J 2006 *Science* **163** 242
[2] Qin Y, Wang X D and Wang Z L 2008 *Nature* **451** 809
[3] Pi Z, Zhang J, Wen C, Zhang Z B and Wu D 2014 *Nano Energy* **7** 33
[4] García-Gutiérrez M C, Linares A, Hernández J J, Rueda D R, Ezquerra T A, Poza P and Davies R J 2010 *Nano Lett* **10** 1472
[5] Lu J, Kim S G., Lee S and Oh I K 2008 *Adv Funct Mater* **18** 1290
[6] Lu J, Kim S G., Lee S and Oh I K 2011 *J Control Release* **152** e229
[7] Zhang D P, Tian P F, Chen X, Lu J, Zhou Z W, Fan X M and Huang R 2013 *Compos Sci Technol* **77** 29
[8] Zhang D P, Tian P F, Chen X, Lu J and Huang R 2013 *J Appl Polym Sci* **130** 1823
[9] Huang W J, Li Z P, Tian P F, Chen X, Lu J, Zhou Z W, Huang R, Liu T X, Zhang C L and Wang X L 2014 *Compos Sci Technol* **90** 110
[10] Huang W J, Li Z P, Chen X, Tian P F, Lu J, Zhou Z W, Huang R, David H, He L P, Zhang C L and Wang X L 2014 *Compos Part B-Eng* **62** 126
[11] Tian P, Xu C, Huang H, Chen X, Zhao L and Lyu J 2016 *Gut* **7** A231
[12] Lu C, Zhang L, Xu C W, Yin Z Z, Zhou S B, Wang J X, Huang R, Zhou X Q, Zhang C L, Yang W Q and Lu J 2016 *RSC Adv* **6** 67400
[13] Ogoshi T, Ichihara Y, Yamagishi T and Nakamoto Y 2010 *Chem Commun* **46** 6087
[14] Yang R S and Wang Z L 2007 *Philos Mag* **87** 2097

An antisymmetric cell structure for high-performance zinc bromine flow battery

Yongbeom Kim[1,a], **Joonhyeon Jeon**[2,b,‡]

[1,2] Division of Electronics & Electrical Engineering, Dongguk University, 30 Pildong-ro 1-gil, Jung-gu, Seoul, Republic of Korea.

Author's e-mail : youngbeom88@dgu.edu[a], memory@dgu.edu[b]

Abstract. Zinc-bromine flow batteries (ZBBs) remain a problem of designing a cell with high coulombic efficiency and stability. This problem is caused intrinsically by different phase transition in each side of the half-cells during charge-discharge process. This paper describes a ZBB with an antisymmetric cell structure, which uses anode and cathode with different surface morphologies, for high-discharge capacity and reliability. The structure of the antisymmetric ZBB cell contains a carbon-surface electrode and a carbon-volume electrode in zinc and bromine half cells, respectively. To demonstrate the effectiveness of this proposed ZBB cell structure, Cyclic Voltammetry measurement is performed on a graphite foil and a carbon felt which are used as the surface and electrodes. Charge and discharge cyclic operations are also carried out with symmetric and antisymmetric ZBB cells combined with the two electrode types. Experimental results show that the arrangement of antisymmetric cell structure in ZBB provides a solution to the high performance and durability.

1. Introduction

Zinc-Bromine flow Battery (ZBB) provide advantages of having cost effectiveness, high energy density, high cell voltage, high long cycle life, low cost of maintenance, for energy storage system [1, 2, 3]. The ZBB electrolyte (composed of an aqueous zinc-bromide salt dissolved in distilled water) is stored in two tanks and circulated through the cell stacks constantly. During charge operation, the zinc bromide salt is divided to zinc ions and bromide ions. Bromide ions which have a low reaction rate [4] convert into bromine molecules and zinc ion deposited on electrode surface as little as reacted bromide ions. On discharge operation, the zinc plated on the negative electrode dissolves in to the electrolyte. But, it can't totally dissolve caused by bromine, which has a low reaction rate. For this reason, low coulombic efficiency is presented. Some researchers had effort for enhancing coulombic efficiency using volume electrode [5]. Also, bromine has a big problem which is high electrochemical corrosion with electrode. For this reason, carbon coated electrode used for enhancing electrochemical resistance with bromine [6]. To whom any correspondence should be addressed.

This paper describes a antisymmetric cell structure of ZBB. The antisymmetric cell structure is described that volume electrode which used for cathode structure can provide large surface area for improving reaction rate and surface electrode which used for anodic structure can secure solid phase deposition area. To verify the electrochemical properties and effectiveness of the structure, Cyclic

‡ To whom any correspondence should be addressed.

IOP Conf. Series: Journal of Physics: Conf. Series **939** (2017) 012021 doi:10.1088/1742-6596/939/1/012021

Voltammetry (CV) measurements and cyclic operation test (charge and discharge) are carried out in 2.0 M $ZnBr_2$ electrolyte solution, respectively.

2. Electrode

Generally, carbon is used for electrode of flow battery. [7, 8] Carbon have high economical profit (cheap), high conductivity, electrochemical resistance against Br_2. [9, 10] this carbon electrode is classified by a surface electrode and volume electrode [6]. As depicted in Fig 1, in case of the surface electrode, the electrolyte flows over only the external electrode. So, the redox reaction occur only surface of the electrode. And as shown right side of the Fig 1, flow of the electrolyte runs into volume electrode and the redox reaction occur entire internal electrode. [11]. Because of above properties of the electrodes, volume electrode has a large reaction area than surface electrode. And surface electrode secure area for zinc deposition. So, because of the bromine that remains liquid phase after charge operation, it can't use surface electrode better than use volume electrode as cathode for enhancing reaction rate of the bromine.

For the purpose of certify reaction rate of the bromine, TF6 (SGL, Germany) is kind of graphite foil is used as a surface electrode and GF-031(JNTG, Korea) which is kind of a carbon felt is used as volume electrode. Each property is shown in Table 1.

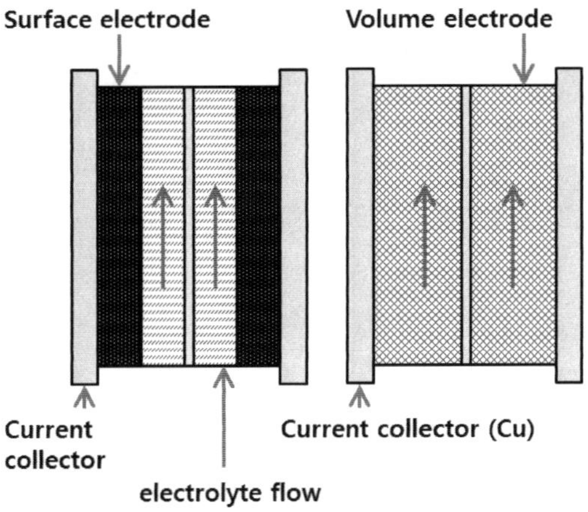

Fig. 1: Structure of the surface electrode, volume electrode.

Table 1: Caption for table goes at the top.

Class	Model Name	Company	Thickness	Electrical Resistivity
			mm	Ω·mm
Surface Electrode	TF6	SGL	0.6	0.13
Volume Electrode	JNT GF-031	JNTG	1	0.03

3. Experiment method

3.1. *Cyclic Voltammetry (CV)*

Cyclic Voltammetry (CV: three electrode system) is a widely used method for the study of electrochemical property of electrode [12, 13]. Surface electrode, volume electrode are used as a working electrode, Pt-wire and standard calomel electrode are used as counter electrode and standard electrode, respectively. Also, 2.0 M of $ZnBr_2$ solution is used for studying electrochemical properties of electrode. 10 mV/S of voltage scan rate and from 1.5 V to -1.5 V of voltage range are injected.

3.1.1. Cell operation test (charge and discharge operation).
Cell operation experiments are carried out for performance comparison of the following five electrode structure : (s) using surface electrode as both side of the electrode, (v) using surface electrode as an anode and using volume electrode as a cathode. In this case thickness of cathode is classified by 1mm (V-1), 2mm (V-2), 3mm (V-3), 4 mm (V-4). 2.0 M of $ZnBr_2$ solution was used to the electrolyte. [14] The charge and discharge current density is 20mA/cm^2. [15, 16] Other operating conditions are described in the table 2, in details. Coulombic efficiency (CE), Voltage efficiency (VE) and Energy efficiency (EE) are calculated by below equations.

Table 2: condition of the unit cell during charge/ discharge

Class	Value [unit]	Remarks
Flow Frame	4x5 [cm^2]	PVDF
Membrane	4x5 [cm^2]	Asahi Kasei / SF600
Active area	6 [cm^2]	
Pump	Masterflex / L/S Pump	
Current Density	20 [mA/cm^2]	
Charge Time	1 [h]	
Operating System	Won-A-Tech / WBCS3000S	

$$Voltage\ Efficiency : \frac{Discharge\ Voltage(V)}{Charge\ Voltage(V)} \times 100[\%] \quad (1)$$

$$\text{Coulombic Efficiency} : \frac{Charge\ Time(\text{sec})}{Discharge\ Time(sec)} \times 100[\%] \quad (2)$$

$$Energy\ Efficiency : \frac{Voltage\ Efficiency \times Coulombic\ Efficiency}{100} [\%] \quad (3)$$

4. Result and discussion

4.1. Cyclic Voltammetry (CV) result
As depicted in Fig 2, oxidation current peak value of surface electrode observed at -80.81 mA. In case of using volume electrode, -171.24mA of oxidation current peak value is observed, respectively. It means that quantity of the redox reaction of bromide cations is enhanced by large reaction surface of the volume electrode. It indicate that using volume electrode as cathode electrode can improve coulombic efficiency And bromine oxidation energy which is above area of the oxidation curve $\int_b^a f(x)\,dx$ dx]is observed that volume electrode oxidation energy is 3 times than surface electrode. And as you can see in the graph, peak current of surface electrode is large than volume electrode. It means that using surface electrode as an anode occur high coulombic efficiency. And as mentioned above, surface electrode can secure deposition area for zinc ion.

For this reason, the antisymmetric structure using surface electrode as an anode and using volume electrode as a cathode is recommended thus result of cyclic voltammetry test. Therefore, cell operation test is performed for demonstrate the result of cyclic voltammetry.

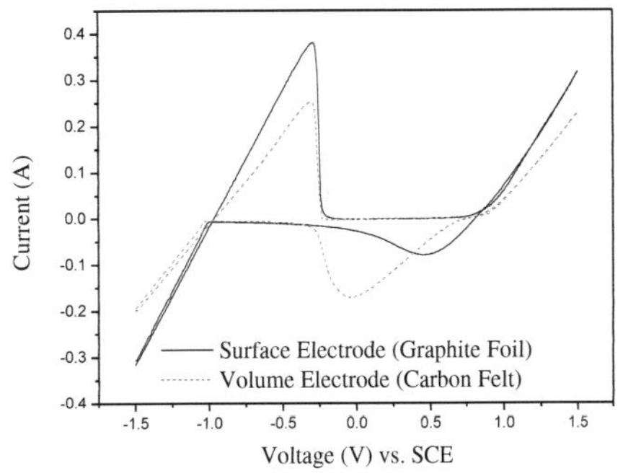

Fig. 2: Cyclic Voltagrams of (a) surface electrode, (b) volume electrode as working electrode

4.2. Cell operation test (charge and discharge operation) result

Cell operation experiments are carried out for compare performance with the following electrode structure: Because of volume electrode has large surface area for reaction, that's why energy efficiency of volume cathode are higher than surface electrode (average 20.49% higher). And discharge capacity using volume electrode as a cathode is longer than using surface electrode as a cathode. As depicted in Fig 3 (a), (b) and Table 3, cell which used surface electrode as a cathode shown low efficiency and low stability, as mentioned above.

(V-1), (V-2), (V-3), (V-4) are tested for optimizing. As a result, (V-3) is the best solution for ZBB. Because, (V-3) has highest energy efficiency and coulombic efficiency (discharge capacity). Because, the thicker volume electrode is used, the higher coulombic efficiency is observed. But, (V-4) has lower coulombic efficiency than (V-3). It is due to the fact that bromine reaction is higher at (V-3). Accordingly, this indicates that (V-3) is used for high energy efficiency.

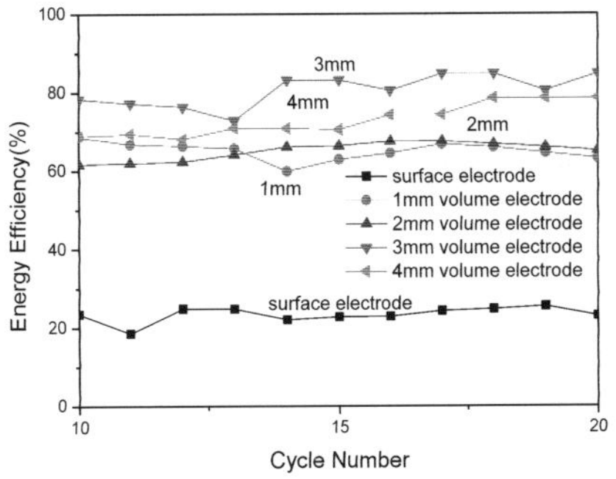

(a)

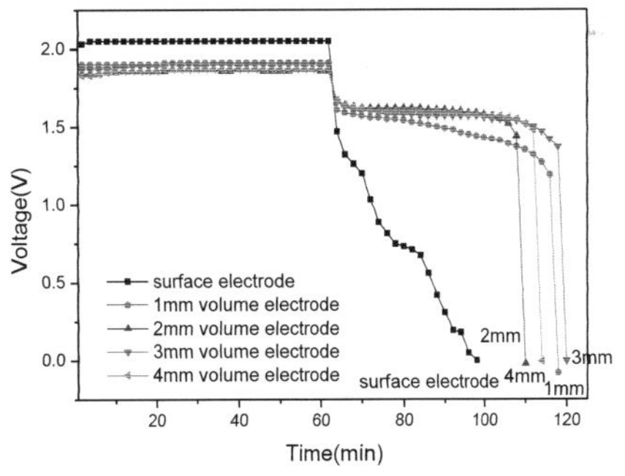

(b)

Fig. 3: (a) energy efficiency, (b) voltage curve(11th) of each cathode

Table. 3: Average efficiencies of each cathode (20cycle)

Efficiency	Surface Electrode	Volume Electrode (thickness)			
		1mm	2mm	3mm	4mm
Coulombic	56.26	72.89	74.60	83.76	81.80
Voltaic	36.84	73.96	79.59	80.93	83.60
Energy	20.49	53.90	59.38	69.67	68.05

5. Conclusion

In this paper, CVs and cell-operation experiments using 2.0 M of $ZnBr_2$ solution are carried out for optimizing proper electrode structure in Zinc Bromine flow Battery (ZBB). The CV experiment has shown that the surface electrode (graphite foil) has the highest zinc oxidation current peak, and volume electrode (carbon felt) has the highest bromide current peak. In addition, cell operations have demonstrated that using volume electrode as a cathode has highest coulombic efficiency by large reaction surface. Hence, a solution to the problem encountered in the ZBBs is found.

References

This work (Grants No. S-2015-A0013-00018) was supported by Business for Cooperative R&D between Industry, Academy, and Research Institute funded Korea Small and Medium Business Administration in 2016

References

[1] Ponce de leon C., Frias-Ferrer A., Gonzalez-Garcia J., Szanto D.A., Walsh F.C. 2006. Power Sources. 160, 716-732
[2] L. Stein, “Random patterns,” in *Computers and You*, J. S. Brake, Ed. New York: Wiley, 1994, pp. 55-70
[3] Wei Wang, 2013, Adv. Funct. Mater. 23, 970–986
[4] Brian Huskinson, 2014. Nature 505, 195–198.
[5] Puiki Leung, Xiaohong Li, Calos Ponce de Leon, Leonard Berlouis, C.T.John Low, Frank C. Walsh. 2012. RSC Publiching. DOI:10.1093/C2RA21342G (Review article) 2. 10125-10156
[6] Guy Rampel, Gainesville, Fla. 1972. U.S. Patent, 3660,170
[7] Hantao Zhou, Huamin Zhang, Ping Zhao, Baolian Yi. 2006. Electrochimica Acta, 51, 6302-6312
[8] Peng Qian, Huamin Zhang, Jian Chen, Yuehua Wen, Qingtao Luo, Zonghao Liu, Dongjiang You, Baolian Yi. 2008. Journal of Power Sources, 175, Issue 1, 613–620
[9] G. J. W. Radford, 2008 Journal of Power Sources, 185 1499-1504
[10] Adam Z. Weber, 2011. J Appl Electrochem 41:1137-1164
[11] Shuhei Maeda, Jun Sugawara, Hiroshi Hayami, 2013, US Patent 0037760A1
[12] United states patent application publication "Layered composite material for use in a redox flow battery" Pub.No : US 2013/0040194 A1
[13] Oswin Öttinger, Rainer Schmitt, Jürgen Bacher, Sylvia Mechen, Bastian Hudler, 2014 US Patent, US 2013/0040194 A1
[14] Donghyeon Kim, Joonhyeon Jeon. 2015. Journal of Physics: Conference Series, 574, 012074
[15] Ralph Zito, Jr,, Durham, N.C.. 1984. U.S.Patent, 4,482,614.
[16] P. Eidler, 1999, Sandia National Laboratories. SAND99-1853
[17] N. Clark, P. Eidler, P. Lex, 1999, Sandia National Laboratories. SAND99-2691. 1999
This reference has two entries but the second one is not numbered (it uses the ‘Reference (no number)’ style.

IOP Conf. Series: Journal of Physics: Conf. Series **939** (2017) 012022 doi:10.1088/1742-6596/939/1/012022

Synthesis, physical and microwave absorption properties of Barium ferrite - P(VDF-TrFE) nanocomposites

Jon Gutiérrez[1,*], Ane Miren Gutiérrez Muto[1], Alazne Peña[2], Maria San Sebastián[3], Ana Catarina Lopes[3], Luis Lezama[2], Izaskun Gil de Muro[2], A. Larrañaga[4] and I. Orue[4]

[1] Departamento Electricidad y Electrónica and [2] Departamento Química Inorgánica, Universidad del País Vasco UPV/EHU, P. Box 644, 48080-Bilbao, Spain
[3] Basque Center for Materials, Applications and Nanostructures (BCMaterials), Ibaizabal Bidea Bndg. 500, Parque Tecnológico de Bizkaia, Derio 48160, Spain
[4] SGIKER, Universidad del País Vasco UPV/EHU, 48940-Leioa, Spain

* email: jon.gutierrez@ehu.eus

Abstract. We present results concerning the synthesis, magnetic, dielectric and microwave absorption properties of $BaFe_{12}O_{19}$ / P(VDF-TrFE) nanocomposites. First, barium ferrite nanopowders were prepared by the hydrothermal synthesis method. Structural and morphological properties of the prepared powders were investigated by X-ray diffraction (XRD) and transmission electron microscopy (TEM), showing as result a good quality spherical plus platelet-like shaped nanoparticles. Afterwards, composite films with filler nanoparticles of barium ferrite $BaFe_{12}O_{19}$ (BaFO, nominal 5-20 wt.%) dispersed within P(VDF-TrFE) acting as polymeric matrix have been prepared by solvent evaporation. Magnetic properties were examined by the vibrating sample magnetometry (VSM), and a direct comparison of results obtained for the composites respect to the pure nanopowder allow us to obtain the true nanofiller content value, different from the initially nominal one. Dielectric properties have been measured up to 2 MHz and show the typical behaviour of Maxwell-Wagner type interfacial polarization. From 10 kHz up, permittivity of the composites remains almost unchanged, with values that show a smooth increase as the %wt. of $BaFe_{12}O_{19}$ does. Finally, microwave absorption properties were analyzed by using ESR technique operating at X-band. All obtained results are discussed in terms of size and quantity of the composites $BaFe_{12}O_{19}$ filler nanoparticles.

1. Introduction

In the last years the fabrication of multifunctional composite materials by combining inorganic constituents with polymeric ones is receiving great attention and efforts [1]. A typical example of this is the inclusion of $BaTiO_3$ in silicon rubber to improve the properties of the dielectric elastomer for electromechanical actuators [2]. Polymers exhibit many advantages as easy processing and forming, thermomechanical stability, high dielectric breakdown strength and low cost. Dielectric and magnetic properties of polymers can be substantially altered by embedding suitable inclusions, and tuned by controlling the amount and the type of the filler particle used [3,4]. More recently, mixtures of piezoelectric polymers with magnetic/magnetostrictive nanoparticles have been proposed to give as result potential magnetoelectric (ME) composites, being a clear example the particulate [poly(vinylidene fluoride)-trifluorethylene, (PVDF-TrFE), - ferrite] nanocomposites [5]. The barium ferrite ($BaFe_{12}O_{19}$) shows particular magnetic properties as high coercive force and hysteresis losses that make this ferrite to be widely used in milli- and microwave devices. It has very high magnetocrystalline anisotropy in the low-symmetry hexagonal *magnetoplumbite* crystal structure with hexagonal unit cell, and it usually appears shaped as thin hexagonal platelets with easy direction of magnetization along the *c* axis, which is perpendicular to the plates.

In a previous work [6] we presented a detailed study on the electric, dielectric and magnetic properties of purchased microsize $BaFe_{12}O_{19}$ (up to a 20% weight) - P(VDF-TrFE) composite films. Here we present new results concerning the behavior of the dielectric and magnetic properties of nanosize $BaFe_{12}O_{19}$ (up to a 10% weight) - P(VDF-TrFE) composite films, but being now the Barium ferrite nanoparticles synthesized at the laboratory by using a hydrothermal method. This study will extend also to microwave absortion properties of these nanocomposite films.

2. Experimental procedures

$BaFe_{12}O_{19}$ ferrite nanoparticles were synthesized following a hydrothermal route [7, 8]. For this purpose, stoichiometric quantities of the starting reagents sodium Hydroxide (NaOH), barium chloride ($BaCl_2.2H_2O$), ferrous chloride ($FeCl_2.4H_2O$), these three from Sigma Aldrich, and potassium nitrate (KNO_3), this last from Panreac, and deionized water were used in the synthesis reaction carried out in a Teflon autoclave without any agitation. The obtained powder was washed repeatedly using deionized water and alcohol, then dried at 240 °C for 10 h, and finally cooled down slowly to room temperature. The structural characterization of the final product was done by using X-ray (Cu-Kα radiation, λ= 1.5418 Å) diffraction patterns analysis. Transmission Electron Microscopy (TEM) micrographs were obtained using a Philips CM200 microscope at an acceleration voltage of 200 kV. To measure particle size, ImageJ software was used.

The P(VDF-TrFE) copolymer in the form of powder was supplied by Piezotech. For the film shaped composites preparation, the desired amount of barium ferrite $BaFe_{12}O_{19}$ (BaFO, 5, 10 and 20 %wt.) was added to Triton surfactant and DMF solvent, and then placed in ultrasound bath for 3 h to faccilitate the particles to be well dispersed in the solution and also to avoid aggregates. Then P(VDF-TrFE) powder was subsequently added. Further, the obtained mixture was placed in a Teflon mechanical stirrer with ultrasound bath for completing dissolution of the polymer. Later, the material was spread on a glass substract and introduced in an oven at 210 °C for 10 min, ensuring the total evaporation of the solvent as well as the melting of the polymer. Finally, the material crystallized at room temperature. In this way composite flexible films of about 85 μm thickness were obtained.

Magnetic hysteresis loops at room temperature were measured using an in house vibrating sample magnetometer up to a maximum field of 1.8 T. Measurements of dielectric constant ε and dielectric loss tangent tgδ were performed with an automatic Keysight E4980A Precision LCR Meter, at frequencies ranging from 20 Hz to 2 MHz. Finally, microwave absorption properties were analyzed by using FMR technique operating at X-band (9.4 GHz).

3. Results and Discussion

3.1. Structure and morphology

From Rietveld analysis of measured X-ray diffraction patterns we infer the existence of two phases in our synthesized nanopowder, the desired $BaFe_{12}O_{19}$ ($\approx$90% of the full content) and also Barium carbonate $BaCO_3$ ($\approx$10% of the full content). From TEM pictures we see two different powder morphologies: majority of spherical nanoparticles agregates with particle diameters ranging about 30-80 nm, plus some hexagonal platelets with size from 100 to 500 nm.

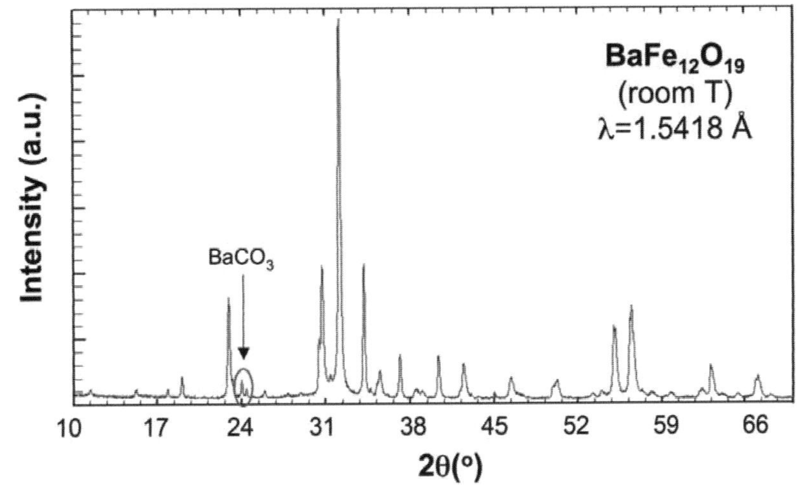

Figure 1. Room temperature measured X-ray measured diffraction pattern of our synthesized barium ferrite nanopowder.

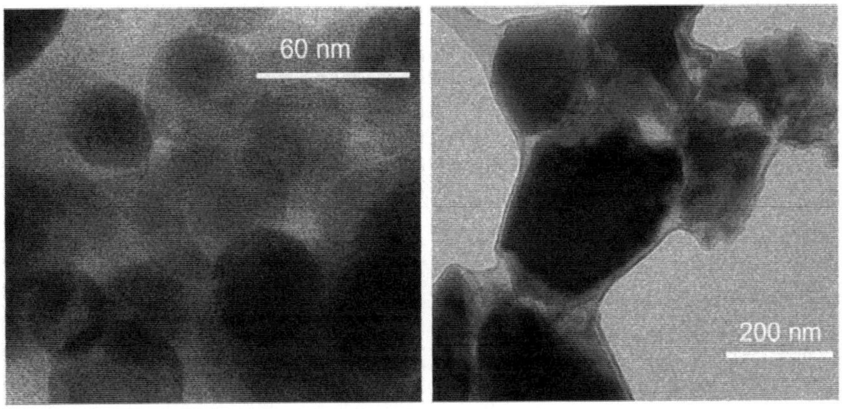

Figure 2. TEM pictures of the obtained BaFO powder showing spherical *(left)* and platelet-like morphologies *(right)*.

3.2. Magnetic characterization
First, we have measured the hysteresis loop at room temperature of the pure Barium ferrite nanopowder, obtaining values of $M_S \approx$ 30 emu/g at 1.8 T, $M_r/M_S \approx$ 0.32 and coercitivity about $H_C \approx$ 800 Oe.

Afterwards hysteresis loops were measured for the different composites. All loops exhibit similar shape with values of $M_r/M_S \approx$ 0.24-0.34 , and $H_C \approx$ 700-800 Oe. As a first consequence of all these measurements, the direct comparison of the M_S values obtained for all the composites respect to the pure nanopowder allowed us to deduce the true nanofiller content value and correct it from the initially nominal one. So, instead of the initially assumed 5, 10 and 20 %wt. contents, we have estimated true values of the BaFeO content within the polimeric matrix of 4.4, 8.1 and 12.5 %wt. content. These true values will be used in the following sections.

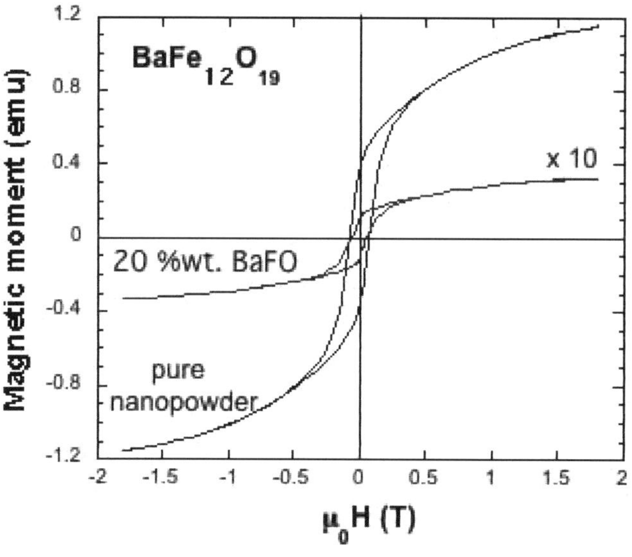

Figure 3. Hysteresis loops at room temperature of the pure Barium ferrite nanopowder and the composite with the 20 %wt. BaFO nominal content.

3.3. Dielectric behaviour

We have observed a quick decrease of the value of the dielectric constant as frequency increases, remaining almost constant for frequencies above 10 kHz. This behavior is explained assuming Maxwell –Wagner type interfacial polarization, in agreement with the Koop's phenomenological theory [9, 10]: the dielectric constant of a composite with heterogeneous structure of ferrite grains embedded in a polymeric matrix can be imagined as consisting of conducting grains separated by highly resistive grain boundaries. Thus, the voltage applied to the sample drops mainly across the Barium ferrite nanoparticle surfaces and space charge polarization is built up at those boundaries.

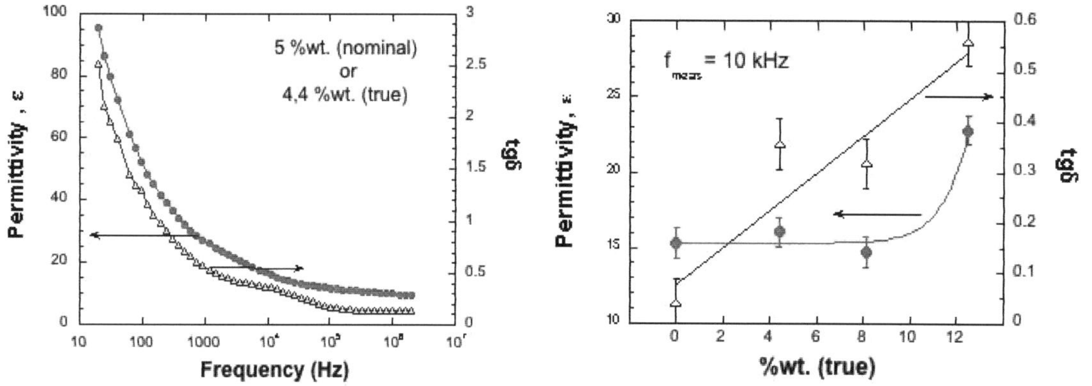

Figure 4. Permittivity and loss tangent against frequency for the 5%wt. nominal content composite (*left*), and values measured at 10 kHz for all the different BaFO content films *(right)*.

From 10 kHz up, permittivity of the composites remains almost unchanged. These measured values show a smooth increasing behaviour as the BaFO filler content increases in the composites that tell us about this nanofiller acting as a reinforcement within the composite.

IOP Conf. Series: Journal of Physics: Conf. Series **939** (2017) 012022 doi:10.1088/1742-6596/939/1/012022

3.4. Microwave absorbing behaviour

All room-temperature recorded spectra exhibit the same essential features, showing complex behavior with rather large line widths ($\Delta H_{pp} \approx 1.8$ kOe) as usually happens with barium ferrite samples [11]. Microwave absorption is maximum around 3 kOe in all studied composites, a field value that is in good agreement with our previous observation of a majority of spherical shaped nanoparticles in our synthesized powder [12].

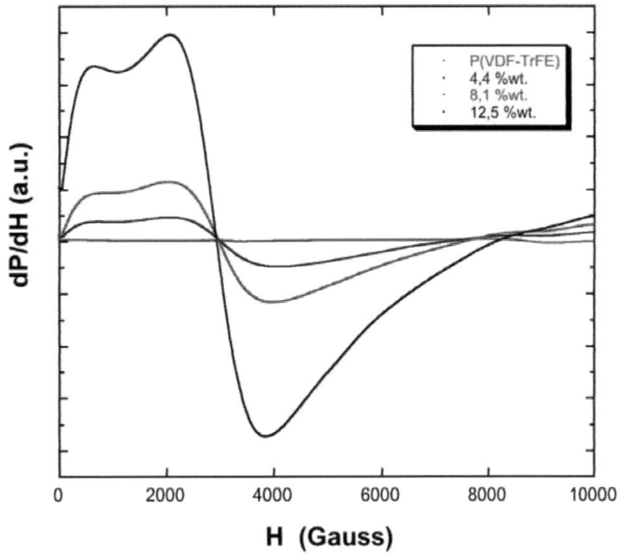

Figure 5. EPR spectra (at 9.4 GHz) obtained for the different composites.

The measured value of H_{res} (field of maximum absorption, when dP/dH=0) smoothly decreases (from 3.01 to 2.95 kOe when changing from true 4.4%wt to 12.5%wt.) with the increase of magnetic filler content, a fact that one can expect when interactions increase inside these composites as Barium ferrite content also does.

In all cases significant absorption near zero-field has been observed, absorption that increases proportionally to the ferrite content in the nanocomposite. This zero field signal is not surprising and arises from the non neglectable electrical conductivity of the BaFO ferrite.

4. Conclusions

We have synthesized $BaFe_{12}O_{19}$ ferrite nanoparticles by using the hydrothermal route. As probed by TEM microscopy, we have obtained a nanopowder with majority of sperical particles with diametres ranging from 30 to 80 nm , plus some bigger platelets.

When these Barium ferrrite nanoparticles are embedded within a P(VDF-TrFE) polymeric matrix, we obtain film-shaped composites in which the dielectric behavior smoothly improves as the filler content increases. Very similar behavior has been observed for microwave absorption characteristics when EPR curves are measured. Thus, we can conclude that the $BaFe_{12}O_{19}$ ferrite nanoparticles act as a reinforcement within the polymeric matrix, making so these nanoparticles promising candidates for the development of new multifunctional composites.

Acknowledgements

J. Gutiérrez wants to thank financial support from the Basque Government Industry Department under Project Actimat, (ELKARTEK Program). A.M. Gutiérrez Muto wants to thank the BCMaterials Center for an expertising stay. Technical and human support for magnetic measurements provided by SGIker (UPV/EHU, MICINN, GV/EJ, ESF) is gratefully acknowledged.

References

[1] Pedro Martins and Senen Lanceros-Méndez, Polymer-based magnetoelectric materials, Advanced Functional Materials 23 (2013) 3371-3385.
[2] H. Zhao, Y.J. Xia, Z.M. Dang, J.W. Zha and G.H. Hu, Composition dependence of dielectric properties, elastic modulus and electroactivity in (carbon black-BaTiO3)/silicone rubber nanocomposites, Journal Applied Polymer Science 127 (2013) 4440-4445.
[3] A. Vavouliotis, E. Fiamegou, P. Karapappas, G.C. Psarras and V. Kostopoulos, DC and AC conductivity in epoxy resin/multiwall carbon nanotubes percolative system, Polymer Composites 31 (2010) 1874–1880.
[4] N. Frickel, A.G. Greenbaum, M. Gottlieb and A.M. Schmidt, Magnetic properties and dielectrical relaxation dynamics in CoFe2O4@PU nanocomposites, The Journal of Physical Chemistry C 115 (2011) 10946–10954.
[5] Pedro Martins, Andoni Lasheras , Jon Gutiérrez , Jose Manuel Barandiarán, Iñaki Orue and Senen Lanceros-Méndez, Optimizing piezoelectric and magnetoelectric responses on $CoFe_2O_4$/P(VDF-TrFE) nanocomposites, Journal of Physics D: Applied Physics 44 (2011) 495303, pp.1-7.
[6] Jon Gutiérrez, Pedro Martins, Renato Gonçalves, Victor Sencadas, Andoni Lasheras, Senen Lanceros-Mendez and Jose Manuel Barandiarán, Synthesis, physical and magnetic properties of $BaFe_{12}O_{19}$/P(VDF-TrFE) multifunctional composites, European Polymer Journal 69 (2015) 224-231.
[7] Kang Li, Hongchen Gu and Qun Wei, A novel hydrothermal synthesis method for barium ferrite, China Particuology 2 (2004) 41-43.
[8] Xuan Shi, Yongping Pu and Dan Liu, Preparation of magnetic barium ferrite powders by microwave hydrothermal method, Journal of Ceramic Processing Research 13 (2012) s414-s417.
[9] C.G. Koops, On the dispersion of resistivity and dielectric constant of some semiconductors at audiofrequencies, Physical Review 83 (1951) 121-124.
[10] K.W. Wagner, Zur theorie der unvoll kommener dielektrika, Annali Physika 40 (1973) 817-855.
[11] Jun Wang, Qianwang Chen and Shan Che, Magnetic properties in $BaFe_{12}O_{19}$ nanoparticles prepared under a magnetic field, Journal of Magnetism and Magnetic Materials 280 (2004) 281-286.
[12] G.V. Kurlyandskaya, Iñaki Madinabeitia, I.V. Beketov, A.I. Medvedev, A. Larrañaga, A.P. Safronov and S.M. Bhagat, Structure, magnetic and microwave properties of FeNi nanoparticles obtained by electric explosion of wire, Journal of Alloys and Compounds 615 (2014) S231–S235.

IOP Conf. Series: Journal of Physics: Conf. Series **939** (2017) 012023 doi:10.1088/1742-6596/939/1/012023

Preparation and evaluation of PdCr thin film resistive strain gauges

Hao Liu, Shuwen Jiang*, Hongchuan Jiang, Xiaohui Zhao and Wanli Zhang

State Key Laboratory of Electronic Thin Films and Integrated Devices, University of Electronic Science and Technology of China, Chengdu, 610054, China

*Email: jiangsw@uestc.edu.cn

Abstract. Thin film resistive strain gauges using PdCr alloy as sensing material were fabricated on nickel-based alloy substrate, and evaluated for strain response, apparent strain and drift strain over a temperature range to 800 °C. The results indicated that the PdCr thin film strain gauge showed an excellent linear response to applied strain and stable gauge factor at different temperatures. The apparent strain can be estimated and compensated due to its repeatability and low value. The drift strain can be neglected until 600 °C. Even at 800 °C the drift strain can also be corrected since it drifted slightly and linearly. The repeatability of PdCr thin film strain gauge was also evaluated. It showed that the repeatable measurement error was 6.4%.

1. Introduction

The engine blades which work in harsh environment of high temperature, heat stress and strong vibration will appear fatigue cracks, lost blocks and other phenomena with the use of time, and thus cause engine failure. In order to meet the needs of the growing demand, the research on more performance of equipment, such as stress, strain, vibration and other mechanical parameters, is urgently needed.

Advanced thin film sensors, which are sputter deposited directly onto the surface of heating components, have thickness on the order of a few micrometers; therefore, they leave intact the surface structural integrity and add negligible mass to the surface and create minimal disturbance of the gas flow over the surface[1]. Thin film strain gauge can provide more effective method of accurate strain and other mechanical parameter measurement at high temperature environment for engine turbine blades and other heating components.

PdCr[2,3] is structurally stable, undergoes no phase transformation or order-disorder transition over temperature range up to 1100 °C, and it forms an adherent, self-protective Cr_2O_3 scale in air[4]. All these properties of PdCr result in repeatable, stable resistance versus temperature relationship which is independent of heating and the cooling rates. As sensing material of thin film strain gauge, the palladium-13w/o chromium alloy (Pd-13Cr) was regarded to be the optimum composition. The alloy of this composition has a lower temperature coefficient of resistance (TCR) than alloys containing less than 13% weight Cr, while alloys with more than 13% weight Cr have poorer resistance to oxidation [5].

2. Experimental

PdCr thin film strain gauges were fabricated on Nickel-alloy (GH3536) flat bar specimens. The test bars were polished with an appropriate degreaser, and then were cleaned by ultrasonic agitation in a bath of acetone, alcohol and deionized (DI) water until the polished surfaces passed the water break test. A NiCrAlY layer was then deposited by sputtering with a thickness of approximately 15 μm. By heating treatment in vacuum and following oxygen atmosphere, on surface of this coating a "graded buffer layer" formed a stable and adherent electrically insulating Al_2O_3 layer. Next, YSZ and Al_2O_3 were deposited as "composite insulation layer" with a thickness of 1μm and 5μm to obtain the required insulation resistance, respectively. The PdCr sensitive thin film was sputtered onto the insulating layer with a controlled resistance and thickness. Finally, the gauges were covered with an Al_2O_3 overcoat to reduce the possible oxidation, followed by vacuum heating treatment for stabilization at 800 °C for 2 hours. The leadwires of platinum (Pt) wire with 100 μm in diameter were connected to each of the two gauge lead pads by high temperature conductive silver paste for tests. Figure 1 shows a sample of PdCr thin film resistive strain gauge on flat bar specimen.

Figure 1. A sample of PdCr thin film resistive strain gauge on flat bar specimen.

The strain-resistance response tests of prepared gauges were performed in Zwick/Roell-Z050 material testing system by step-by-step loading at different temperatures. The applied strain to flat bar was determined by laser extensometer with a resolution of 0.11μm; and the resistance and temperature of the thin film gauges were acquired through a HBM QuantumX MX840B amplifier. Four-wire measurement of resistance was used to minimize the lead wires effect. During test, all parameters were recorded by a computer, simultaneously.

The gauge factor (GF) of PdCr thin film resistive strain gauge, which is determined by measuring the fractional change in resistance at applied strain, can be expressed as [1]

$$GF = \frac{\Delta R}{\varepsilon R_0} \tag{1}$$

where, ΔR is the change in gauge resistance; R_0 is unstrained resistance of strain-sensing element; and ε is applied strain.

For elevated temperatures, the gauge resistance changes with temperature T and with time τ at high temperature, and the measured strain ε_m results from true strain, apparent strain, and drift strain, expressed as

$$\varepsilon_m = \varepsilon_t + \varepsilon_a + \varepsilon_d \tag{2}$$

where ε_t is true strain, ε_a apparent strain, and ε_d drift strain.

The apparent strain, due to temperature change and thermal mismatch rather than applied strain is described by the following equation:

$$\varepsilon_a = \frac{\Delta R_T/R_0}{GF_T} \tag{3}$$

And drift strain, caused by the resistance change of the sensing material during holding time at elevated temperature, is defined by:

$$\varepsilon_d = \frac{\Delta R_{T,\tau}/R_T}{GF_T} \tag{4}$$

In the equations, R_0 and R_T are unstrained resistance of gauge resistance at room temperature and temperature T, respectively, and ΔR_T is their difference as $R_T - R_0$; $\Delta R_{T,\tau}$ is the drift of unstrained resistance during holding time at temperature T, and GF_T is the gauge factor at temperature T.

For strain gauge system, careful attention must be given to minimize or compensate the apparent strain and drift strain to obtain applied true strain. In this work, the apparent strain and drift strain of PdCr thin film resistive gauge was experimentally evaluated.

3. Results and Discussion

The strain-resistance response for PdCr thin film strain gauge at room temperature is presented in figure 2. Figure 2(a) shows the applied strain and the corresponding resistance value versus loading time. Note that the trends of the two curves were exactly the same. The step-by-step applied strain results in the step change of gauge resistance. Figure 2(b) presents the change in relative resistance versus applied strain during the test. As can we seen, the relationship between the variation of resistance and strain was linear and positive, from which the gauge factor can be obtained to be 1.41 according to formula (1).

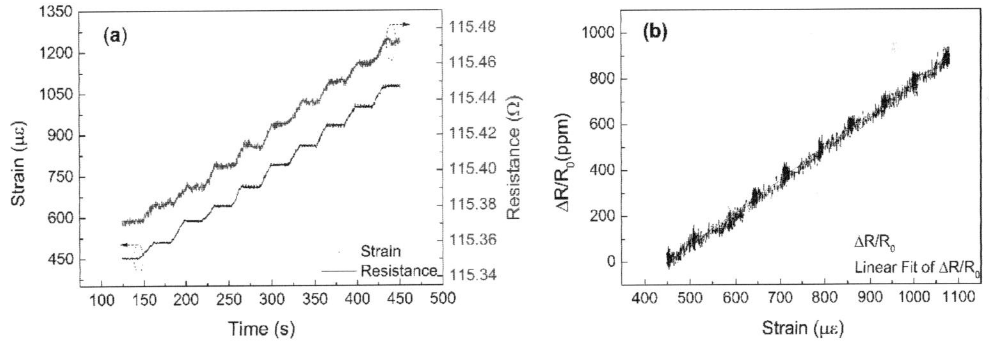

Figure 2. The strain-resistance response for PdCr thin film strain gauge at room temperature. (a) The strain and corresponding resistance versus loading time; (b) The relative change in resistance versus applied strain.

After the room temperature test, the PdCr thin film strain gauge was subjected to elevated temperature tests at 600 °C and 800 °C, respectively. For each circle of tests, the specimen was first cooled down to room temperature to evaluate repeatability, then raised to next target temperature for test.

The test results for 600 °C were shown in figure 3. Although the obvious fluctuations of resistance, the step change can be clearly observed. Resistance variations were less than 0.02 Ω, resulting in a

IOP Conf. Series: Journal of Physics: Conf. Series **939** (2017) 012023 doi:10.1088/1742-6596/939/1/012023

error of ±50 με. The linear curve of the resistance change with applied strain was presented in figure 3 (b), the gauge factor is fitted to be 1.55.

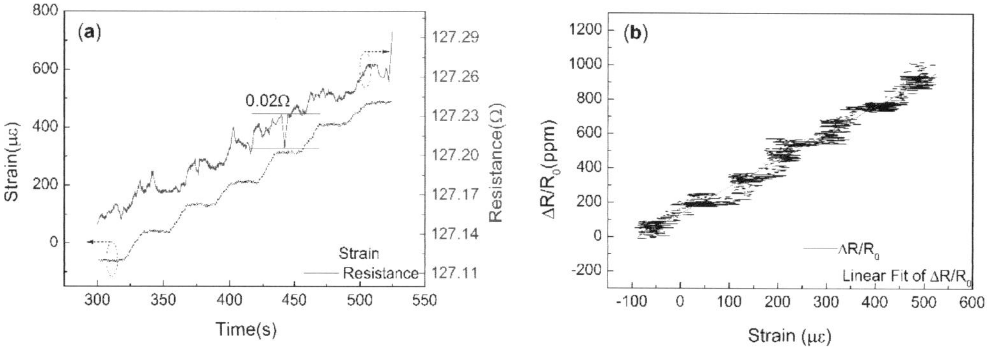

Figure 3. The strain-resistance response for PdCr thin film strain gauge at 600 °C. (a) The strain and corresponding resistance versus loading time; (b) The relative change in resistance versus applied strain.

The apparent strain and drift strain of PdCr thin film strain gauge can be determined by heating and soaking the gauge in high temperature environment. The curve of the resistance with heating and holding time for 600 °C was presented in figure 4, as well as the resistance as a function of temperature showed in the inset, from which we can see that the resistance increased linearly with temperature. The apparent strain sensitivity, defined as $\delta\varepsilon_a/\delta T$, was 115με/°C. The drift strain of PdCr thin film gauge at 400 °C can be neglected since the resistance remained almost unchanged at holding stage.

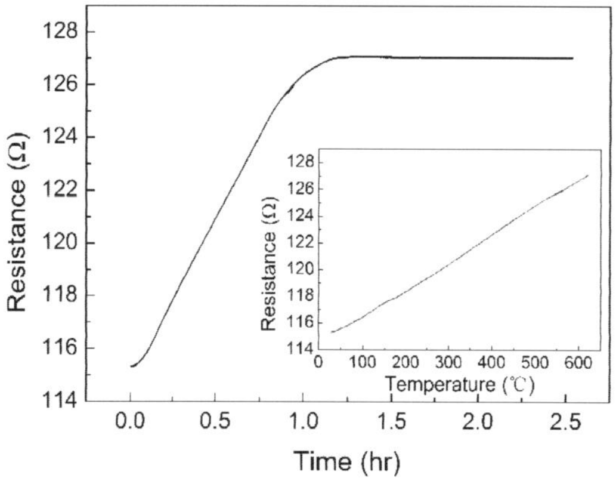

Figure 4. The resistance changes of the PdCr gauge heating up to and soaking at 600 °C.

Figure 5 presented the results of the PdCr thin film strain gauge tested at 800 °C. It was observed from figure 5(a) that the variation in resistance was significantly increased to 0.05 Ω, about an error of ±100 με. This error is believed to be acceptable in 800 °C harsh environment. The increased fluctuations of resistance might be attributed to the thermal noise of the electronic motion in high temperature environment. From figure 5(b), the gauge factor of PdCr strain gauge at 800 °C was determined to be 1.41.

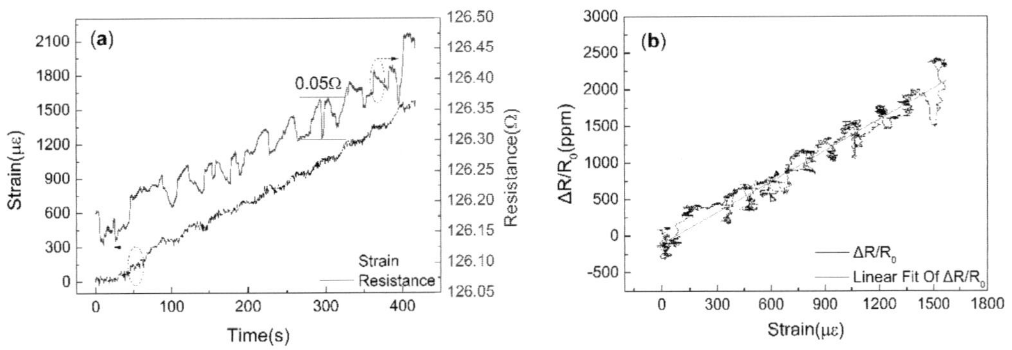

Figure 5. The strain-resistance response for PdCr thin film strain gauge at 800 °C. (a) The strain and corresponding resistance versus loading time; (b) The relative change in resistance versus applied strain.

The heating and holding curve of PdCr thin film strain gauge for 800 °C was presented in figure 6. The linear resistance change with temperature at heating phase suggested that the apparent strain sensitivity remained constant, which was found to be 127με/°C. The resistance drift of the gauge at 800 °C during 4 hours soak was recorded to decrease about 1.1 Ω. Notice that the resistance drifts slightly and linearly, indicating no serious oxidation in PdCr thin film at 800 °C. The results show that PdCr thin film remains stable at high temperature. According to equation (4), the drift strain is 1800 με/hr at 800 °C. Since the resistance of PdCr thin film strain gauge decrease linearly with time, the drift strain can be easily corrected.

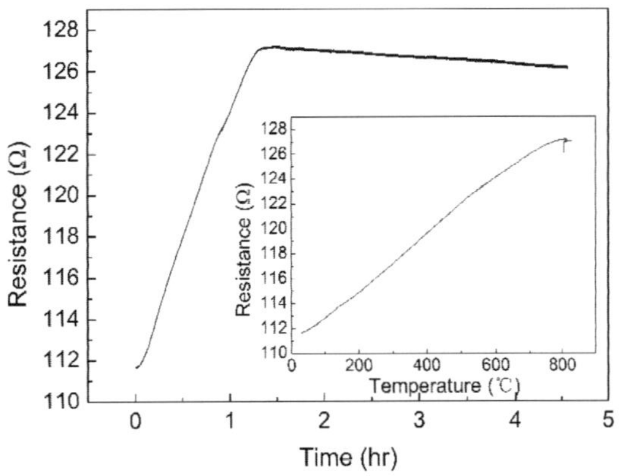

Figure 6. The resistance change of the PdCr gauge heating up to and soaking at 800 °C.

The repeatability of prepared PdCr thin film strain gauges was evaluated simultaneously. The room temperature test results after cooling down from 600 °C and 800 °C, respectively, were presented in figure 7. The results indicated that the resistance change of the PdCr thin film gauge remains linear under applied strain after several high temperature tests. However, the calculated gauge factor was decreased from original 1.41 at room temperature to 1.34 and 1.32, respectively. That is, the repeatable error after several high temperature tests was 6.4%. And the total lifetime of the PdCr gauge was more than 10 hours.

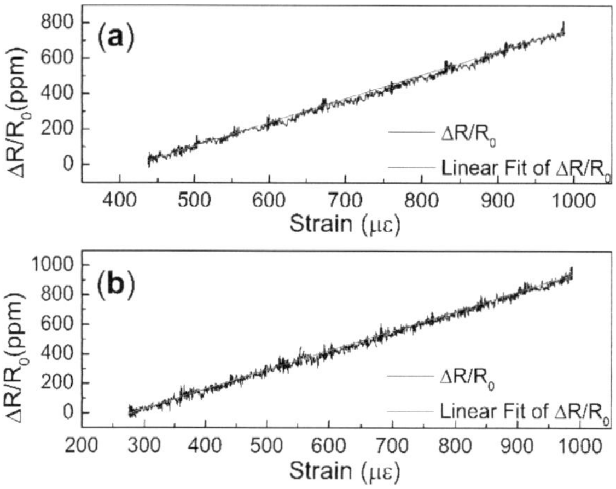

Figure 7. The repeatability of PdCr thin film strain gauge.

(a) cooled down from 600°C; (b) cooled down from 800°C.

4. Conclusion

High temperature thin film resistive strain gauges using PdCr alloy as sensing material were fabricated on nickel-based alloy substrate. The test results showed that the PdCr thin film strain gauge had an excellent linear response to applied strain and stable gauge factor at different temperatures, as well as repeatable low apparent strain and neglectable, correctable drift stain. The repeatability and lifetime of PdCr thin film strain gauge were also evaluated by several thermal circles at different temperatures. It showed that the repeatable measurement error was 6.4% at different circles, and lifetime was over 10 hours.

Reference

[1] Lei J F, Martin L C, et al 1997 *Advances in thin film sensor technologies for engine applications* (Cleveland, OH United States: NASA Technical Report) NASA-TM-107418

[2] Liu J F 1991 *Platinum Met. Rev.* **35** 65

[3] Kayser P, Godefroy C J and Leca L 1993 *Sens. Actuators, Phys.* **37** 328

[4] Hulse C O, Bailey S R, et al 1987 *High temperature static strain gage alloy development program* (Cleveland, OH United States: NASA Technical Report) NASA-CR-174833

[5] Hulse C O, Bailey S R, et al 1987 *High temperature static strain GAGE development contract* (Cleveland, OH United States: NASA Technical Report) NASA-CR-180811

IOP Conf. Series: Journal of Physics: Conf. Series **939** (2017) 012024 doi:10.1088/1742-6596/939/1/012024

Design and development of a new magnetic sensor for stress monitoring and annihilation /control in steels

E Mangiorou[1]
Dpt. of Mining & Metallurgical Engineering,
National Technical University of Athens, Zografou Campus, Athens, 15780, Greece

E-mail: mangiorou.eleni@gmail.com

Abstract. This paper describes a new method and apparatus for monitoring the distribution of the hydraulic stresses (type stresses I) and the residual stresses (type stresses II and III) in ferromagnetic steels within the elastic region and plastic deformation region, determining the deforming step them and use them for elimination and / or control of the stresses in the steel.

1. Introduction

Ferromagnetic steels have a wide range of applications, such as in transformers, aerospace and automotive gas industries. The mechanical behaviour of these construction components are strongly related with the distribution of the internal stresses, which exist within a material. The residual stresses are caused by inhomogeneous deformations, which result from mechanical and thermal stresses and from the stress in the crystal lattice. These stresses lead to a rearrangement of the material's magnetic domains [1-5].

When a mechanical field applied in a ferromagnetic material, the total stress tensor is increased. This increment influences the microstructural configuration, as well as the magnetic and mechanical properties of the material. Therefore, the knowledge of the existence of the residual stress state in a material should be evaluated, in order to effectively monitor the health of a ferromagnetic steel structure.

In many studies, have been described the exploitation of several semi- and non-destructive techniques in measuring residual stresses, such as X-ray and neutron diffraction and quasi-dc magnetic permeability measurement [4-10].

This papers describes the design and the development of a new magnetic sensor for stress measurements using the magnetic permeability and magnetostriction techniques in ferromagnetic steels.

2. Measuring procedure

The magnetic non-destructive methods use the combined knowledge of the structural and the magnetic configuration for testing and evaluating metallurgical, microstructure, mechanical properties of ferromagnetic steels grades [11-22].

[1] Tel: +30 2107723737, E-mail: mangiorou.eleni@gmail.com

The proposed sensor (figure 1) is capable of measuring the component of the algebraic sum of the residual stresses (type I and II stresses) and hydraulic stress (type I stresses) to geometries of steel, whose surface and / or cross-section comparable to the length them, counting the tensor of magnetic permeability, (μ), and the magnetostriction, (λ), as a function of the applied field.

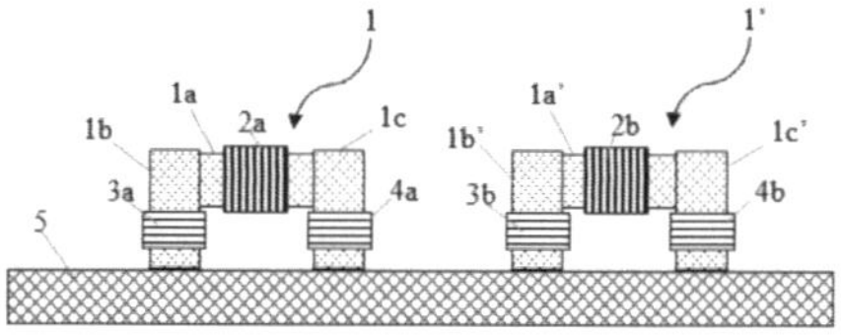

Figure 1. The apparatus of the sensor.

The sensor device consists of two U-shaped cores (1 and 1'), which are a certain distance from apart and arranged in series on the surface of the test ferromagnetic steel (5). In one leg of the core (1b and 1b' respectively) is wrapped around the excitation/ receiving coil (3a and 3b respectively), while the respective biasing coils (2a and 2b) are wrapped around the core's body, which is parallel to examined steel. Considering the in series positioning of the electromagnets (1 and 1') at the surface of the under-measurement steel, the output fields of the biasing coils are collinear.

Furthermore, the electromagnets include each of them a further modulation coil (4a and 4b respectively) wound around the section of the other leg (1c, 1c' respectively). The modulation coil produces an alternating magnetic field in the steel area under the electromagnet. Therefore, when the excitation and generation of the elastic waves, which produced by the excitation / pick-up coils (3a, 3b respectively), combined with the alternating magnetic field, which created by the modulation coil (4b or 4a) result in the optimizing configuration of output voltage, (V_o).

In the excitation coil is applied a triangular signal with a frequency of 10Hz, and the pulse width from peak to peak 14Volt. The magnetic noise is obtained by the receiving coil of the electromagnet. The signal in the receiving coil is amplified and obtained from a data acquisition card (Figure 2).

(a)
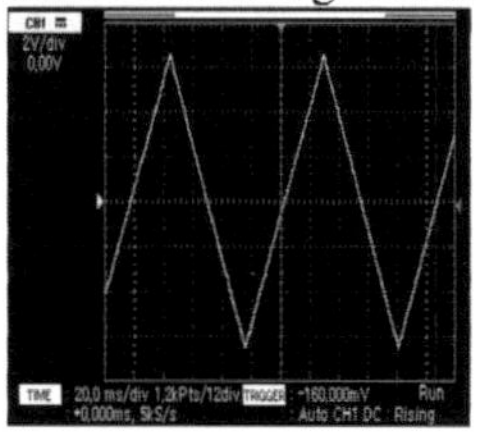

(b)
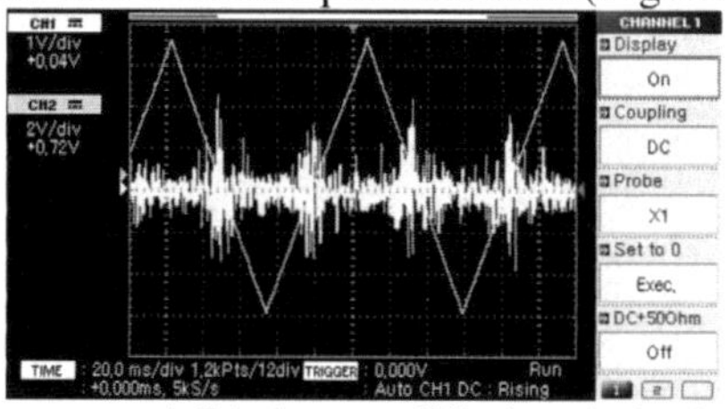

Figure 2. (a) The applied triangular signal to the sensor and **(b)** the amplified and obtained from a data acquisition card signal.

The supply of adjustable intensity bias current (H) in the biasing coil (2a) located around the excitation/receiving coil (3a) and thus the biasing field in the magnetizable area of the examined steel (5), results in the modulation of the output voltage, (V_o), of the excitation/receiving coil (3b). The maximum value of output voltage is proportional to the mean value of the magnetostriction (λ) of the examined steel's volume located beneath the excitation/receiving coil (3a), according to the type:

$$V_o = C_1 \frac{d\lambda}{dH} \quad (1)$$

where C_1 is a constant, determined from the experimental data.

Similarly, the generation of a biasing field from the biasing coil (2b) located around the excitation/receiving coil (3b) leads to the modulation of the peak of the output voltage (V_o) at the excitation/receiving coil (3b), the peak of which is proportional to the mean magnetic permeability (μ) of the volume of the steel, located beneath the excitation/receiving coil (3b):

$$V_o = C_2 \frac{dM}{dH} = C_2 \mu \quad (2)$$

where C_2 is a constant, determined from experimental data.

By this way, the magnetostrictive function λ(H) of the steel volume, located beneath the excitation/receiving coil, (3a), and the magnetic permeability function μ(H) of the steel located beneath the excitation/receiving coil, (3b), are determined.
Reversing the role and the function of the two excitation/receiving coils (3a and 3b), using the (3b) coil as the excitation coil and the (3a) coil as the receiving coil, the measurement of the magnetostrictive function (λ) in the steel volume, enclosed by the (3b) coil and the magnetic permeability function of the steel volume, enclosed by the (3a) coil is achieved. Thus, the measurement of the λ(H) and μ(H) functions of the steel volume located beneath the two, (3a) and (3b) excitation/receiving coils are obtained.
By this way and by the movement of the coil device (2a)–(3a) and (2b)–(3b) along the steel, the measurement of the distribution of the magnetic permeability μ(H) and the magnetostriction λ(H) along the steel and hence the distribution of the mechanical stress along the under-measurement steel are obtained.

3. Discussion
Considering that λ(H) functions of the mentioned device, follow monotonic response until to the Villari point of the under-measurement steel, where their slope change, maintaining again the monotonicity of their response after this point, then the selection of the appropriate part of the μ(H) function is accomplished by monitoring the λ(H) function. In Figure 3 appears a typical response of the μ(σ) function in relation with the applied positive mechanical stress and the waveform of the λ(H) function in relation with the stress of the examined steel with positive magnetostriction coefficient in the elastic region, is shown. The reverse response is observed for negative values (compressive) of stress, where the μ(H) function will be decreasing until the negative Villari point and subsequently it will be increasing until to the UTS point. The exactly opposite behavior will be observed on a steel with a negative magnetostrictive function in the elastic region.

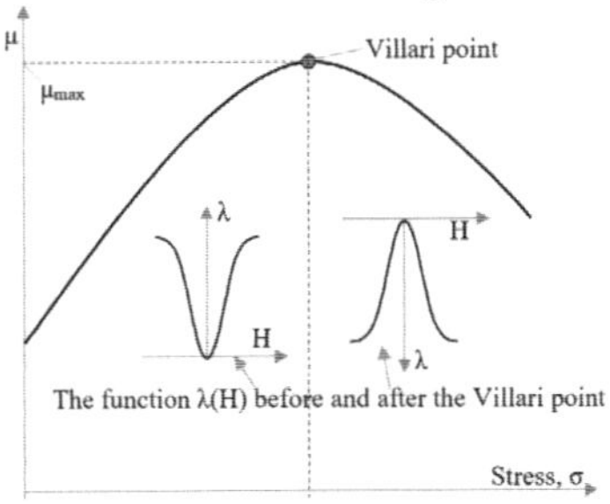

Figure 3. Diagram of magnetic permeability μ(H) and magnetostriction λ(H) at the Villari point.

Consequently, the monotonic correlation of the local value of the magnetic permeability, μ, with the local value of the algebraic sum of the residual stresses (type II and III stresses) and the hydraulic stress (type I stresses) to the measuring axis, in the region of the excitation and reception of the elastic waves, becomes possible, by selecting one of the two monotonic μ(H) functions, based on the measured value of the λ(H) function: for each one of the two ranges of values of the μ function, corresponding one and only one sign of the λ function. Accordingly, the μ(σ) function is transformed into a monotonic function, not only until to the positive and negative Villari point, but also to the positive and negative UTS point, with the simultaneous monitoring of the λ(H) function. Thus, the Magnetic Stress Calibration Curve (MASC) to the UTS point, is obtained. Such a universal MASC curve is illustrated in the Figure 4.
Having calibrated the appropriate MASC, correlating the magnetic permeability with the mechanical stress on the same axis for the examined steel, it becomes possible to measure the algebraic sum of the residual stresses and the hydraulic stress along the two electromagnets (1 and 1') in the volume of the steel beneath the regions of excitation and reception of the elastic waves.

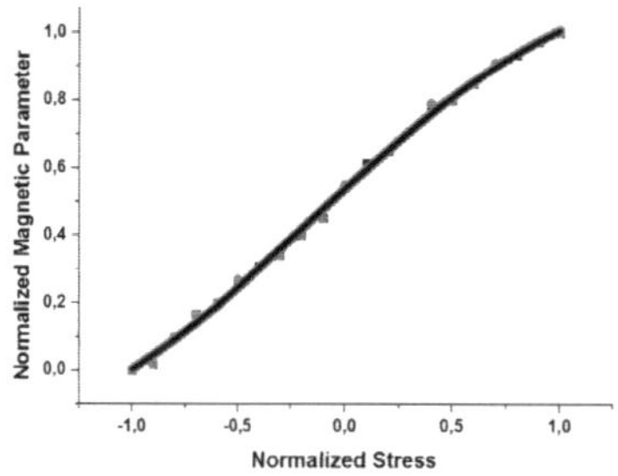

Figure 4. The normalized universal MASC curve: measurement of two specific magnetic parameters permit the stress component along the direction of magnetic measurement.

In this way, there is no need of knowledge of the stoichiometry and the phase of each steel: the measurement of the hysteretic permeability function μ(H) and the hysteretic function λ(H), allows the finding of the normalized value of the magnetic parameters and hence, the value of the component of the stress in the measuring axis.

4. Conclusions

The results mentioned above indicate that the developed sensor can be used effectively to measure local magnetic permeability and local rate of magnetoelasticity or magnetostriction correlated with the externally imposed magnetic field, μ(H) and λ(H) respectively. This measurement is able to determine the local constituents of the stresses tensor of the steel, including the algebraic sum of the residual stresses tensor and the hydraulic stresses tensor to the under-control steel.

5. References

[1] Iordache E, Hug E and Buiron N 2003 *Mater. Sci. Eng. A* **359** 62-74
[2] Perevertov O. 2007 J. *Phys. D: Appl. Phys.* **40** 949-954
[3] Cullity BD and Graham CD. 1972 *2nd ed. London: Addison-Wesley.*
[4] Hristoforou E 2002 *J. Opt. Adv. Mat.* **4** 245-260
[5] Hristoforou E and Ktena A 2007 *J. Magn. Magn. Mater.* **316** 372-378
[6] Vourna P, Ktena A, Tsakiridis P E and Hristoforou E 2015 *Measurements* **71** 31
[7] Vourna P, Ktena A and Hristoforou E 2014 *IEEE Trans. Mag.* **50** 1
[8] Ktena A, Hristoforou E, Gerhardt GJL, Missell FP, Landgraf FJG, Rodrigues Jr DL and Alberteris-Campos M 2014 *Physica B: Coned Matter* **435** 109
[9] Vourna P, Ktena A, Tsakiridis P E and Hristoforou E 2015 *NDT & E Int.* 71 33
[10] Vourna P, Hervoches C, Vrána M, Ktena A and Hristoforou E 2015 *IEEE Trans.* 51 7029219
[11] Rossini S, Dassisti M, Benyounis andY, Olabi AG 2012 *Mater. Des* **35** 572
[12] Vourna P, Ktena A, Svec P and Hristoforou E, 2016 *IEEE Trans. Magn.* **52** 5
[13] Martinez-de-Guerenu A, Gurruchaga K and Arizti F, 2007 *J. Magn. Magn. Mater.* **316** e842
[14] Kobayashi S, Kikuchi N, Takahashi S, Kamada Y, Kikuchi H, 2010 *J. Appl. Phys.* **108** 043904-1
[15] Vértesy G, Mészáros I and Tomáš I, 2013 *NDT&E Int.* **54** 107
[16] Stupakov O., 2012 *IEEE Trans. Magn.* **48** 4718
[17] Hristoforou E and Reilly R E, 1993 *J. Magn. Magn. Mat.* **119** 247-253
[18] Hristoforou E, Kosmas K and Kollar M, 2008 *Journal of Electrical Engineering* **59** 90-93
[19] Ioakeimidis E, Kytopoulos V and Hristoforou E, 2013, Materials Science and Engineering A, **MSEA-D-13-00774**, 254-260
[20] E. Hristoforou and R.E. Reilly, New Mechanical Stress Transducers Based on Amorphous Alloys, IEEE Trans. Mag., 26, p. 1563-1565, 1990.
[21] Christopoulos A, Hristoforou E and Tsamasphyros G, 2012 *Smart Materials and Structures* **21** 8
[22] Christopoulos A, Hristoforou E, Koulalis I and Tsamasphyros 2014 *Smart Materials and Structures* **23** 8

Flexible microstrip antenna based on carbon nanotubes/ (ethylene–octene copolymer) thin composite layer deposited on PET substrate

J Matyas, R Olejnik, P Slobodian

Centre of Polymer Systems, University Institute, Tomas Bata University in Zlin, tr. T. Bati 5678, 760 01 Zlin, Czech Republic

matyas@cps.utb.cz

Abstract. A most of portable devices, such as mobile phones, tablets, uses antennas made of cupper. In this paper we demonstrate possible use of electrically conductive polymer composite material for such antenna application. Here we describe the method of preparation and properties of the carbon nanotubes (CNTs)/ (ethylene–octene copolymer) as flexible microstrip antenna. Carbon nanotubes dispersion in (ethylene–octene copolymer) toluene solution was prepared by ultrasound finally coating PET substrate by method of dip-coating. Main advantages of PET substrate are low weight and also flexibility. The final size of flexible microstrip antenna was 5 x 50 mm with thickness of 0.48 mm (PET substrate 0.25 mm) with the weight of only 0.402 g. Antenna operates at three frequencies 1.66 GHz (–6.51 dB), 2.3 GHz (–13 dB) and 2.98 GHz (–33.59 dB).

1. Introduction

Electronic devices use mainly copper based microstrip antenna for communication between each portable devices in network [1-3]. In many fields of science, carbon nanotubes are one of the most promising materials. Gradually, there are increasing possibilities for the use of carbon nanotubes in electronics. The main reasons for employing these materials are their characteristics, which are, for instance, good conductivity (after complex adjustments), or great potential for miniaturization. As the composite materials can perform several functions simultaneously, they can be applied to various products and surfaces not only in the ICT field. This contribution is focused on the use of carbon nanotubes in the passive antenna, which consists of a layer of carbon nanotubes and (ethylene–octene copolymer), applied to PET substrate. The antenna described in this article operates at three frequencies 1.66 GHz (–6.51 dB), 2.3 GHz (–13 dB) and 2.98 GHz (–33.59 dB).

Nevertheless, the aim of the research is to produce the layer with the lowest possible resistivity. The low resistivity ensures a better suitability for the material for the construction of the antenna. This characteristic can be influenced by the use of specific chemicals and by the type of carbon nanotubes. The produced specimen of the antenna employs multi-walled carbon nanotubes (MWCNTs) (Figure 3). The main reason for choosing MWCNTs is the fact that the composite does not have to be perfectly pure; for the pure composite it would be more appropriate to use the single-walled carbon nanotubes (SWCNTs). In addition, the final price of the subsequent applications to the devices plays an important role. Therefore, the price is an important criterion. However, this would lead to the price of the final product being much higher than in the case of multi-walled carbon nanotubes (MWCNTs).

2. Experimental methods

Carbon nanotubes and (ethylene–octene copolymer) EOC was used as an electrically conductive polymer base for preparation of conductive antenna layer. Engage (ethylene–octene copolymer) is a new elastomer produced by DuPont Dow Elastomers. By bridging the gap between rubber and plastic, versatile engage polymers inspire new design possibilities. The flexibility and mechanical properties of synthetic rubbers combined with the process ability of plastics make engage a great material. Low hardness and low viscosity are the advantages for easy thermoplastic processing as well as in solution state. EOC copolymer was dissolved in toluene to reach 16% wt polymer solution. Multiwall carbon nanotubes supplied by Sunnano china was treated by sonication in toluene by using sonication horn UZ Sonopuls HD 2070 kit for 15 minutes to disperse carbon nanotubes. Obtained dispersion was mixed together with (ethylene–octene copolymer) solution and stirred for two hours. The resulting dispersion contains 30% wt% of carbon nanotubes. PET substrate was dipped in the carbon nanotubes/ (ethylene–octene copolymer) dispersion in toluene for 10 s to reach the resulting active layer. The composite was kept at room temperature for 48 hours [4-6].

The PET substrate strip (5 × 50 mm) was coated by dip coating method on the following scheme (Figure 1) is demonstrate the process of dip coating of PET substrate by CNTs and (ethylene–octene copolymer) dispersion. First step: Dip a specimen into Dip Coat dispersion; second step: Wait for a dispersion surface to become the stationary state after dipped a specimen perpendicularly; third step: Raise a specimen up perpendicularly by optional speed; forth step: Available to control film thickness with raise-up speed for specimen and by the density of particle in solution.

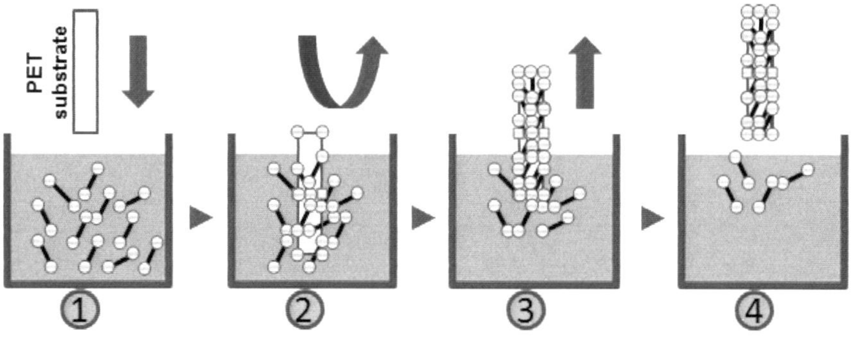

Figure 1. Preparation of electrically conductive layer by dip coating process.

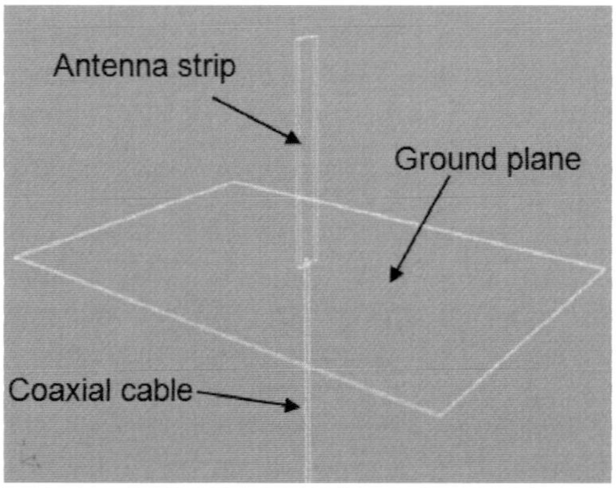

Figure 2. Construction of the antenna with antenna strip (coated PET), ground plane and coaxial cable.

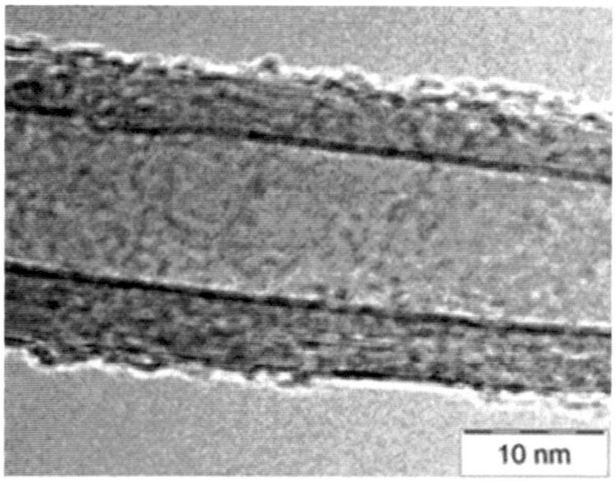

Figure 3. HRTEM detailed view of the structure of the MWCNT nanotube (scale of 10 nm).

IOP Conf. Series: Journal of Physics: Conf. Series **939** (2017) 012025 doi:10.1088/1742-6596/939/1/012025

3. Results

All measurements of the prepared antenna were performed in the anechoic chamber using the N9912A FieldFox Handheld RF spectrum analyzer with a measuring range within 2 MHz to 4 GHz. By means of this spectrum analyzer the parameter S11 was measured; this parameter determines the best frequencies of impedance matching for the antenna. The dimensions of the coated PET substrate are 5 × 50 mm. The total weight of the coated antenna is 0.402 g. The final size of flexible microstrip antenna was 5 x 50 mm with thickness of 0.48 mm (PET substrate 0.25 mm) The ground plane of the antenna is made of FR-4 copper substrate sized 65 × 75 mm (Figure 2). All measured values (Figure 4) show that the PET substrate and carbon nanotubes/ (ethylene–octene copolymer) in combination with dip coating technology proved to be very promising materials. By combining these materials a very efficient design of the microstrip antenna was achieved. Micrographs (Figure 5) obtained by scanning electron microscopy (FEI Nova NanoSEM 450) showed MWCNTs/ (ethylene–octene copolymer) conductive structure on the PET substrate.

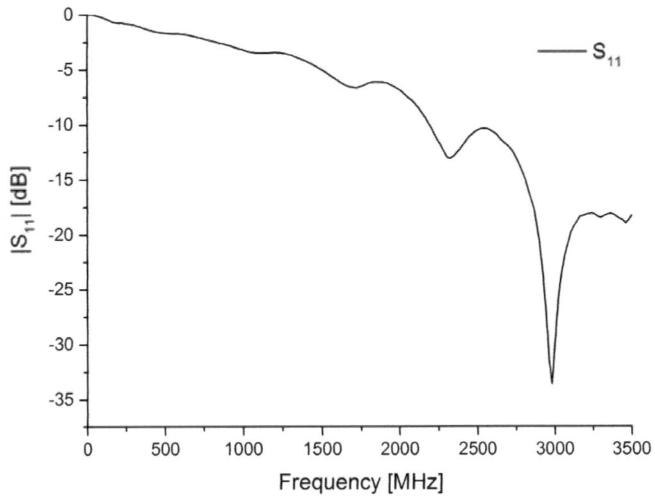

Figure 4. This graph represents the reflection coefficient S11 (expressed in dB) with dependence on the frequency with impedances to 1.66 GHz (–6.51 dB), 2.3 GHz (–13 dB) and 2.98 GHz (–33.59 dB).

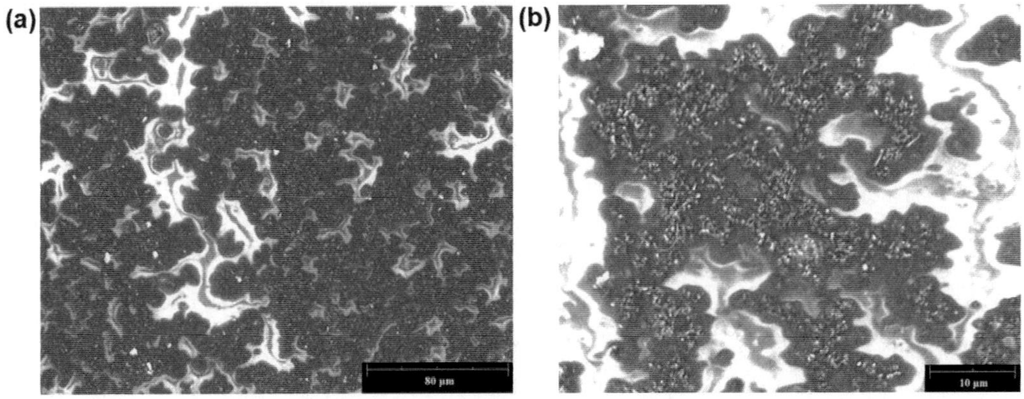

Figure 5. (a) SEM microscopy of the conductive structure on the PET substrate (scale of 80 µm), (b) SEM microscopy of the conductive structure on the PET substrate (scale of 10 µm).

IOP Conf. Series: Journal of Physics: Conf. Series **939** (2017) 012025 doi:10.1088/1742-6596/939/1/012025

4. Conclusion

Multiwall carbon nanotubes and (ethylene–octene copolymer) were used to prepare conductive antenna layer. The antenna works at three frequencies 1.66 GHz (–6.51 dB), 2.3 GHz (–13 dB) and 2.98 GHz (–33.59 dB) and could be used in telecommunication field. Main advantages are light weight and the possibility to be incorporated to a mobile phone cover. The carbon nanotubes and (ethylene–octene copolymer) network as an antenna could replace commonly used materials in the future.

The main advantages of the antenna are low weight and the possibility for applying to plastic covers of portable telecommunication devices and wearable electronics. Portable telecommunication devices, such as mobile phones, use for communication antennas made of copper foils and strips, which could be replaced by the developed antenna. The majority of widespread technologies, e.g. GSM, UMTS, or LTE operate at frequencies for which it is possible to produce a customized antenna made of carbon nanotubes and (ethylene–octene copolymer) deposited on the PET substrate, because PET substrate is mostly used for next generation of the flexible antenna [7-15].

Acknowledgments

This article was supported by the Ministry of Education, Youth and Sports of the Czech Republic – Program NPU I (LO1504) and with the support of Operational Program Research and Development for Innovations co-funded by the European Regional Development Fund (ERDF) and national budget of the Czech Republic, within the framework of the project CPS - strengthening research capacity (reg. number: CZ.1.05/2.1.00/19.0409).

References

[1] K. Gupta, R. Garg, I. Bahl, P. Bhartia, Microstrip Lines and Slotlines, second ed., Artech House, Norwood, MA, 1996.
[2] R. Garg, P. Bhartia, I. Bahl, A. Ittipiboon, Microstrip Antenna Design Handbook, first ed., Artech House, Norwood, MA, 2001.
[3] M.C. Dang, T.M.D. Dang, E. Fribourg-Blanc, Adv. Nat. Sci. Nanosci. Nanotechnol. 4 (2013) 015009.
[4] M. Maiti, S. Sadhu, A. K. Bhowmick. J. Appl. Polym. Sci. 2006, 101(1), 603-610
[5] P. Slobodian, P. Riha, R. Olejnik, M. Kovar, P. Svoboda. J. Nanomater.2013, 1-7
[6] R. Theravalappi, P. Svoboda, J. Vilcakova, S. Poongavalappil, P. Slobodian, D. Svobodova. Mater. Design. 2014, 60, 458-467
[7] L. Yang., A. Rida, M. Tentzeris, Design and Development of RFID and RFID-Enabled Sensors on Flexible Low Cost Substrates, first ed., Morgan & Claypool Publishers, San Rafael, CA, 2009.A. Georgiadis, A. Collado, S. Kim, H. Lee, M. Tentzeris. M. IEEE MTT-S Int. (2012) 1-3.
[8] H.K. Yoon, Y. Yoon, H. Kim, C.H. Lee, IET Microw Antenna P. 5 (2011) 1463-1470.
[9] D. Betancourt, J. Castan, Prog. Electromagn. Res. C 38 (2013) 129-140.
[10] S.R. Forrest, Nature 428 (2004) 911-918.
[11] L. Wang, Y. Guo, B. Salam, C.W. Lu, IEEE Asia–Pacific Conf. Antennas and Propagation, (2012) 239-240.
[12] J. Siden, M.K. Fein, A. Koptyug, H. Nilsson, IET Microwaves Antennas Propag. 1 (2007) 401-407.
[13] J. Perelaer, M. Klokkenburg, C.E. Hendriks, U. Schubert, Adv. Mater. 21(2009) 4830-4834.
[14] J. Matyas, R. Olejnik, K. Vlcek, P. Slobodian, P. Urbanek, P. Krcmar, L. Munster, Adv. Mater. Res. 1101 (2015) 245-248.
[15] J. Matyas, L. Munster, R. Olejnik, K. Vlcek, P. Slobodian, P. Krcmar, P. Urbanek, I. Kuritka. Jpn J Appl Phys. 55 (2016) 02BB13-4.

IOP Conf. Series: Journal of Physics: Conf. Series **939** (2017) 012026 doi:10.1088/1742-6596/939/1/012026

Advances in Magnetic Field Sensors

Kapsalis VC

Laboratory of Electronic Sensors, School of Electrical and Computer Engineers, National TU of Athens

bkapsal@mail.ntua.gr

Abstract. The important procedures in the field of magnetic measurements, due to the versatile and numerous applications in different sciences and the advantage of accuracy, non-destruction and contact less techniques are discussed here. Due to the universality of the methods and the prototype techniques which used in the magnetic field measurements, a review on the latest advances in the topic is essential. This paper deals with the determinant factors of the design and fabrication process of the magnetic sensors and especially the techniques and the materials used in the industry of the transducers. Following a brief examination on the magnetic sensors and main operating issues, specific focus is given to the critical factors for obtaining a high quality output. We focus our discussion on those effects that influence the efficiency and provide the information towards an accreditation system establishment.

1. Introduction

A sensor system consists of transducers input, output and intermediate signal processing. The collection of information to the input (sensor) is converted into an electrical signal, is processed (e.g., amplified, purified, etc.) and converted into a format, perceptible by the senses or trigger some event (actuator). The sensing mechanism is derived from different physical phenomena which also define the specific properties of the inputs, signal processing and the outputs. They are manufactured in different sizes of the active area of the order of microns or below. Recent developments refer to nano sensors with picotesla resolution, using metals such as gold, aluminium, metal alloys, ferromagnetic and nonmagnetic multilayers, semiconductors, graphene, etc. On the other hand another necessity derives from the most precise applications where other parameters like temperature coefficient, linearity and cross field sensitivity become more important, beside the signal to noise ratio coefficient.

Magnetic field sensors has been extensively researched in terms of their principles of operation and their properties behavior under different application [1-15]. In the evolution of the magnetic field sensors, two critical properties are critical in order to obtain high fidelity outputs, namely the sensitivity and selectivity. The increasing of effective sensing area and the modification of the planar surface with various nano porous materials enhance the former while the utilization of several materials and techniques to optimize the inter facial interactions result in the latter improvement. In other words, the sensitivity enhancement may induce a negative effect to the electrons transportation but the proper selectivity procedure could compensate the outputs towards an efficient, reliable and cost-effective result.

High surface areas can be attained by fabricating small particles or clusters where the surface to volume ratio is large or by creating materials where the void surface area is high compared with the bulk support material. Moreover, the development of several methods to gain information of the functional interactions and signalling nano impacts [16-19] the interfacial conductivity and the efficient energy dissipation also contribute to the magnetic field sensors cost effective development.

2. Sensors design and fabrication steps

The design of a magnetic sensor according to the specific application is the more significant stage for an optimized configuration but also a multi criteria decision procedure. Therefore, the best results comes from each step optimization process. Typically, the development of a sensor consists of several steps. We group them here in four main stages which include the optimized steps to be followed. The sensitivity definition, the selectivity including the electronic portion and the calibration control of proficiency testing un er interlaboratory comparison.

S1. Sensitivity definition derives from the design stage. We define the problem and the scale of the measurements, the principle that should be used, and the specific characteristics of the physical properties to be measured (e.g. operation range, sensitivity, uncertainty and dependence of the response with parametric phenomena such as the field environment, temperature and humidity). In this stage the granularity of the sensor should be analyzed, which means the smallest period which can be detected and measured. Calibration processes and standards of the level of uncertainty define the time and the calibration technique.

S2. Selectivity defines the adaptation and characterization of the materials and the manufacturing technique. Adaptation of the materials properties includes annealing by heating and magnetic field, application of current and mechanical stress, heat treatment and stretching. Another way is to introduce impurities in the magnetic material to be hardened through the anchoring of the magnetic dipoles in specific orientations, either by diffusion or by ion implantation. The manufacturing techniques adaptation may include thin films, thick films and rapid quenching techniques. The characterization of the sensor includes several methods such as electronic scanning microscopes (SEM, TEM), atomic microscopy and magnetic force and cross scanning (AFM, MFM and STM respectively), x-ray diffraction (XRD) and differential thermal analysis (DTA). In this stage we consider here the electronic portion of the sensor which used for the excitation circuit, signal configuration and self-calibration methods widely use amplifiers and integrated circuits.

S3. Proficiency and interlaboratory comparison provide accreditation of the calibration, testing and measurement procedure. The control of parametric phenomena, such as electrical, magnetic, temperature, humidity and time (which denotes the degradation of properties) provide the responses of the sensor to the desired application.

Latest integrated circuit technologies and packaging techniques allow for the compensation of the temperature dependence not only of the sensor itself but also of the magnetic circuit enclosing the sensor such as the coefficient of the permeability and the changes in the air gap. Moreover, integrated flat coil simplifies the design of magnetometers but also causes problems such as linearity error. Negative feedback with sufficient gain may adjust the temperature coefficient in a desirable range. Since the output voltage depends on the measuring field direction as well as the perpendicular one there is a cross field error. Strain sensing materials fabricated with CNTs polymer hybrids succeeded in large range of resistance – strain linearity and reproducing sensitivity highlighting the promising role of the materials. The type of output configuration plays an important role due to the presence of offset voltage, amplitude imbalance and additional harmonics. For example in recent study the use of two Wheatstone bridges, due to the non-ideal properties of sensor, the output sinusoidal signal presented undesired characteristics. A trade-off between the desired properties and the specific application usually defines the selection of the sensor characteristics.

3. Types of magnetic sensors

Generally speaking, the detection of changes in the eddy current magnetic field and the variation of the magnetic polarization have been extensively using in the evaluation of non – destructive testing (NDT). The resolution and the sensitivity of the sensors over the samples reveal the magnetic leakage on the defects and the flux distribution is detected through a data acquisition system. The sensors Hall are solid state devices that convert the magnetic field into an electrical signal for use in electronic circuits. Apart from the accuracy in measurements, strength in manufacturing and non-contact measurement, the biggest advantage is their incorporation into integrated circuits with different technologies. They suffer from the piezo effect and are more appropriate for in-plane sensing. The single domain ferromagnetic thin films exhibit anisotropic magneto restrictive (AMR) resistivity. In non-magnetic metals the phenomenon is weaker. Anisotropy in electric resistance between the direction of magnetization and the perpendicular one is a feature which makes the AMR sensor more sensitive in sensing direction from the Hall one, they exhibit a better stability but they cannot integrate the driving and the signal processing electronics in the same chip like the later does. For example the AMR sensors the periodical flipping or re magnetization of sensor structure is a good technique to improve the accuracy through pulses into a coil, usually integrated in the chip. Magnetic field sensors using Giant magneto restrictive (GMR) structures with high sensitive multilayers consisting of alternating ferromagnetic (f) and nonmagnetic (n) metals. The magnetic moment of an electron charge, typically the opposite quantum spin ½, can lead in large changes in electric resistance with the changes in the moment ordering of adjacent f-layers.

Recent developments of GMR with Current flow Perpendicular to the layers Planes (CPP) quantify the scattering of spin flipping asymmetries and the gained knowledge of the f and n metals and alloys. The

importance of antiparallel and parallel flow on the electron transportation through the interface between he f and n layers is a challenging topic for investigation to improve the performance.
Both GMR and Spin – dependent tunneling (SDT) devices are used in in nondestructive evaluation and bio assay applications and can be integrated in silicon on chip with isolator small packaging technology improving the speed performance and the ground noise. Fluxgate sensors are based on the modulation of permeability of the soft magnetic core, which reaches the maximum and minimum value twice in each excitation frequency. The low permanence can be achieved by the deep saturation of the core of the sensor. This can be done, with the lower power consumption, by nonlinear excitation tuning. The feedback has been succeeded with three orthogonal windings and the coil wounded on the top of a sphere to homogenize the magnetic field. The development of thin film based superconducting quantum interference device magnetometer (SQUID) has presented superior noise performance due to the niobium thin film pick up coil which is connected with the input coil and the low capacitance cross type Josephson junctions. Beside the high sensitivity several modifications are needed to enhance the modest spatial resolution. Giant magneto impedance sensors (GMI) are based in the large changes in the real and imaginery part of the impedance upon the static magnetic field of soft materials. The advances in soft materials play a crucial role in sensitivity, accuracy and stability of the demanded applications.
What primarily we are interested in the dimensional reduction of the active regions of the sensor without the loss of sensitivity of the magnetic field. Sensitivity and accuracy are the main properties of sensors we are looking for to improve. Integrated magnetic flux concentrators at the surface of magnetic sensors dramatically improve their performance. They need to be made of high permeability materials to increase the sensitivity. The results are better to the AMR and GMR sensors than the Hall ones. At the same time they have to lower the temperature dependence of the magnetic gain factor, to handle with the possibility of saturation and the non-linearity behaviour and of course to deduce the remanence.
A typical technique to shift the signal frequency out of the noise range is the modulation of the permeability of the material by the excitation of a magnetic field, using long wire concentrators or the periodically moving concentrators using Micro Electromechanical Systems (MEMS). Moreover, the development of resonators, which operate with the Lorenz force, provide low energy consumption and high sensitivity and reduced fabrication costs. In Hall sensors a low noise promising fabrication was achieved with a two dimension quantum – well multilayer heterojunction based on GaAs. The external spinning current electronic could be further improved with the usage of leakage free switches. The temperature increase above the Curie point can also modulate the permeability but it is impractical because of the limited operation of such devices. Super conductivity loop has been also proposed but they need low temperature to maintain a good sensitivity. Advances in the knowledge of Meissner currents and the induced vortex like dipoles or the cooling field effect beside the magnetic hysteresis loop and the phonon electron interactions on the lattice are to be investigated in order to achieve the most demanded applications.

4. Conclusions
In this paper a review on the advances in the range of the field is presented. Typical properties of the magnetic sensors, design and fabrication issues and critical steps in real life applications support the efforts for the consideration of the sensitivity and selectivity which remain the primary issues for an efficient and cost effective process. The analysis reveals more challenging factors to be investigated such as surface porous cycles, interfacial interactions, thermal noise and energy dissipation which affect advanced magnetic field sensors application and the establishment of an accreditation system.

References
1. Hristoforou, E. and R.E. Reilly, *New mechanical stress transducers based on amorphous alloys*. IEEE Transactions on Magnetics, 1990. **26**(5): p. 1563-1565.
2. Hristoforou, E. and D. Niarchos, *Mechanical sensors based on reentrant flux reversal*. IEEE Transactions on Magnetics, 1992. **28**(5): p. 2190-2192.
3. Hristoforou, E. and R.E. Reilly, *Tensile stress distribution sensors based on amorphous alloys*. Journal of Magnetism and Magnetic Materials, 1993. **119**(3): p. 247-253.
4. Hristoforou, E., et al., *Non-destructive evaluation distribution sensors based on magnetostrictive delay lines*. Sensors and Actuators A: Physical, 2001. **92**(1–3): p. 132-136.
5. Dimitropoulos, P.D., J.N. Avaritsiotis, and E. Hristoforou, *A novel micro-Fluxgate sensor based on the AMR effect of ferromagnetic film-resistors*. Sensors and Actuators A: Physical, 2003. **107**(3): p. 238-247.

6. Kosmas, K., et al., *Non-destructive evaluation of magnetic metallic materials using Hall sensors.* Journal of Materials Processing Technology, 2005. **161**(1–2): p. 359-362.
7. Piotrowski, L., et al., *Evaluation of Barkhausen Noise and Magnetoacoustic Emission Signals Properties for Plastically Deformed Armco Iron.* IEEE Transactions on Magnetics, 2010. **46**(2): p. 239-242.
8. Augustyniak, B., et al., *Barkhausen Noise Properties Measured by Different Methods for Deformed Armco Samples.* IEEE Transactions on Magnetics, 2010. **46**(2): p. 544-547.
9. Ktena, A., et al., *Barkhausen noise as a microstructure characterization tool.* Physica B: Condensed Matter, 2014. **435**: p. 109-112.
10. Rossini, N.S., et al., *Methods of measuring residual stresses in components.* Materials & Design, 2012. **35**: p. 572-588.
11. Vourna, P., A. Ktena, and E. Hristoforou, *Residual Stress Analysis in Nonoriented Electrical Steel Sheets by Barkhausen Noise Measurements.* IEEE Transactions on Magnetics, 2014. **50**(4): p. 1-4.
12. Vourna, P., et al., *A novel approach of accurately evaluating residual stress and microstructure of welded electrical steels.* NDT & E International, 2015. **71**: p. 33-42.
13. Vourna, P., et al., *Correlation of Magnetic Properties and Residual Stress Distribution Monitored by X-Ray and Neutron Diffraction in Welded AISI 1008 Steel Sheets.* IEEE Transactions on Magnetics, 2015. **51**(1): p. 1-4.
14. Vourna, P., et al., *An accurate evaluation of the residual stress of welded electrical steels with magnetic Barkhausen noise.* Measurement, 2015. **71**: p. 31-45.
15. Vourna, P., *An accurate method for determining residual stresses with magnetic non-destructive techniques in welded ferromagnetic steels.* IOP Conference Series: Materials Science and Engineering, 2016. **108**(1): p. 012017.
16. Di Bari, C., et al., *Fabrication of high surface area graphene electrodes with high performance towards enzymatic oxygen reduction.* Electrochimica Acta, 2016. **191**: p. 500-509.
17. Berkesi, K., et al., *Comparative study of the adsorption of and Cl− anions on smooth and rough surfaces of noble metal electrodes by in situ radiotracer thin gap method.* Journal of Electroanalytical Chemistry, 2014. **712**: p. 151-160.
18. Berkesi, K., et al., *Development and application of the in situ radiotracer thin gap method for the investigation of corrosion processes II. Validation of the thin gap method adapted for the application of porous surfaces.* Electrochimica Acta, 2013. **109**: p. 790-797.
19. Kätelhön, E., et al., *Destructive nano-impacts: What information can be extracted from spike shapes?* Electrochimica Acta, 2016. **199**: p. 297-304.

IOP Conf. Series: Journal of Physics: Conf. Series **939** (2017) 012027 doi:10.1088/1742-6596/939/1/012027

The hardness of the hydroxyapatite-titania bilayer coatings by microindentation and nanoindentation testing

Djahida SIDANE, Hafit KHIREDDINE, Sabeha YALA

Laboratoire de Génie de l'Environnement, Faculté de Technologie, Université Abderrahmane Mira, Bejaia, Algérie

dj.sidane@yahoo.fr

Abstract. The aim of this paper is to investigate the effect of the addition of titania (TiO_2) inner-layer on the morphological and mechanical properties of hydroxyapatite (HAP) bioceramic coatings deposited on 316L stainless steel (316L SS) by sol-gel method in order to improve the properties of hydroxyapatite and expand its clinical application. The addition of TiO_2 as sub-layer of a hydroxyapatite coating results in changes in surface morphology as well as an increase of the microhardness. The deposition of the inner-layer provides the formation of new types of hydroxyapatite coatings at the same condition of annealing. This represents an advantage for the various applications of the hydroxyapatite bioceramic in the medical field. Classical hardness measurements conducted on the coated systems under the same indentation load (10g) indicated that the microhardness of the HAP coating is improved by the addition of TiO_2 inner-layer on the 316L stainless steel substrate. The hardness values obtained from both classical tests in microindentation and the continuous stiffness measurement mode in nanoindentation are slightly different. This is because nanoindentation is more sensitive to the surface roughness and the influence of defects that could be present into the material. Moreover, nanoindentation is the most useful method to separate the contribution of each layer in the bilayer coatings. In this study, the hardness is comparable with those reported previously for pure HAP ceramics (1.0–5.5 GPa) which are close to the properties of natural teeth.

1. Introduction

Hydroxyapatite ($Ca_{10}(PO_4)_6(OH)_2$, HAP) is widely used for hard tissues repair due to its chemical and structural similarities with the mineral phase of bone and teeth. In this paper, we proposed a strategy of coating 316L stainless steel (316L SS) substrate with HAP ceramic. The sol–gel dip coating process is easily applicable to surface coating, and it allows the preparation of high quality HAP thin films on metal substrates [1, 2]. Many recent studies have investigated the variation of the microstructural and morphological properties of the hydroxyapatite coatings with the influence of the reinforcing process such as the addition of titania (TiO_2) phase in order to improve the bioactivity, mechanical, and bonding strength of the nanocomposite coatings developed [3–6]. It is recognized that thin films of TiO_2 on 316L SS possess combined advantages of biocompatibility and corrosion-resistant properties [7]. As a simple and direct measurement, indentation hardness is often used as an initial guideline for the qualification of a coating for any application requiring wear resistance. As a rule of thumb, intrinsic film hardness can be directly measured when the penetration depth of the indenter is lower than 10% of the coating thickness. In this case, the substrate is supposed to not interfere in the hardness measurement. For the recent generation of coatings that are becoming thinner and thinner, very often below than 4-5 μm, this rule of thumb is sometimes difficult to be respected mainly in microindentation [8, 9].

2. Experimental details

Defined amounts of phosphorus pentoxide (P_2O_5, Prolabo 100 %) and calcium nitrate tetrahydrate ($Ca(NO_3)_2.4H_2O$, Fluka 98 %) were dissolved in absolute ethanol, to form solutions with concentrations of 0.5 and 1.67 mol/l, respectively. The two solutions were mixed and this results in the formation of HAP sol having Ca/P molar ratio of 1.67 [10]. The mixture was continuously stirred at ambient temperature for 24 hours. This produces translucent sol. Titanium isopropoxide (TIP, Fluka 100 %) was used as a titania precursor in the sol–gel process. The reactivity toward water is modified by acetic acid (HOAc) (molar ratio of TIP/HOAc = 1/10) which is also used as catalyst. 2-Methoxy ethanol was added to adjust the degree of viscosity of the solution. This solution with titanium molar concentration of approximately 0.47 M was vigorously stirred under ambient temperature conditions [11]. 316L SS substrates were mechanically polished using different silicon carbide grit papers up to #1200 grades. Final polishing was done using diamond paste (2 μm) and (0.7 μm) to produce scratch-free mirror finish surface. The substrates were ultrasonically degreased with acetone and washed with running double distilled water. Finally, they were dried at 150 °C for 10 minutes. The TiO_2 coatings were obtained by dipping the polished, washed, and dried metal substrates in the suspension with dipping rate of 20 mm/min and annealed at the temperature of 450 °C for 60 minutes. The HAP-TiO_2 bilayer coatings were obtained by the deposition of the hydroxyapatite on the outer surface of the TiO_2 single coatings, with dipping rate of 80 mm/min and annealed at the temperature of 500 °C for 60 minutes.

The morphology and the elemental analysis of the surface of the coated specimen were performed using scanning electron microscopy (SEM QUANTA 200, detector SUTW-Sapphire). The thickness of the coating layer was measured using a profilometry analysis ''DEKTAK 150 SURFACE PROFILER,'' by scanning the surface of the coating at an interval of 1000-8000 μm. Three different areas were scanned and measured to determine the mean thickness.

Vickers microhardness indentations were performed using a Zwick microhardness machine (ZHV10) with a dwell time of 15 seconds applied at the maximum indentation load. Hardness of the coatings is determined under the lowest indentation load available on the instrument, i.e., 10 g, to minimize the influence of the substrate on the hardness measurement. The hardness value was the result of 10 indents measurements to be representative.

Nanoindentation experiments were performed with a Nano Indenter XP ™ (MTS Nano Instruments) with Berkovich diamond indenter. The samples were fixed on a metallic support using the heat softening glue crystalbond 509. 25 indentation tests were conducted randomly on the surface of the material with the same indentation testing conditions. The maximum indentation depth reached by the indenter was fixed at 2000 nm and the strain rate was equal to 0.05 s^{-1}. The instrument was operated in the continuous stiffness measurement (CSM) mode allowing the computation of the hardness continuously during the indentation loading. The harmonic displacement was 2 nm and the frequency was 45 Hz.

3. Results and discussion

3.1. Morphological properties

Figure 1 shows the SEM images related to the HAP single coating (Figure 1a) and bilayer coating (Figure 1b).

The HAP single coating exhibited a porous surface composed of spherical agglomerates, whereas the surface of the HAP-TiO_2 bilayer appears dense and uniform. SEM surface examination reveals no detectable cracks for these coatings.

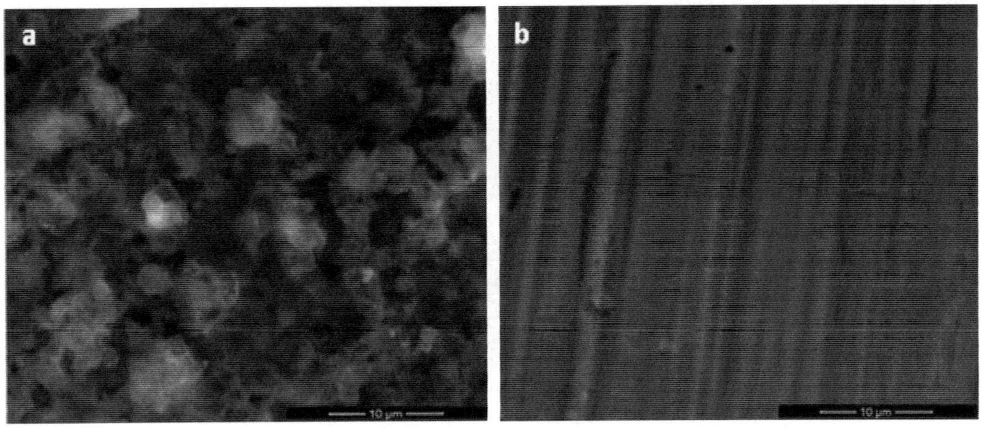

Figure 1. SEM observations of the (a) HAP single coating and (b) HAP-TiO_2 bilayer coating.

3.2. Vickers microhardness

To analyze the homogeneity and the indentation size effect (hardness-load dependence) of the substrate, a series of indentation loads ranging between 10 and 200 g were performed. Table I gives the hardness number computation for each indentation load. As it can be seen, the microhardness of the substrate varies according to the applied load between 133 HV and 168 HV, the hardness values are close enough to consider a mean value, i.e., 150 HV for the substrate hardness.

Table I. Experimental data of the 316L SS substrate as a function of the applied load.

Load (g)	10	20	50	150	200
Hardness (HV)	155	168	150	149	133

For characterizing the hardness of the coatings, only indentation experiments using the indentation load of 10 g have been performed. The comparison between the HAP and HAP-TiO_2 coatings is acceptable since their coating thicknesses are quite similar, around 2 µm. The hardness value indicated in Table II is a composite hardness taking into account a part of the elasto-plastic response of the substrate [12]. In order to estimate the apparent hardness of the coating, the model of Jönsson and Hogmark [13] is the only model applicable because the calculation has been performed under only one indentation load. This model, which is predictive and not descriptive since all the involved parameters are known, expresses the composite hardness, H_C, as a function of the hardness of the substrate, H_S, and that of the film, H_F, as follows:

$$H_C = H_S + a.(H_F - H_S) \qquad [1]$$

Where a is the parameter representing the influence of the film in the hardness measurement, this parameter can be expressed as a function of the indent diagonal, d, and the coating thickness, t, as follows:

$$a = \frac{2Ct}{d} - \left(\frac{Ct}{d}\right)^2 \qquad [2]$$

Where C is a constant equals to 0.5 for brittle materials which is the case here since cracks have been observed around the indents in the coatings (data not showed).

Considering t = 2 µm, the average diagonal value of 9 µm for d and the constant C equals to 0.5, the computation of the parameter a leads to 0.21. Applied to Eq. [1] using the hardness value of the substrate of 150 HV and the hardness value of the composite of 235 HV for the bilayer HAP-TiO_2 coating and

215 HV for the HAP coating, the approximate Vickers hardness number of the separate coating is found to be close to 555 HV for the bilayer HAP-TiO_2 coating, and 459 HV for the HAP coating. This result is interesting but must be verified by nanoindentation employing the continuous stiffness measurement method. Indeed, this methodology will be necessary to separate the contribution of each layer in the composite coatings [14–16].

Table II. Indent diagonals and Vickers hardness number obtained on the different coatings by applying the same indentation load of 10 g.

Coating	HAP	HAP-TiO_2
Diagonal (µm)	9.3 ± 0.2	8.9 ± 0.3
Hardness (HV)	215 ± 10	235± 16

3.3. Nanoindentation hardness

Figure 2 represents the hardness variation for the different situations. As it is clearly shown on this figure, the uncoated substrate presents the highest hardness value with a hardness variation, which decreases from 7 GPa, close to the external surface, to 3 GPa toward the core of the material for the highest indenter displacements. The hardness gap is due to the presence of the iron oxide on the surface of the substrate [17].

For the TiO_2 coating, we note that the hardness is very similar to that of the substrate except for indenter displacements very close to the outer surface, typically over the first 150 nm in depth. In this region, we applied the model of Jönsson and Hogmark (Eq. [1]) for determining the hardness of the film only. Here d is the indent diagonal:

$$d = 2tg74h \quad [3]$$

Where h is the indentation depth.

Applied to the hardness variation of the TiO_2 films and using the hardness values of the substrate, we obtained a value of 3.5 GPa for the hardness of the titania oxide. For information, Fu et al. [18] indicated that the hardness of titania films measured by nanoindentation, can vary to a great extent, between 2 and 18 GPa depending on crystalline nature and microstructure of the films [19,20].

For the HAP coating, we can note that the hardness value is constant for the lowest indenter displacements over 200 nm in depth. In this region, the hardness is close to 1 GPa. After 200 nm in depth, the hardness value increases toward a value of 3 GPa which corresponds to the hardness of the steel substrate. In this zone, the substrate interferes with the hardness measurement. For this coating, no application of model is required since the film hardness has been determined for the lowest displacements of the indenter. Indeed, the hardness has a constant value of 1 GPa for indenter displacements between 100 and 200 nm.

For the HAP–TiO_2 coating, it is interesting to note that the hardness value varies between two limits corresponding to the HAP hardness and the substrate hardness, respectively for the lowest and highest indenter displacements. Between these two limits, the benefit of the sub-layer of TiO_2 in terms of mechanical properties is clearly visible since the hardness variation is observed between 2.5 GPa at the extreme surface, as compared to only 1 GPa for the HAP coating, until 6 GPa corresponding to the TiO_2 coating or steel substrate hardness at only 300 nm. As a conclusion, we showed that the TiO_2 sub-layer significantly improves the hardness of the HAP coating. In this study, the hardness is comparable with those reported previously for pure HAP ceramics (1.0–5.5 GPa), which are close to those properties of natural teeth [21].

Above, it is shown that the microhardness of the HAP coating, which is of 459 HV (4.50 GPa) increases with the addition of the TiO_2 inner layers, to reach approximately 530 HV (5.19 GPa) for the global HAP-TiO_2 bilayer coatings.

IOP Conf. Series: Journal of Physics: Conf. Series **939** (2017) 012027 doi:10.1088/1742-6596/939/1/012027

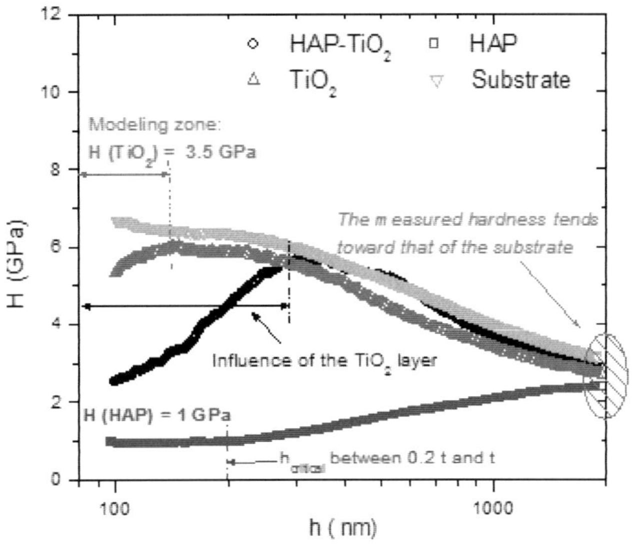

Figure 2. Hardness versus the indenter displacement obtained by using the CSM mode of nanoindentation performed on the uncoated substrate, the HAP, TiO_2, and HAP–TiO_2 coatings.

4. Conclusion

In addition, the hardness values obtained from both classic tests in microindentation and the continuous stiffness measurement mode in nanoindentation are slightly different. This is due to the fact that nanoindentation is more sensitive to the surface roughness and the influence of defects, which could be present in the material. Moreover, nanoindentation is the most useful method to separate the contribution of each layer in the bilayer coatings.

References

[1] D.M. Liu, T. Troczynski, and W.J. Tseng: Biomaterials, 2001, vol. 22, pp. 1721–30.
[2] A. Montenero, G. Gnappi, F. Ferrari, M. Cesari, E. Salvioli, L. Mattogno, S. Kaciulis, and M. Fini: Mater. Sci., 2000, vol. 35, pp. 2791–97.
[3] A.J. Nathanael, N.S. Arul, N. Ponpandian, D. Mangalaraj, and P.C. Chen: Thin Solid Films, 2010, vol. 518, pp. 7333–38.
[4] A.J. Nathanael, D. Mangalaraj, and N. Ponpandian: Compos Scie Technol, 2010, vol. 70, pp. 1645-51.
[5] C.C. Yang, C.Y. Huang, C.C. Lin, and S.K. Yen: J. Electrochem. Soc., 2011, vol. 158, pp. 13–20.
[6] A.J. Nathanael, J.H. Lee, and S.I. Hong: Adv. Sci. Lett., 2012, vol. 15, pp. 285–90.
[7] J. Blumn, K.L. Eckert, A. Schroeder, M. Petitmermet, S.W. Ha, and E. Wintermantel: in Bioceramics, T. Kokubo, T. Nakamura, and F. Miyaji, eds., Pergamon, Otsu, 1996, pp. 89–92.
[8] D.E. Kramer, A.A. Volinsky, N.R. Moody, and W.W. Gerberich: J. Mater. Res., 2001, vol. 16, pp. 3150–57
[9] A.C. Fischer-Cripps: Vacuum, 2000, vol. 58, pp. 569–85.
[10] S. Kim and P.N. Kumta: Mater. Sci. Eng., B, 2004, vol. 111, pp. 232–36.
[11] A. Balamurugan, S. Kannan, and S. Rajeswari: Mater. Lett., 2005, vol. 59, pp. 3138–43.
[12] D. Chicot, E. Bemporad, G. Galtieri, F. Roudet, M. Alvisi, and J. Lesage: Thin Solid Films, 2008, vol. 516, pp. 1964–71.
[13] B. Jönsson and S. Hogmark: Thin Solid Films, 1984, vol. 114, pp. 257–69.
[14] K. Zeng and C.H. Chiu: Acta Mater., 2001, vol. 49, pp. 3539–51.
[15] B. Wolf, A. Richter, and V. Weihnacht: Surf. Coat. Technol., 2004, vol. 183, pp. 141–50.

[16] J. Musil, F. Kunc, H. Zeman, and H. Polakova: Surf. Coat. Technol., 2002, vol. 154, pp. 304–13.
[17] D. Chicot, J. Mendoza, A. Zaoui, G. Louis, V. Lepingle, F. Roudet, J. Lesage: Mater. Chem. Phys. 129 (2011) 862–70.
[18] T. Fu, C.S. Wen, J. Lu, Y.M. Zhou, S.G. Ma, B.H. Dong, B.G. Liu: Vacuum, 2012, vol. 86, pp. 1402–07.
[19] O. Borrero-López, M. Hoffman, A. Bendavid, P.J. Martin: Thin Solid Films, 2011, vol. 51, pp. 7925–31.
[20] T. Modes, B. Scheffel, C. Metzner, O. Zywitzki, E. Reinhold: Surf. Coat. Technol. 2005, vol. 200, pp. 306–9.
[21] J. Song, Y. Liu, Y. Zhang, L. Jiao: Mater. Sci. Eng., A., 2011, vol. 528, pp. 5421–27.

IOP Conf. Series: Journal of Physics: Conf. Series **939** (2017) 012028 doi:10.1088/1742-6596/939/1/012028

Characterization of various Si-photodiode junction combinations and layout specialities in 0.18μm CMOS and HV-CMOS technologies

I.Jonak-Auer, O. Synooka, A. Kraxner and F. Roger

ams AG, Tobelbaderstraße 30, A-8141 Premstätten, Austria

ingrid.jonak-auer@ams.com

Abstract. With the ongoing miniaturization of CMOS technologies the need for integrated optical sensors on smaller scale CMOS nodes arises. In this paper we report on the development and implementation of different optical sensor concepts in high performance 0.18μm CMOS and high voltage (HV) CMOS technologies on three different substrate materials. The integration process is such that complete modularity of the CMOS processes remains untouched and no additional masks or ion implantation steps are necessary for the sensor integration. The investigated processes support 1.8V and 3V standard CMOS functionality as well as HV transistors capable of operating voltages of 20V and 50V. These processes intrinsically offer a wide variety of junction combinations, which can be exploited for optical sensing purposes. The availability of junction depths from submicron to several microns enables the selection of spectral range from blue to infrared wavelengths. By appropriate layout the contributions of photo-generated carriers outside the target spectral range can be kept to a minimum. Furthermore by making use of other features intrinsically available in 0.18μm CMOS and HV-CMOS processes dark current rates of optoelectronic devices can be minimized. We present TCAD simulations as well as spectral responsivity, dark current and capacitance data measured for various photodiode layouts and the influence of different EPI and Bulk substrate materials thereon. We show examples of spectral responsivity of junction combinations optimized for peak sensitivity in the ranges of 400-500nm, 550-650nm and 700-900nm. Appropriate junction combination enables good spectral resolution for colour sensing applications even without any additional filter implementation. We also show that by appropriate use of shallow trenches dark current values of photodiodes can further be reduced.

1. Introduction

The selection of Si as the starting material for photodiode (PD) fabrication, gives the possibility to monolithically integrate the sensor with the reading circuitry, lowering device costs, and enhancing the sensor capabilities [1]. With the ongoing miniaturization of CMOS technologies the need for integrated optical sensors on smaller scale CMOS nodes arises. In this paper we report on the development and implementation of different optical sensor concepts in high performance 0.18μm CMOS and HV-CMOS technologies. The 0.18μm CMOS and HV-CMOS process technologies offer several junctions with different doping levels and depths which have been investigated as single junction and double junction PDs and have been characterized with respect to their electrical and optical performance.

IOP Conf. Series: Journal of Physics: Conf. Series **939** (2017) 012028 doi:10.1088/1742-6596/939/1/012028

2. Structures and Measurements

In this section PD layouts and measurement setup are described. Test chips containing various junction combinations were processed in the 0.18µm CMOS and HV-CMOS process options of ams AG [2]. Only standard process steps were used. The only modification was the opening of the passivation nitride layer on top of the PDs in order to minimize multiple reflections and interferences in the illumination path in the dielectric layer stack between top wafer surface and the silicon based diodes.

2.1. Layouts

Figure 1 shows schematics of the single junction PDs processed and measured in the course of this work. For the electrical contact, an anode (A) and a cathode (C) contact are used as indicated in Figure 1a. For the cathode, a contact with the active area of the PD is established by an n+ source/drain doped region. The anode contact is build-up of a p+ source/drain-doping in the p-substrate. Figure 1(c) shows a PD with narrow shallow trench (ST) between cathode and anode. The size of all PDs is 100µmx100µm.

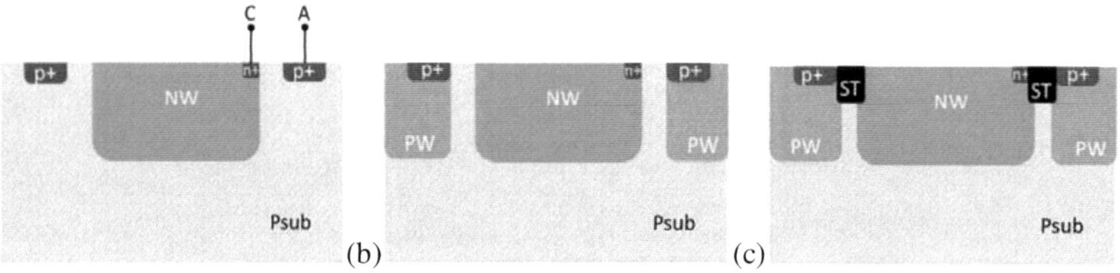

(a) (b) (c)

Figure 1. Schematic layout of single junction photodiodes.

Figure 2 shows schematics of the investigated double junction PDs (a and b) as well as a triple junction PD (c). They consist of a combination of p-wells in n-wells in p-substrate, thus forming two or three possible junctions, respectively. The wells used here differ in depth as well as in doping concentration.

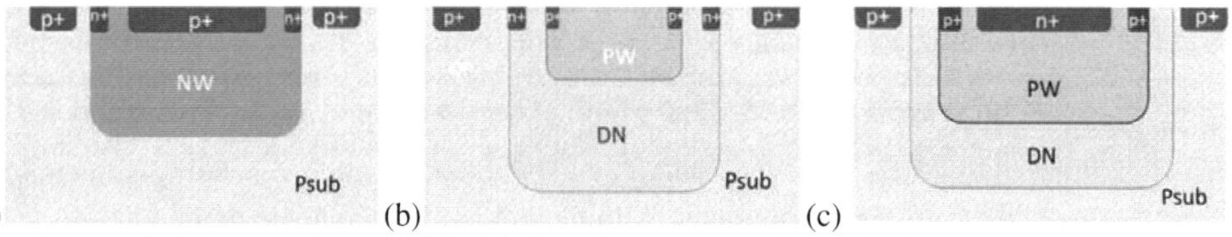

a) (b) (c)

Figure 2. Schematic layout of double junction PDs (a and b) and triple junction PD (c).

Three different substrate materials were investigated in the course of this study including lightly p-doped high quality bulk silicon as well as 14µm and 20µm thick lightly p-doped EPI layers on highly p-doped substrates.

2.2. Measurement Setup

To characterize the performance of the PDs we did capacitance (C) and dark current (DC) measurements as well as spectral responsivity (SR) measurements. Measurements have been performed with a Cascade Micromanipulator equipped with thermo chuck. Additionally a HP4284A precision LCR meter was used for C measurements and a Parameter Analyzer HP4155B for DC measurements. DC and C measurements have been performed at a temperature of 60°C. All optical measurements were performed according to the measurement setup depicted in Figure 3. An almost white Xenon lamp is used as light source, which is discrete band filtered in 1nm wavelength steps via a spectrometer. The monochromatic light beam is split, one half goes into a reference detector for

IOP Conf. Series: Journal of Physics: Conf. Series **939** (2017) 012028 doi:10.1088/1742-6596/939/1/012028

actual power measurement, while the other half is fed onto the optically active area of the device under test. To determine the responsivity of the device under test, the photocurrent generated by the incident light power is measured with a Parameter Analyzer HP4155B. Measurement data recordings as well as filter steering, calibration and reference light power measurements are controlled via serial bus interfaces and personal computers. The optical measurements were done at 0V bias in a wavelength region from 400nm to 900nm.

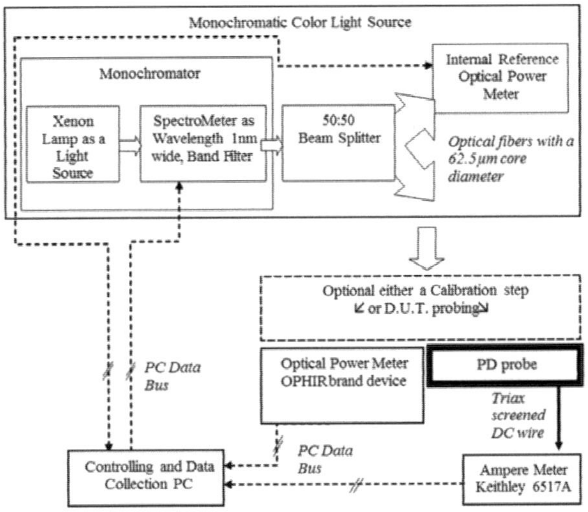

Figure 3. Experimental setup for spectral responsivity measurements.

3. Results

This section covers spectral responsivity (SR) results as well as dark current (DC) and capacitance (C) values for several PD layouts. The influence of the base material and substrate contact design on these parameters is shown on single junction diodes. On double and triple junction diodes we focus on spectral selection by different shorting methods.

3.1. Results on Single Junction PDs

Measurements have been performed on PDs as depicted in Figure 1.

3.1.1. Influence of base material. Figure 4a is a summary of SR data comparing 3 different base materials. The passivation nitride layer was removed on top of all PDs. Calculated external quantum efficiency (EQE) values of 90% and 100% are also included.

The 0.18μm n-well PDs show a very good EQE about 90% for the spectral range between 480nm to 720nm. For longer wavelengths the influence of the base material can clearly be seen. The bulk material has the highest SR followed by 20μm EPI and 14μm EPI. With increasing wavelength the penetration depth of light in silicon also increases according to Lambert-Beer's law [1].

Table 1 shows depth values for some selected wavelength values, at which 70% of the incoming light intensity is absorbed in Si. At a wavelength of 850nm 30% of the incoming light is transmitted to a depth of 15μm. So eventually photo generation happens very close or even inside the highly doped EPI substrate where the generated carriers recombine thus decreasing the photocurrent.

DC measurement results on the three base materials are plotted in Figure 4b. DC on EPI wafers is about 4 times lower than on bulk wafers indicating higher lifetime and less defects in the EPI material. Capacitance measurements show no significant difference between different base materials.

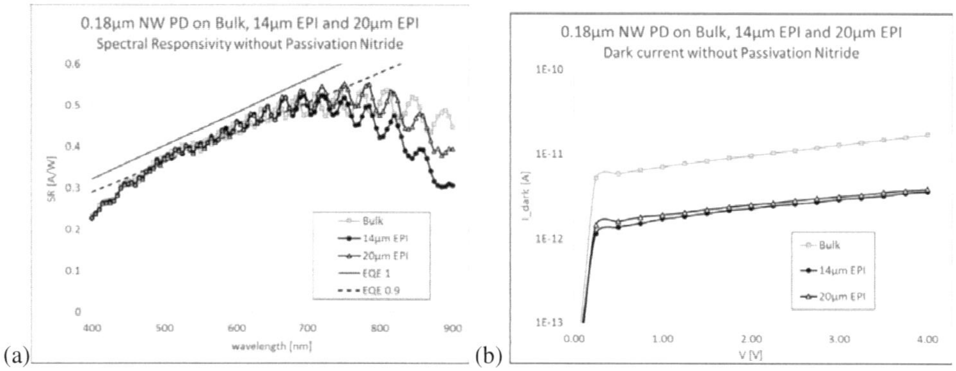

Figure 4. SR and DC results of 0.18μm n-well PDs on three different base materials

Table 1. Penetration depth of light in silicon.

Wavelength (nm)	Depth of 70% absorption (μm)
410	0.14
660	3.9
780	10
850	15

3.1.2. Influence of substrate contact layout. This section shows DC measurements at 60°C on PDs with differently designed substrate contacts shown in Figures 1a to 1c. Different depths of p-doped substrate contacts have been studied and also distances between substrate contacts and PD cathode have been varied. In the case of Figure 1c an additional narrow shallow trench (ST) guardring has been added. For a better understanding of the electrical and optical behaviour of the photodiodes presented in this section, two-dimensional TCAD simulations [3] have been performed following our 0,18μm HV process technology and PD layout. DC simulation results presented below utilize standard recombination models including Shockley-Read-Hall and Auger models combined with standard mobility models dependent on the doping concentration, temperature, normal and saturation electrical fields and scattering effects. As DC simulations are uncalibrated the current scale has been omitted in the plots.

Figure 5 shows the influence of p+ substrate contact distance on DC for PDs processed on 20μm EPI substrates together with TCAD simulation results. A smaller distance of the substrate contact to the PD cathode shows smaller DC values, a trend which has also been confirmed by TCAD simulations. The further the substrate contact is away from the PD the larger is the space charge region (SCR) and the higher is the electron generation current and therefore the dark current of the PD.

In Figure 6a we compare DC measurement results of 3 different substrate contact layouts drawn in Figures 1a to 1c. It shows that having a deeper p-well connection to the substrate results in lower DC values compared to shallow p+ substrate contacts. If an additional narrow ST is placed between anode and cathode of the PD dark current values can be reduced further. In Figure 6b we provide the corresponding TCAD simulations which show the same trend as the measurement results.

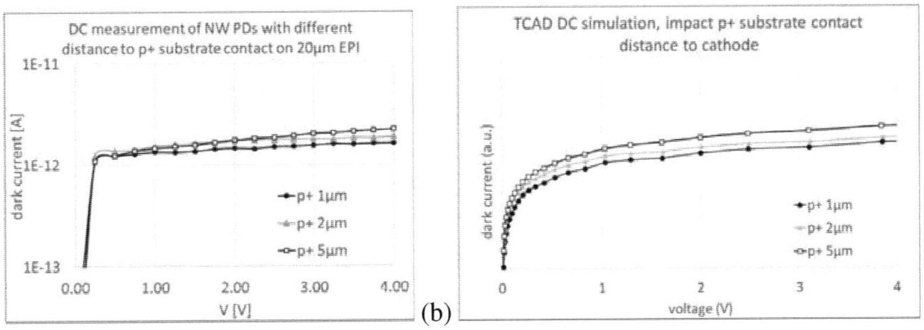

(a) (b)

Figure 5. DC measurement (a) and TCAD simulation results (b) of n-well PDs with p+ substrate contacts at distances of 1µm, 2µm and 5µm from the cathode.

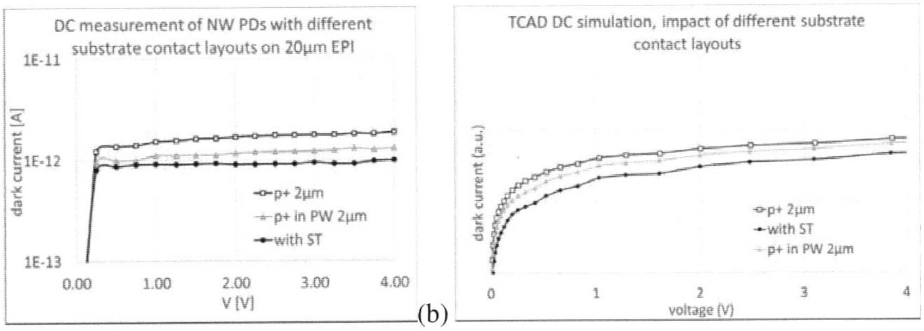

(a) (b)
(b)

Figure 6. DC measurement (a) and TCAD simulation results (b) of n-well PDs with different substrate contact layouts: p+ contact at a distance of 2µm from the cathode (p+ 2µm), p+ in p-well contact at a distance of 2µm from the cathode (p+ in PW 2µm) and narrow ST between p-well contact and cathode (with ST).

Figure 7 shows the corresponding capacitance results. PDs with the lowest DC values show the highest C values and vice versa. This tendency of increasing capacitance with increasing guard-ring depth is due to narrowing of the SCR and therefore increasing the sidewall capacitance of the structure.

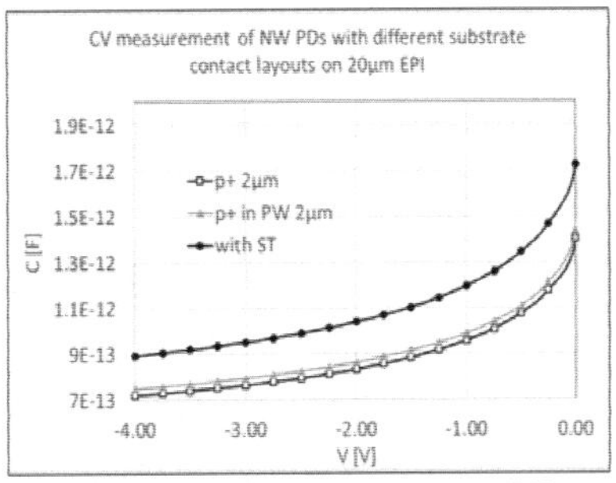

Figure 7. Capacitance results of 0.18µm n-well PDs with different substrate contact layouts.

3.2. Results on Double- and Triple Junction PDs

This section provides an overview of spectral response of PDs consisting of two or three pn-junctions each, as schematically shown in Figure 2. The structures differ in junction-depth and doping concentration as indicated by the shading. We show that appropriate junction combinations enable good spectral resolution for colour sensing applications even without any additional filter implementation. By shorting one of the two available junctions as indicated in Figure 8 the photocurrent contribution from this junction can be suppressed [4]. Shorting the p+ to the n-well (Figure 8a) blocks the upper junction and carriers generated in this region are forced to recombine and therefore do not contribute to the photocurrent while the lower junction (marked in Figure 8a) is active. In the same way the lower junction can be blocked and the upper junction set active as indicated in Figure 8b.

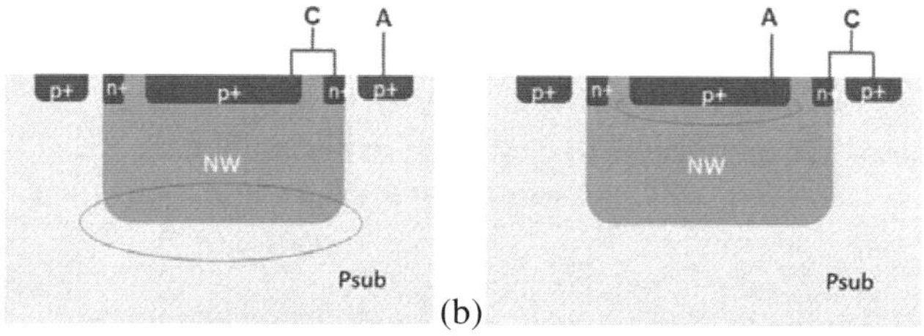

Figure 8. Shorting modes of double junction PDs: (a) n-well to p-substrate junction is activated; (b) shallow p+ to n-well junction is activated.

Figure 9 shows a comparison of the SR data as a function of wavelength for p+ in n-well in p-substrate PDs (a) and p-well in deep n-well in p-substrate PDs (b). Applied shortings are indicated in the legend.

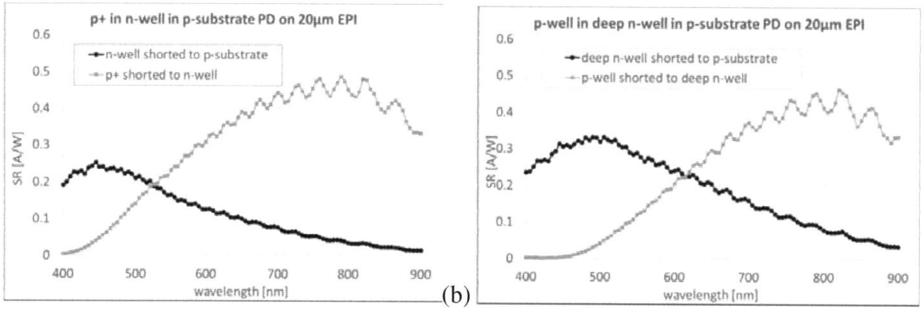

Figure 9. SR of double junction PDs under different shorting conditions: p+ in n-well in p-substrate (a) and p-well in deep n-well in p-substrate (b).

For the n+ in p-well in deep n-well in p-substrate triple junction PD the applied shorting variations are shown in Figure 10. Active pn-junctions are again marked.

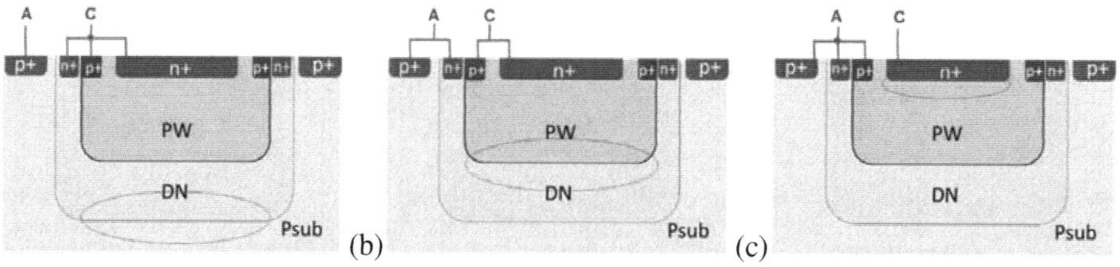

Figure 10. Shorting modes of triple junction PD: (a) deep n-well to p-substrate junction is activated; (b) p-well to deep n-well junction is activated; (c) n+ to p-well junction is activated.

Corresponding SR data for the three different shorting variations of the triple junction PD are shown in Figure 11.

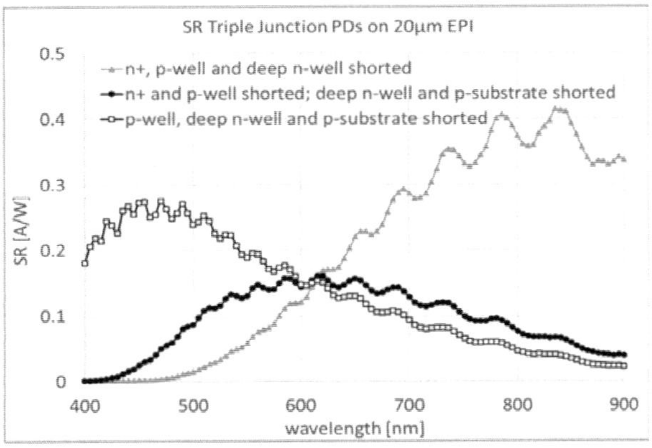

Figure 11. SR of n+ in p-well in deep n-well in p-substrate triple junction PD under different shorting conditions.

As can be seen in Figures 9 and 11 spectral response of junction combinations can be optimized for peak sensitivities in the range of 400-500nm, 550-650nm or 700-900nm. By appropriate junction choice the ratios of the generated photo currents in their respective peak zones can exhibit more than a factor of 10 compared to the other photo diode combinations. This enables a fairly good spectral resolution for colour sensing applications without any filters.

4. Conclusion

It was shown by TCAD simulations as well as measurement data that different substrate contact designs available in 0.18μm CMOS and HV-CMOS processes help to lower the dark current values of PDs. Moreover by exploiting the wide variety of junctions available in these processes, the spectral responsivity of PDs can be optimized for peak sensitivities in the ranges of 400-500nm, 550-650nm or 700-900nm, thus offering the possibility to provide a very good spectral resolution for colour sensing applications without filters.

References

[1] Zimmermann H., *Silicon Optoelectronic Integrated Circuits*, Springer Series in Advanced Microelectronics, 2004.
[2] Minixhofer R. et al., *A 120V 180nm High Voltage CMOS smart power technology for System-on-chip integration,* Proceedings of The 22nd International Symposium on Power Semiconductor Devices & ICs, Hiroshima, 2010.
[3] Sentaurus Device User Guide, Ver. L-2016.03.
[4] Kraxner A. et al., *Characterization of spectral optical responsivity of Si-photodiode junction combinations available in a 0.35μm HV-CMOS technology*, SPIE Integrated Photonics: Materials, Devices, and Applications II, Grenoble; 04/2013.

IOP Conf. Series: Journal of Physics: Conf. Series **939** (2017) 012029 doi:10.1088/1742-6596/939/1/012029

New Developments for Radiation Enhancements from Metal Surfaces by Using Nanoscale Materials in the Membrane

Koji Yamada[1,2] and Masami Matsuda[1]

[1]Future Product Co., Kita-Asaka, Saitama 351-0031, Japan

[2]Saitama Universiy, Shimo-ohkubo, Sakura-ku, Saitama 338-8570, Japan
msmqx323@yahoo.co.jp, Koji Yamada

Abstract. The enhancements of thermal radiations from the surfaces of devices are very important for electric machines to prevent from heating up and/or efficiency degradations. In this investigation, new applications of micro-scale membrane of Si, SiO_2 etc. on the metal surfaces have been studied to cool down the temperature without breaking insulations of the devices by selecting materials. The modified black-body radiations were sensitively detected by thermisters with sub-second responses. The optimum membrane thicknesses were successfully determined by subtractions a of radiation intensities between those at membranes with and without membrane, respectively. We obtained the best cooling condition in SiO_2 membrane with 20μm' for an Al-plate of 10cmx10cmx1mm'. Further, we observed the detaching/attaching processes of massive molecule clusters from the metal surface as a sudden change in temperature changes just like the noises in the detectors. A characteristic pattern of temperature change was observed in diatomite membranes during the cooling process in a temperature range between 200-50°C. These radiation phenomena as a function of temperature might be available as a molecular analysis on the metal surface.

1. Introduction

The easy cooling methods of materials and devices are of great interest without any additional energy during cooling. In this study, we investigated on the enhancements of thermal radiations from metal surface by using attached membranes of nano-scale materials of Si, SiO2 etc [1]. Here we made a mixture of membrane material and small amount of water as the solvent and also for the adhesive, and the baking temperatures were as low as 150-200℃. The observations of sample temperatures as a function of time elapses have been technically performed by using CCD cameras. However, the real time temperatures with "high precision" with an error less than 0.1% might be very difficult, mainly due to refresh time of CCD elements or frame rates are slow as 10ms in general. Therefore, in this study we have been investigated on the improvements of temporal change in the thermister by solving an equation of temporally retarded propagation of the thermister resistances. As will be mentioned after, the sub-second resolutions of phenomena were attained.

On the material side of preparations of this investigation, the nano-scale powders for the membranes were obtained and from markets [2] which were usually made of the following manners so called as hydrolysis method of nano-scale SiO_2: The hydrophobic fumed silica was obtained by a hydrophobization treatment of fumed silica with an organosilane compound. After, the drying processes, the humed particle are obtained from the diffusion of original material in a furnace with a high temperature range more than1000-1500℃ and collect powders from the dropping stream of the material. We used the these starting materials of Si (20-400nm), SiO_2 (Dia.<200nm) and Ag+SiO_2 (particle diameters<400nm) etc. The raw materials of nano-scale-powders were mixed with small amount of high purity water with 20MOhm or larger. The outside appearances of the mixed powder were just paste-like and these could be diluted with small amount of organic liquid, e.g. ethanol. The most important feature of the paste was the full mixing of the original material with the smallest amount of water to avoid residual H_2O molecules after baking process. The residual H_2O molecule might cause the breaking of bonding between the nano-particles. In the conventional method, the formed material might include the full of imperfections at the surface. To avoid this situation, the membrane might be better quality by limiting the thicknesses less than 20-40μm. All these processes were performed in a clean box with immersed clean air. The prepared mixture of pasts were coated on a metal by using a precision rolling bar with 20cm long, $1.5cm^D$ and with a constant dips with 5cm along the bar direction to determine the coating thickness on the substrate of glass or metal.

The prepared substrates were baked at the low temperature of 150℃-200℃, 1hr in the ordinary atmospheric pressure. Here, the baking temperature must be performed at a temperature as low as possible, not to injure the device. The prepared samples in this experiment showed transparent in samples with thicknesses of 2-30 μm and showed a little bit opaque in those thicker samples [1]. The strengths of the membranes were strong enough and attached tightly on the substrate of glasses and metals, which were tested by scratching by using sound-papers [3].

2. The sample preparations and experimental procedures

We first performed the experiment on the transparencies of Si membrane and of SiO_2 in the visible light wavelength range. It was shown that the reflection become better at the metal surface with SiO_2 membrane than those without. Here we used conventional spectrometer with a tungsten lamp as the light source which was driven in a low voltage to obtain longer wavelength lights, and detected by a photomultiplier with S1 curve.

The enhancements of infrared light emission were observed by constructing the experimental apparatus as shown in Figure 1. The sample was set at the central position in between the thermistors (S1 and S2 in Figure1) with the activation energy of E_A=0.290eV (=3380K) to observe the blackbody radiation which is modified by the coated membrane. One more thermistor (S) with E_A=0.343eV (=3995K) [2] was attached tightly at the sample surface with a very small size as *1*mm$^{\Phi}$ to minimise the heat capacity in comparison with those of S1 and S2. It will be shown later that the improvement of the temporal responses as short as 0.2s was achieved by solving the an equation of the retarded propagations in thermistors. The precise observations of the resistances and temperatures in the thermister were realized by using a 16-bits ADC and recorded up to 100kword*s* up to 200s (2ms/w). Note here that the sample was supported all by the edges of glass substrates to minimise the contacting cross section between the sample and glass supporters. Here, it must be also noted that the incidental electric power to detect the resistances in thermisters, must be less than 10μW which was inevitably limited by the observing incidental small radiation power.

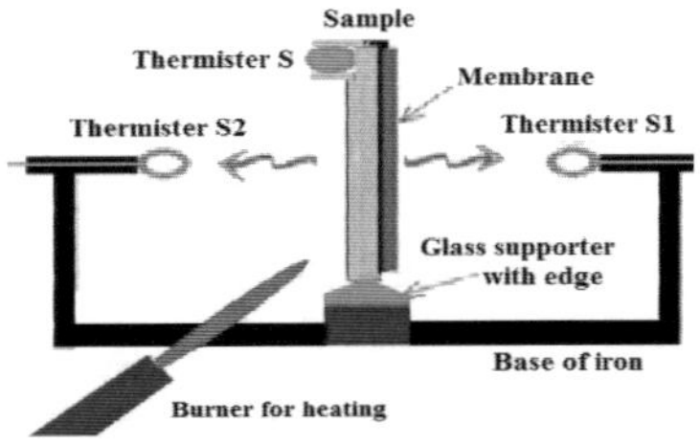

Figure 1. The experimental setup of the radiation detection

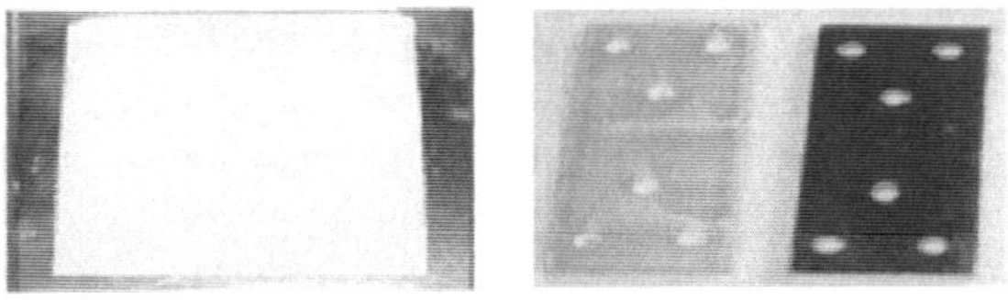

Figure 2. Sample shapes of an Al plate with 10cmx10x0.3mm^t (left) and fittings of soft iron with a shape of 30mmx70mmx3.1mm^t (right) and the membrane thickness of 20μmand 75μm, respectively.

Figure 2 shows examples of samples of Al-plate with SiO_2 coatings and conventional fittings of soft iron with and without coating of SiO_2.

3. Experimental Results

The observable temperature range was in between 0 and 200℃, which was limited by the thermisters. The time lag of the response was corrected by a method of retarded propagation of temperature which will be explained later in discussions. As explained before, the membranes are made of nano-scale powder of SiO_2 etc. with the solvent of smallest amount of water. The strength of attached membrane was very tight and attached directly with the metal surface within nano-meter distance between membrane and metal without any adhesive material as shown in Figure 2. The experiments on the cooling effects of membranes as shown in Figure 2 were performed by using a device in Figure 1. All the samples showed the monotonic temporal decreases with time elapses as expected before. The sample temperature (T) and the first derivative with time t ($-dT/dt$) followed exactly by exponential functions of (T-T_0). However, precise curves of ($-dT/dt$) with high temporal resolution include turbulent noises around at the function of T-T_0.as shown in Figure 4.

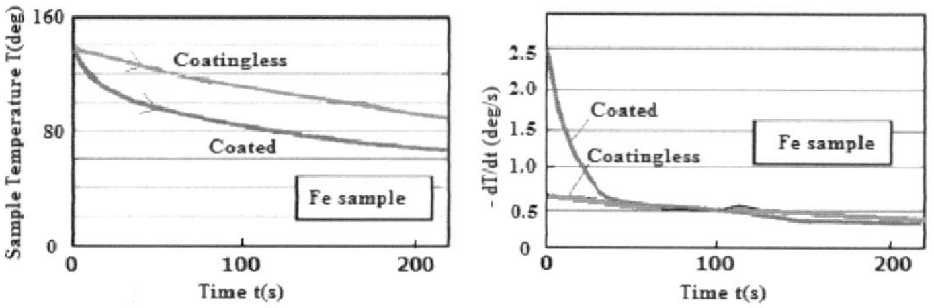

Figure 3. Experimental results of. the cooling effects in soft iron fittings with and without cover by SiO_2 membrane with 100µm in a temporal range up to 220s in the l.h.s. figure. The r.h.s. figure shows the first derivative of sample temperature with respect to time t.

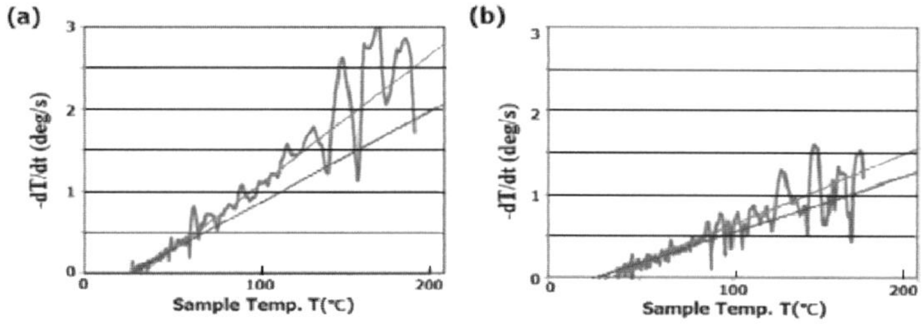

Figure 4. The cooling effects of membranes at Cu surface with membrane with Ag nano-powder in SiO_2 with thicknesses of (a) 22µm and (b) 8µm, one side coated respectively

It is apparent to observe the enhancements of cooling effects by larger radiations. The time delay caused by the heat capacity of the thermistor is corrected in this investigation which will be also discussed later. It must be noted here that the coating of SiO_2 was performed both sides of the surfaces, or only one side for simplicity. The temperature decreasing rate became 2 times than that without coating in this experimental condition.

Figure 4 shows two different samples of Cu-plates with Ag powder+SiO_2 coating with the different coating thicknesses. Here, the experiment of radiation efficiency was performed as a function of the coating thickness. It will be also shown together with the experimental results in the discussion. In Figigure4(a) and (b), the inserted lines with red and blue, stand for the approximated two lines of responses in a low temperature range between 40-80℃(red line) and that in a higher temperature than 80℃(blue line), respectively. It must be noticed here that the turbulences of the curves were not reproducible and the physical origins of the phenomena will be discussed later. Further, the temporal behavior of $-dT/dt$ shows a proportional function of (T-T_0) decay in this temperature range with some turbulences around at the function of T-T_0.as shown in Figure 4. For the improvement of responses, the retarded propagations of sequential equations at each point were developed.

4. Discussions

The cooling effects by radiation enhancements at the metal surface were theoretically analysed, where phonons played an important role [4] and the nonlinear effects on temperature might be important. Here, the effects might be expressed by classical equation of thermal conduction as .

IOP Conf. Series: Journal of Physics: Conf. Series **939** (2017) 012029 doi:10.1088/1742-6596/939/1/012029

$$dT(t)/dt = c(T - T_0) + ho \tag{1}$$

Here, T and t stand for the sample temperature and time t after heating the sample, respectively. This coefficient c stands for all cooling mechanism as convections, radiations and etc. T_0 stands for the ambient temperature to be equal after the thermal equilibrium. The most simple case, the coefficient c might be a constant or it expresses some function of T or/and t. Eq.(1) is precisely solved by the method of "variable separable" with the initial condition of $T=T_i$ as

$$T(t) = (T_i - T_0)exp[-\int c(t)dt] + T_0. \tag{2}$$

Here, $c(t)$ stands for a nonlinear dependence of thermal conductions in the wider temperature range. For this analysis, the temperature dependence of dT/dt as a function of T were well measured experimentally as shown in Figure 4, except the large fluctuations which might be caused by some molecular attaching/detaching processes at the surfaces. Here, we discuss on the cooling effects of the membrane without the effects of convections. For this purpose, we calculated the difference of the output voltages as defined by $V_{12}=V_{S1}-V_{S2}$. Now, we adopt the first order approximation to consider the effects of both side convections are almost same. Figure 5 shows the derived voltages of The output voltages V_{12}.as a function of membrane thicknesses d.

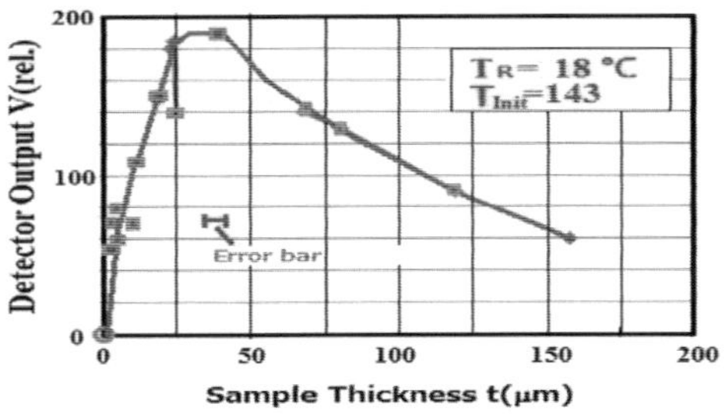

Figure 5. The cooling effects of membrane of SiO_2 on the metal as a function of different thickness

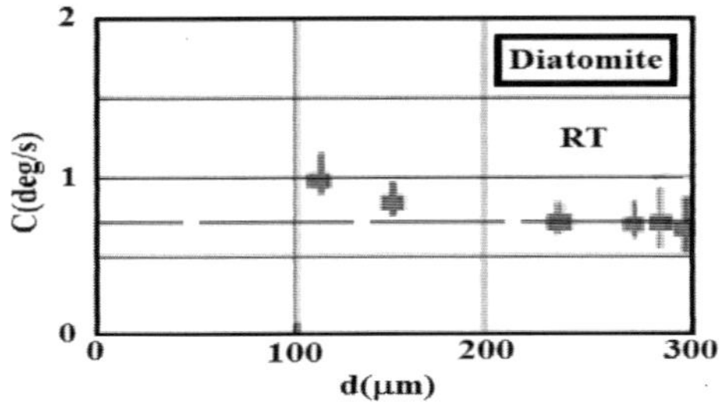

Figure 6. The coefficients c as a function of different membrane thicknesses with T_i=200℃

Now, we discuss on the time delay of the thermistor, which is caused by the heat capacity of the material inside of the thermistor with about 3 second of order. To solve this problem, we tried the corrected temperatures T'_j at $t=t_j$ which derived by the old T_j ($j=0,1,2,\ ,N$) as

$$T'_j=5T_{j-1} - (T_{j-1}+T_{j-2}+ T_{j-3}+ T_{j-4}), \tag{3}$$

$$=T_{j-1} +c_1(T_{j-1}-T_{j-2})+c_2(T_{j-1}-T_{j-3})+c_3(T_{j-1} -T_{j-4}) \tag{4}$$

$$[\ t=t_j\ \ (j=0,1,2,3,,,N),\ (c_1=c_2=c_3=1)\ ]$$

where *j*'s stand for the time t_j defined by $j\Delta t$ ($N\Delta t=t_N$, N=10^5, Δt=2ms, t_N=200s) in analogue digital converter with clock frequency of $1/\Delta t$. T_j , T_j' stand for an apparent and superficial temperature in the thermistor S and the corrected temperature at $t=t_j$, respectively. Note here that. Eq.3 and Eq.4 express the Markovian process with 4th folds with the coefficients of $c_0(=1)>c_1>c_2>c_3>0, c_{j>3}=0$, where the extension of the effective time span is determined by the causality extent. Here, we set $c_0=c_1=c_2=c_3=1$ for discussions. By using smoothing process, the data number was reduced to $N=10^4$ and ΔT=0.1s. We obtained the results as shown in Figure 7. It shows the temperatures T_j'(red) after corrections and T_j(blue) those before. The time response were best fitted with 2.4s delay in the algorism as shown in Eq.3. After the corrections, the superficial delay was improved by the total delay of 0.2s, which must be probed by independent other devices. Figure 8 shows the characteristic "spectra" of diatomite membranes which appeared in different samples and showed in the similar pattern.

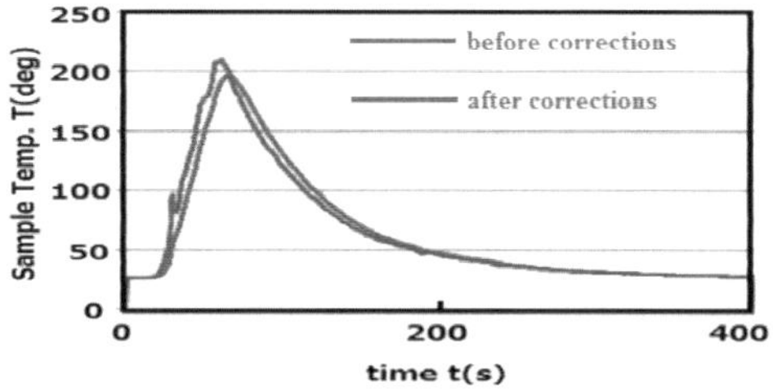

Figure 7. The effects of corrections of temporal lelays due to heat capacity of thermister

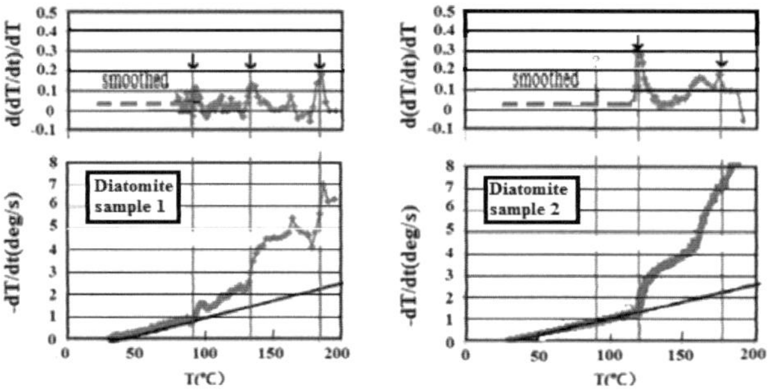

Figure 8. The characteristic features of "spectra" in different Al-plate samples with diatomite membrane with the thickness of 20μm of just an ingradient expression except "diatomyte"

5. Conclusions

In this investigation, we first presented the direct attachment of nano-material on the metal surfaces which show the large enhancements of radiations or cooling effects. The physical origin of the effect of direct coupling of nanomaterial on the metal surface must be examined in the future [4]. However, the first candidate of this mechnism might be the coupling by inter-atomic force or inter-molecule force as Lenard-Jones potential. The cooling effects are the conventional phenomena as those in characole membrane. However, this material is electrically conductive and dangerous when the carbon membrane is exfoliated by the damage. The method and the technology might be avalable to use these in electrical machines to improbe the effciency by cooling down.

References

[1] KYamada et.al. Materials Science Forum , Vol. 856(2016) pp 53-57
[2]http://http://www.aist.go.jp/aistj/press_release/pr2015/pr20151001/pr20151001.html (in Japanese)
[3]Uzhe Ding, et.al, Adv. Mat. Vol. 23, Issue 46, Dec 8, 2011, pp. 5551–5556
[4] A.J.Siebert, J. Optical Soc. America, Vol 68 Issue11, 1978, pp. 1505-1516

IOP Conf. Series: Journal of Physics: Conf. Series **939** (2017) 012030 doi:10.1088/1742-6596/939/1/012030

Characterisation of electrical resistance for CMC Materials up to 1200 °C

T. Stäbler, H. Böhrk and H. Voggenreiter

German Aerospace Center, Institute of Structures and Design, Germany

E-mail:tina.staebler@dlr.de

Abstract. Damage to thermal protection systems (TPS) during atmospheric re-entry is a severe safety issue, especially when considering re-usability of space transportation systems. There is a need for structural health monitoring systems and non-destructive inspection methods. However, damages are hard to detect. When ceramic matrix composites, in this case carbon fibre reinforced silicon carbide (C/C-SiC), are used as a TPS, the electrical properties of the present semiconductor material can be used for health monitoring, since the resistivity changes with damage, strain and temperature. In this work the electrical resistivity as a function of the material temperature is analysed eliminating effects of thermal electricity and the thermal coefficient of electrical resistance is determined. A sensor network is applied for locally and time resolved monitoring of the 300 mm x 120 mm x 3 mm panel shaped samples. Since the material is used for atmospheric re-entry it needs to be characterised for a wide range of temperatures, in this case as high as 1200 °C. Therefore, experiments in an inductively heated test bench were conducted. Firstly, a reference sample was used with thermocouples for characterising the temperature distribution across the sample surface. Secondly, electrical resistance under heat load was measured, time and spatially resolved. Results will be shown and discussed in terms of resistance dependence on temperature, thermal coefficient of electrical resistance, thermal electricity and electrical path orientation including an analysis on effective conducting cross section. Conversely, the thermal coefficient can also be used to determine the material temperature as a function of electrical resistance.

1. Introduction

In a time in which the ecological aspect of space travel is becoming increasingly important, re-usability of space-vehicles is a major concern. Building and optimizing space vehicles for re-entry is not a trivial matter as, depending on the flown trajectory, temperatures of over 3000 K can be reached during re-entry. It is a severe challenge for scientists to develop materials and sensors which can withstand these extreme conditions. For gaining better knowledge of the re-entry phase and for subsequently optimizing structural safety margins, sensors are needed which can collect data and monitor the heat shield during an actual re-entry flight. This is not only important for developing better TPS, but also for detecting defects during flight. In case of minor or major defects, re-entry trajectories can be adapted and might not only be able to save a spacecraft but also save lives.

Several methods are available to monitor the condition of a heat shield in ground testing. During actual re-entry flight; however, most of these methods cannot be used as most of them need external optical access. Common heat shield instrumentation comprises temperature, heat flux and pressure sensors. There is also a range of recession sensors available or in development e. g. the Hollow aErothermal Ablation and Temperature (HEAT) sensor from Santos et. al. [1].

IOP Conf. Series: Journal of Physics: Conf. Series **939** (2017) 012030 doi:10.1088/1742-6596/939/1/012030

Two isolated wires of platinum and tungsten are wound on a kapton tube which is filled with heat shield material. During re-entry an electrical circuit between the wires is closed when the kapton chars at a known temperature.

Other nondestructive monitoring techniques for TPS include ultrasonic scanning and thermography [2]. Fibre optic sensors can be used for measuring strain and temperature by means of wavelength shifts. One such sensor was developed for the structural core of the X-38 spacecraft [2]. Other concepts use acoustic emissions and ultrasonic systems, such as timed echos of sound waves on an ablator surface [3, 4]. Carbon fiber reinforced polymers (CFRP) and ceramic matrix composites (CMC) material have been investigated for in-situ monitoring of laminar failure using a passive method of resistance measurement [5, 6, 7, 8, 9, 10]. Special emphasis of this study is that the sensor should be reusable. Therefore the approach is that the material itself, which is reusable, is being used as the sensor. It is important for health monitoring systems to know certain properties of the materials used, to determine whether the reason of a change in resistance is due to a change in temperature, strain or to damage. In this work the main focus is on the influence of the sample geometry and the temperature on the resistance, including compensation for thermally induced electric current.

2. Material

The material used is a CMC material, carbon fibre reinforced silicon carbide (C/C-SiC) as can be seen in Fig. 1. The lay-up are stacked 0/90°plies of a 2/2 twill weave. The material is highly stressable and unlike monolithic ceramics not brittle. It contains the positive properties of a ceramic, such as corrosion, abrasion and oxidation resistance, high temperature tolerance and adds to them a higher damage tolerance with a quasi ductile fracture behaviour. C/C-SiC has a long history, it has been developed since the 1980's at the German Aerospace Center (DLR) [11] and has already seen many different fields of use, including TPS on re-entry vehicles and in combustion chambers. The C/C-SiC material is manufactured at the DLR Institute of Structures and Design with the liquid silicon infiltration (LSI) process [11].

The material has been qualified in plasma wind tunnel testing and during real re-entry flight such as the SHEFEX missions [12]. For use in this work the plates are cut into rectangular samples of geometry 120 mm x 320 mm x 3 mm (width x length x thickness). A C/C-SiC sample with a polished surface can be seen in Fig. 1. Further information about the production process can be found in [11] and [13]. Measurements at room temperature in previous works have shown a specific electrical resistance for C/C-SiC of approximately 38 µΩm [5].

Figure 1: C/C-SiC sample with polished surface.

3. Theoretical approach

Change in electrical resistance can be due to a multitude of reasons and is therefore a useful indicator for a range of effects that should be detected in a sample. Different sample sizes, geometries or contact layouts result in different electrical resistances. The electrical resistance is a function of the sample temperature. Damage in a sample from small fractures to complete destruction of a sample can be measured with electrical resistance [14]. To be able to determine the cause of a change in electrical resistance different approaches must be taken.

3.1. Effective cross sectional area

The related cross sectional area A^* of the sample is important, as the panel shaped samples have a much bigger cross sectional area A than a usual conducting wire has. The relation for resitance R

$$R = \rho \cdot \frac{l}{A} \tag{1}$$

to specific resistivity ρ is valid for electrical conductors with a much bigger length l than width. The area A needs to be replaced by the effectively conducting area A^* for use with the wide samples in this work. The related cross sectional area A^* needs to be experimentally determined. Method and results are described below.

3.2. Thermal electricity

Adding to the temperature dependence of the electric resistance for some material combinations, a nonhomogeneous temperature distribution results in thermoelectric charge. Therefore, between two measurement points a vectored voltage is measurable depending on the temperature distribution and materials used. This voltage has a large influence on the resistance measurements, as the resistance of the material is measured by measuring the voltage drop over the sample. As an example, the voltage drop of 10 V over a sample at a current of 10 A according to Ohm's law

$$U = R \cdot I \tag{2}$$

would result in a measured resistance of 1 Ω. With an induced thermoelectric voltage of -5 V, i.e. antipodal to the measurement direction as depicted in Fig. 2 case (1), the measurement according to

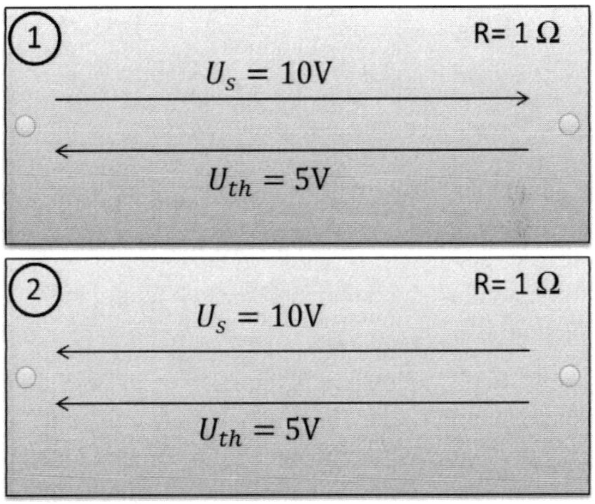

Figure 2: Cases of different electrical resistances due to the thermoelectric effect.

$$U_{measured} = U_{sample} - U_{thermic} \tag{3}$$

yields 15 V resulting in a falsely evaluated resistance of 1.5 Ω. With an induced thermoelectric voltage of +5 V, in measurement direction as is shown in Fig. 2 case (2), the measurement yields 5 V resulting in the falsely evaluated resistance of 0.5 Ω.

To derive the real resistance from measurements and to subtract the thermoelectric voltage, both resistance and voltage measurements are conducted. With those measurements the values can then be corrected mathematically, since the measuring amperage is known and the measured resistance is known. Using Eq. (2) the resulting voltage can be calculated, the vectored voltage can then be found by Eq. (3).

3.3. Temperature dependence

Furthermore, it is known that the electrical resistivity ρ is the reciprocal quantity of conductivity $\sigma \sim e^{-\alpha/T}$, with $\alpha = E_g/2k$ representing energy gap over Boltzman's constant. The resistivity of a semiconductor is thus dependent of temperature. With T_0 as reference temperature and T as material temperature it can be described by

$$\rho(T) = \frac{1}{\sigma} = C\left(\rho, \rho(T_0)\right) e^{\alpha(T)}. \tag{4}$$

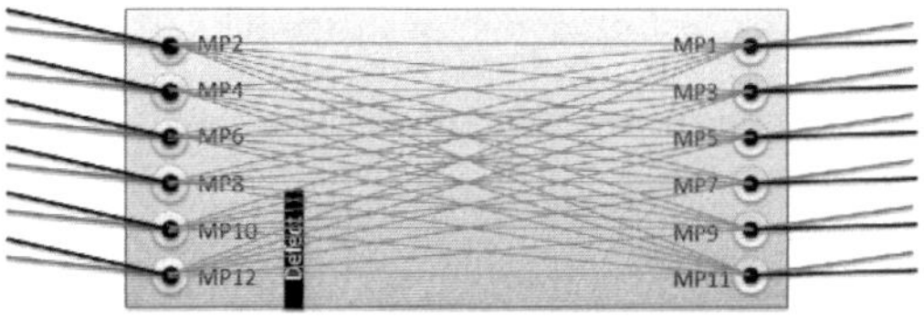

Figure 3: Schematic of measuring principle.

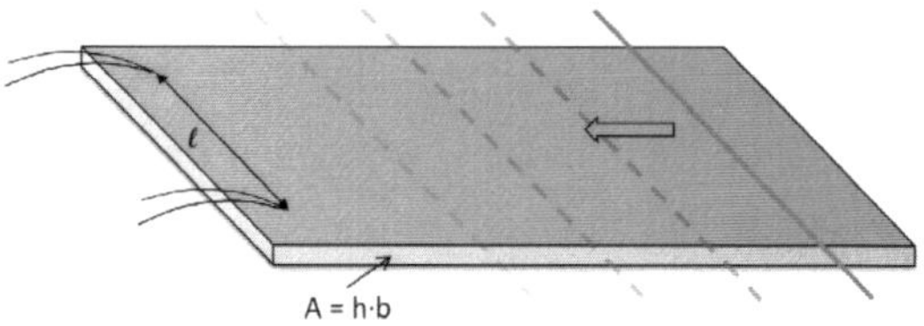

Figure 4: Sample PH1877P to show changing effect of electrical resistance near the edges in C/C-SiC.

4. Experimental set-up

Three different experimental set-ups are used in this work. All of them use the test bench for resistance measurements (WISTAM), which measures the electrical resistance of a sample time and spatially resolved. Samples with up to twelve contacts on either side can be used. WISTAM is clocking the measurements, measuring first one contact on one side with all of the contacts on the other side, one after the other, then the next contact and so on, resulting in 144 single measurements. A schematic of the measured virtual current paths is shown in Fig. 3. A detailed description of the test bench and the measuring principle can be found in [14]. The individual test set-ups are described in the following section.

4.1. Mesh Experiments

For a fundamental analysis of the measurement principle a hexagon mesh was used. On each side 24 wires for twelve contacts were soldered and connected to the test bench for four-wire-measurements of electrical resistance. The hexagon mesh was chosen because of its distinctive feature being that all distances from one side to the other are the same regardless of the path taken, as long as the length is bigger than the width. As all the possibilities of current distribution have the same distance from measuring point A to measuring point B, it can be said that a difference in the measured resistance can only be a result of a difference in the related cross sectional area. As a mesh has very distinctive possible paths it is also a very good demonstrator for damaged sections, as different paths can easily be cut to show a change due to cut fibres.

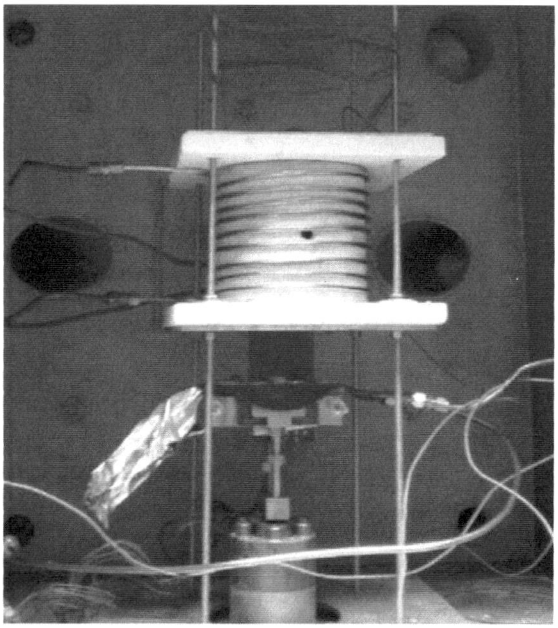

Figure 5: Experimental set-up in the Indutherm test-bench.

4.2. Edge Experiments

This mesh is also used for experiments showing that electrical resistance near the edge of a mesh is higher than in the center of a mesh. Section 3.1 described why this investigation is needed. This experiment was also conducted with C/C-SiC material. A rectangular sample was installed with two contacts, and two wires were soldered to each contact for a four-wire-resistance-measurement. The sample layout is shown in Fig. 4. As pictured the contacts were on one side of a rectangular sample and the measurement section marked with l is seen as the sample length. Therefore, the cross section A is bigger as it would be in a rod. In the experiment the sample was measured and then shortened, indicated in Fig. 4 by the arrow along the red line, and afterwards measured again, repeating this process until the sample was too small to be shortened further.

4.3. High Temperature Experiments
Electrical resistivity of material samples is measured under high temperature. The measurements are carried out in the temperature range from room temperature to 1200°C. The samples are heated inductively by a hot test facility [5]. The samples are placed inside the susceptor ring during the experiment. To avoid oxidation of the carbon in the sample material, all high temperature experiments are carried out under vacuum conditions at 100 Pa with the whole experimental set-up, as can be seen in Fig. 5, being placed inside a vacuum chamber. The heating process takes less than 3 minutes. Thermocouples, a pyrometer and an infra-red camera are used to monitor the temperatures and a pressure sensor for monitoring of the chamber pressure is installed. To make sure that the sample is heated homogeneously, the highest temperature is held for 2 minutes before cutting the power. The induction coil induces current in the susceptor but also in the measurement devices. Therefore, measurements are only taken during the cooling period when the generator and therewith the magnetic field are shut off. Depending on the starting temperature this lasts for one to five hours.

Figure 6: Sample with partly riveted contacts.

4.4. Sample Layout and Test Procedure for High Temperature Experiments
The panel shaped samples are made of C/C-SiC and have 12 small drilled holes on either side, as can be seen in Fig. 6. In the holes, wires are riveted with copper rivets to connect the sample electrically. The contacts are connected to WISTAM to monitor voltage and electrical resistance of the sample. The temperature of the sample surface and the riveted contacts, which are in water-cooled clamps situated outside the graphite cylinder, are monitored during cool down, as well as electrical resistance and voltage on the sample, as there is a thermoelectric potential due to different temperatures over the sample surface.

5. Results
Preparatory measurements are mandatory that correct calculations of the high temperature experiments can be made, with all effects mentioned in the theory section being considered. One of the important measurements is to know the effective relating cross section A^* as it was described previously in Section 3.1. It was observed in wide samples that near sample edges the measured resistance is higher than the theory according to Eq. (1) indicates. This is accounted to the measurement current having a limited cross section to pass across the sample. In turn, in wide sample centres, the current is not using all of the space provided. For compensation of this effect a materially and geometrically dependent correlation factor is used. With it, a measured value on the edge of a sample can be compared to a measured value in the center of the sample.

5.1. Edge effects in a mesh
To show the edge effect, a hexagon mesh was used as was shown in the experimental set up in Section 4.1. Measurements of one thread of 494 mm length gave a resistance of 323 mΩ. Eq. (1) yields with the thread diameter 0.514 mm a specific resistivity of 0.1359 $\mu\Omega$m of the material.

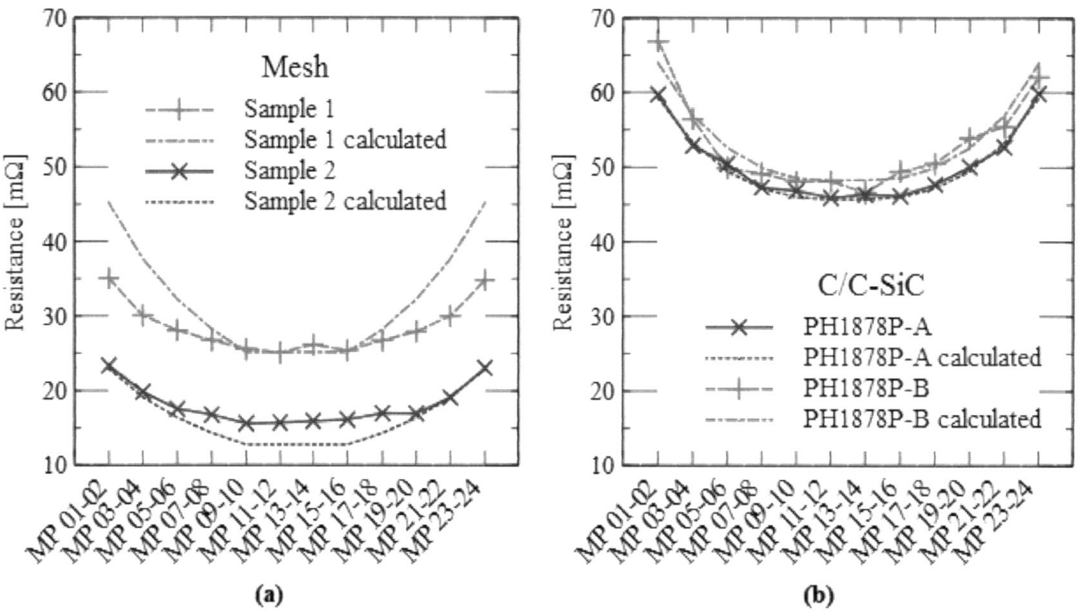

Figure 7: (a) Electrical resistance over path position of the mesh sample 1 (590 mm) and mesh sample 2 (350 mm). (b) Electrical resistance over path position of panel shaped C/C-SiC samples PH1878P-A and PH1878P-B.

Measurement of electric resistance between directly opposite contacts across the sample yields the results shown in Fig. 7 (a). The figure shows that the outer three to four measuring points have a higher resistance than the inner measuring points. This is due to the propagation of the electrical field in the mesh, which is restrained by the edge. Since the resistance is constant at the measuring points in the center of the sample, it is assumed that the resistance between measuring point 7-8 and measuring point 15-16 has reached its maximum effective related cross sectional area. As it is assumed that the current distribution in the grid is homogeneous the cross sectional area can be calculated as A*

$$A^* = 2(1 + x + n) \cdot A_0 \,, \tag{5}$$

where A_0 is the cross sectional area of one wire, n is the number of wires from the measuring point mentioned above to the nearer edge of the sample and x the number of maximum wires effectively used. In this example with this material $n_{max} = x = 4$. As for every contact two paths can be taken in the grid, due to the nature of the hexagon grid, which has a double wound wire on both longer sides of the hexagon, the number of paths is multiplied by two. The results from Eq. (1) using the effective cross sectional area from Eq. (5) above are also given in Fig. 7 (a). The concurrence is good with deviations at the edge for Sample 1.

5.2. Edge effects in C/C-SiC

Further experiments in C/C-SiC samples also show this edge effect, as shown in Fig. 7 (b). Therefore, experiments were conducted to determine A^* in C/C SiC as described in Section 4.2.

In Fig. 8 the measured electrical resistance is shown over the sample width b, a function of which is the cross sectional area $A = b \cdot h$ indicated by the blue line. The calculated resistance with Eq. (1) is shown as the red line in Fig. 8. It can be seen that while the sample is wider than 40 mm the measured resistance is constant. This indicates that, as mentioned above, the current is not using all of the space provided. With width smaller than 40 mm the measured resistance

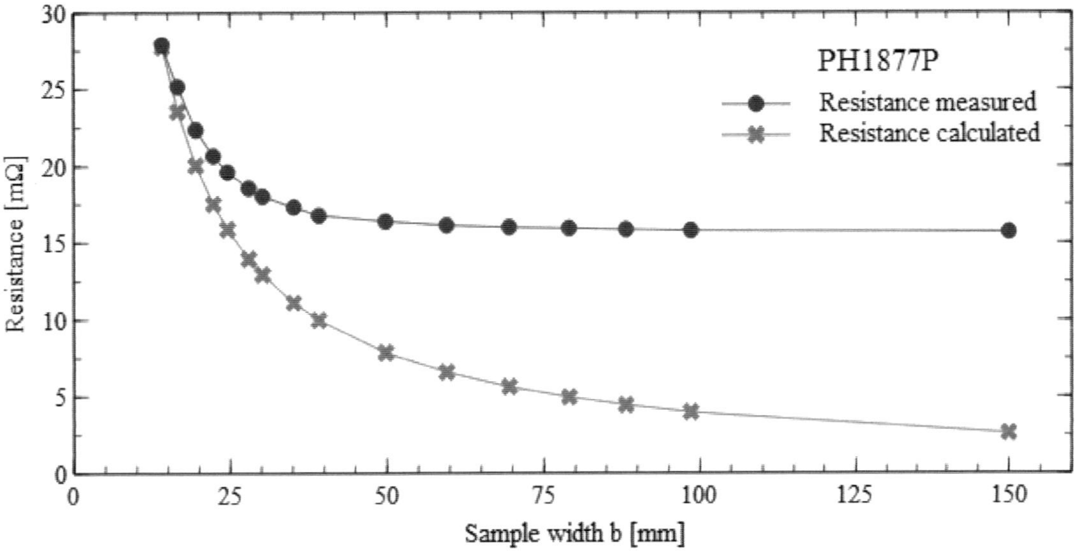

Figure 8: Measured and calculated resistance of a sample which was shortened for each measuring point over the sample length.

starts rising and below a width of approximately 25 mm it is nearly coincident with the theory from Eq. (1).

This effect of a rising resistance near an edge, as shown in this section and Section 5.1, can now be used to calculate an effective cross sectional Area A^* in a C/C-SiC sample. A^* can be calculated with the specific resistance of the material and the lowest measured resistance in the middle of the sample using Eq. (1). The effective sample width b^* can be calculated using

$$A^* = b^* \cdot h \tag{6}$$

with the sample height h. b^* consists of two parts

$$b^* = b_1^* + b_2^* \tag{7}$$

whereas b_1^* is a fixed value independent of the contact location, only dependent of the sample and the sample material, and b_2^* is dependent on the contact position. b_2^* can be calculated with an experimentally determined quadratic formula

$$b_2^* = -8.282x^2 + 0.119x + 0.041 \tag{8}$$

whereas x is the distance of the contact to to middle of the sample for samples with a width of 0.12 m as used here. The calculated values for the electrical resistance for each contact pair using Eq. (8) is shown in Fig. 7 (b) and show a very good correlation of measured and calculated values.

5.3. Damage in a mesh

When all paths in a mesh can be compared, irrespective of their location, damage to a mesh can be both detected and located. The hexagon mesh from Sections 4.1 and 5.1, is used. Areas are cut out during measurements, as can be seen in schematic pictures of Fig. 9. Green and violet colours indicate current lines along the upper and lower edges of the sample. The violet line

IOP Conf. Series: Journal of Physics: Conf. Series **939** (2017) 012030 doi:10.1088/1742-6596/939/1/012030

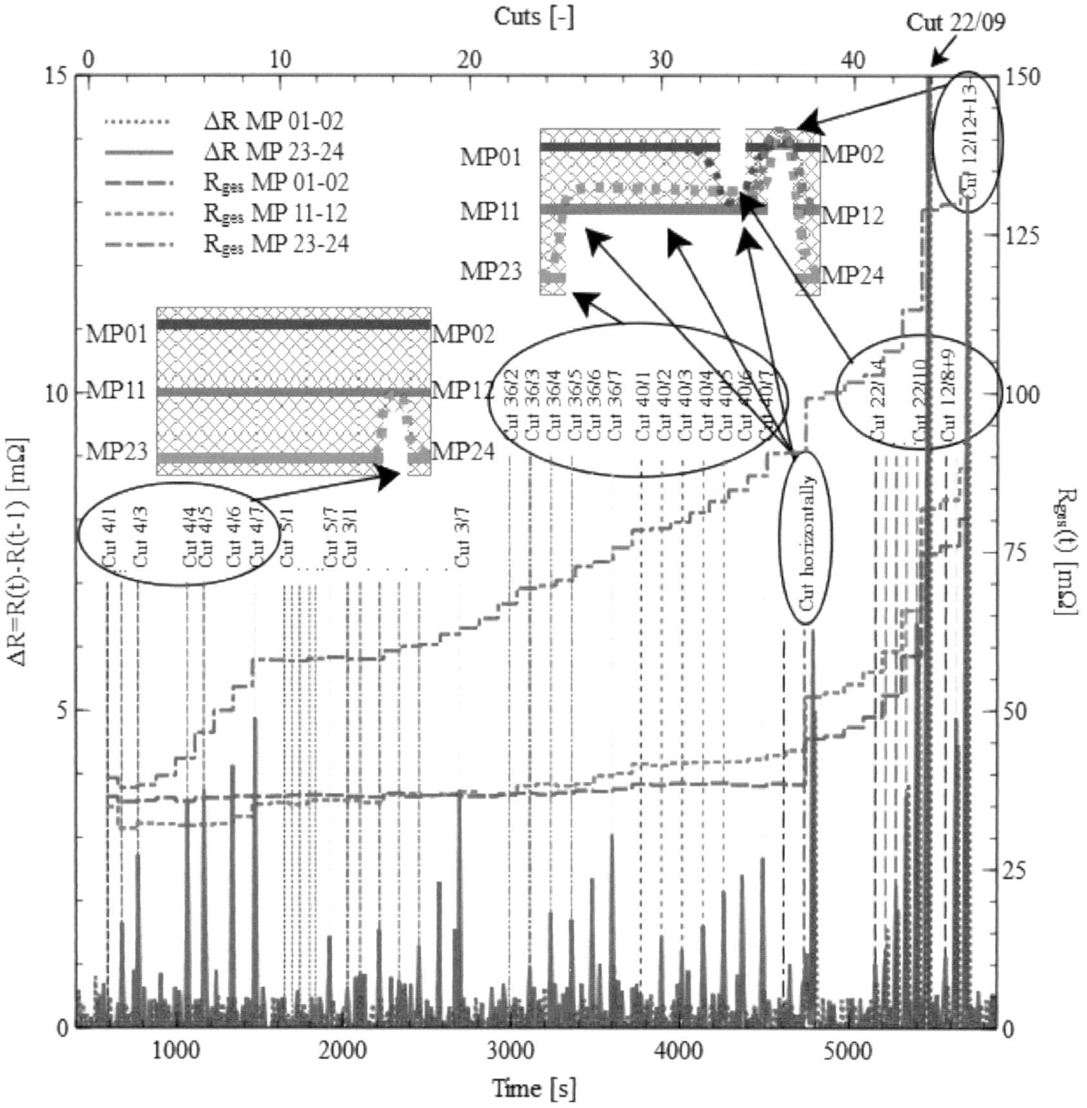

Figure 9: Plot of resistance change and plot of resistance during cutting of connections in a grid, which is shown schematically in the upper pictures.

corresponds to the measurement points 1 and 2, which is a straight line from left to right at the top of the grid. The green line corresponds to the measurement points 23 and 24, which is a straight line from left to right at the bottom of the grid. The experiment is conducted by measuring the electrical resistance across the sample while cuts are being made along the regions indicated by the white space and the arrows in the schematic pictures.

The diagram of Fig. 9, shows resistance R and its deviation $\Delta R = R(t) - R(t-1)$ over cuts and respectively time. When a cut is made the different measurement pole permutations correspond to the cut depending on their location. This can be seen in the different peaks in the lower section of Fig. 9. The first cut is at the right of the grid, far away from the violet line. Therefore, for the cuts 4/1 to 4/7 the measurement at points 1 and 2 (solid green curve) responds with peaks, which means a rise in the electrical resistance, while the violet line does not respond at all. Apart from their last incisions, the subsequent cuts at 5/1 to 5/7 and 3/1 to 3/7, right next to the location of the first cut, do not induce much change of resistance. This is explained by the current path between the measurement points not being strongly influenced. The cuts at locations 36/1-7 and 40/1-7, at the left edge of the mesh, in turn do receive a resistance response.

On the upper right hand side, in a dashed green line, a potential current flow at the end of the experiment is shown in the pictogram. The lower middle area of the grid is completely removed, in the upper part are two long cuts, which each then only leave one connection in the grid. The cutting of the first of those mentioned two long cuts, cut 22/14 to cut 22/09 in the diagram in Fig. 9, can be seen from the 5000th second. The resistance response to the cuts can be seen immediately and rises with each cut. Both, the violet and the green, are affected and therefore both are rising with each cut. The last cut, 22/9 leaves only one connection and has therefore the biggest effect with the highest difference in resistance before and after the cut.

This shows that the nearer a virtual current path is to a damaged section the more it will affect the measured electrical resistance. It is not necessary for the damaged section to be directly in the virtual current path. A bigger impact, like the cut that removed the lower part of the grid, shown as 'Cut horizontally' in Fig. 9, also affected the measurements from measuring point 1 and 2. Even though the virtual current line, the violet line in the pictures of Fig. 9, is not directly affected, a rise in resistance can still be measured. However, it is lower than the directly affected green line.

In Fig. 9 the cumulated resistance change of the different cuts can also be seen in dashed and dash-dotted lines. An additional small dashed line is shown (red), which corresponds to the measurement points 11 and 12. From 700th second, the red line shows less resistance (32 mΩ)

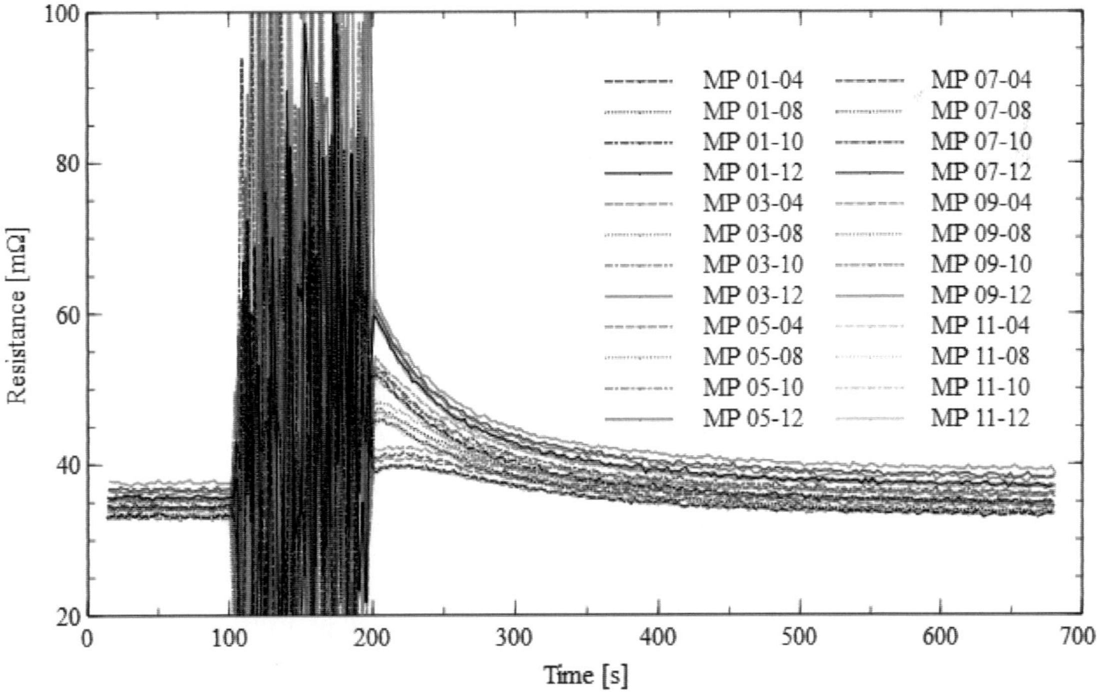

Figure 10: Electrical resistance for all current paths during testing.

than the green, 35 mΩ, and violet, 38 mΩ, lines. This is due to the edge effect, described in Section 5.1, as the virtual current line of the measuring points 11 and 12 is in the center of the grid. During the experiment, the green line, which is more deeply affected by the cuts starts rising, which can also be seen by the corresponding peaks of the resistance change ΔR. The red line starts rising earlier than the violet line, as it is nearer to the cuts and therefore affected more by them. As all of the 144 virtual current paths are affected differently, depending on the distance between the area of defect and the virtual current path as well as the impact of the defect (e. g. size), the location of the defects can be traced to the appropriate areas.

5.4. Defects in C/C-SiC

Damage can also be detected in C/C-SiC, as is shown in Fig. 10. A defect was inserted near measurement point 12 of Fig. 3. All connections with measurement point 12 were deeply affected and group together after the damage was inserted between the 100th and 200th second in the upper area. Going from measurement point 10 to 8 and 4 the effect shows less in the data and the measurement areas group clearly together. This shows that the behaviour of the material can be compared to the behaviour of the mesh. This experiment is described in detail in [14] and shall not be discussed here further.

5.5. High Temperature Experiments

Further characterizations are carried out in an inductively heated test bench, described in Section 4.3. In Section 3.2 the effect of thermal electricity and its influence on resistance measurements was described. Fig. 11 shows the experimental data during the cooling process of a C/C-SiC sample PH1878P-A. The data recording started at a temperature of 1154 °C and ended at a temperature of 400 °C. The potential difference of the center measuring points 7 to 8 was reverse to measuring points 9 and 10, resulting in mirrored data as shown by the voltage curves in Fig. 11. The red and purple lines show the measured resistance of both connections. Resistance and voltage measurements were done alternately. The measuring device is using a fixed current of 10 mA, and the voltage drop over the sample can therefore be calculated using Ohm's law, $U = R \cdot I$. Using this voltage with Eq. (3) and the measured voltage yield the voltage drop over the sample without thermal electricity. This can now be converted again using Eq. (2) into resistance values. The result represents the true material resistance and is shown as the dark blue and turquoise lines in Fig.11. The curves show a linear behaviour of increasing resistance over decreasing sample temperature.

With the measured data the temperature coefficient of the sample PH1878P-A can be calculated with Eq. (4) and α=207 [5] to $C = 2.843 \cdot 10^{-5} \pm 1.118 \cdot 10^{-6}$.

6. Conclusion

Theories for determination of the related cross sectional area as well as for behaviour near sample edges were presented. With a representative hexagonal mesh these effects were further analysed and calculations were shown. The findings are transferred to the heat shield material C/C-SiC. Thermal electricity in hot samples was discussed and data processing to eliminate this effect was shown. With this knowledge, experiments in a inductive heated test bench were conducted up to 1200 °C showing this thermal electric effect. Measurements of electric resistance and voltage were performed, discussed and analysed. The true material resistance was calculated and shown and the thermal coefficient of the electrical resistance for the used sample was determined.

The results lead to a better understanding of electrical effects in samples and allow the test bench to be enhanced for more accurate measurements, such as double measurements in reversed directions to eliminate thermal electricity effects. They also confirm established theories and give a better insight into the used material for characterisation of material properties.

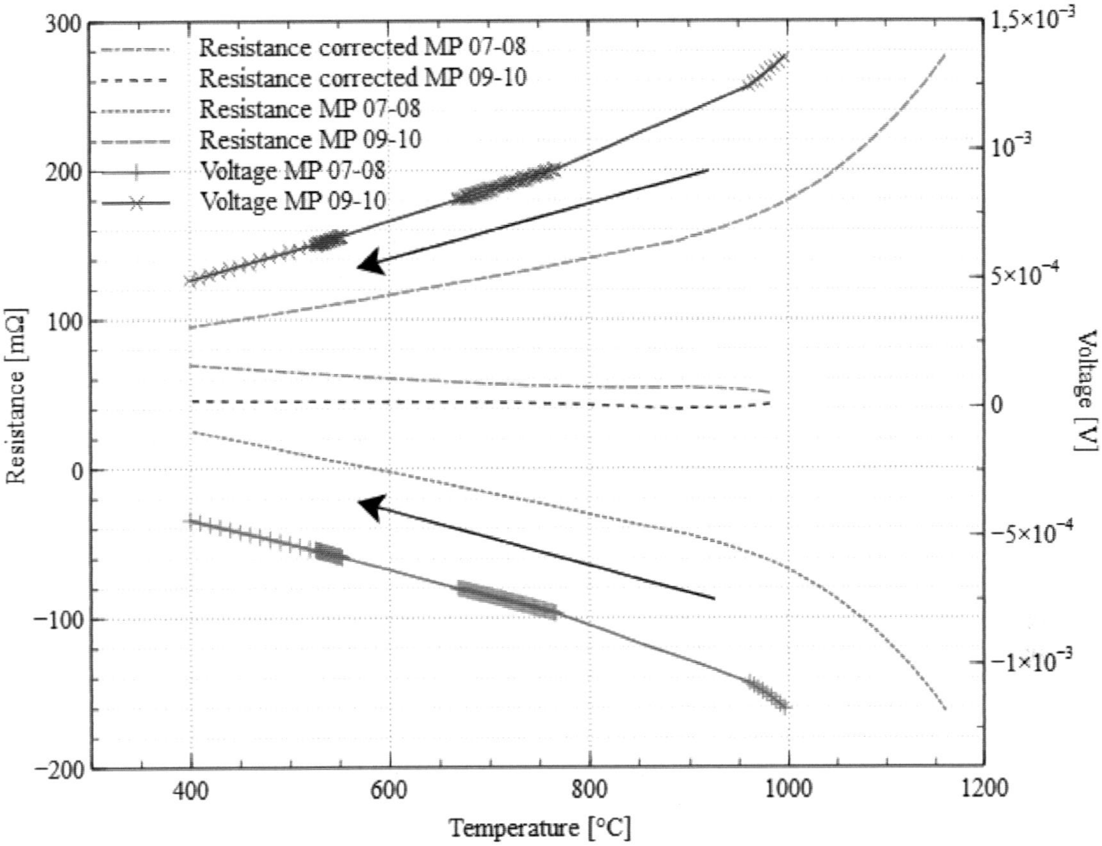

Figure 11: Voltage, resistance measured and corrected over maximum temperature of sample PH1878P-A.

Acknowledgments

This work is financed by the Helmholtz Alliance as the Helmholtz Young Investigator's Group VH-NG-909 'High Temperature Management in Hypersonic Flight'. Thanks for help with experiments, mechanical work or advice go to R. Bähnisch, S. Bletsch, F. Hofmeister, M. Ross, O. Schatz and H. Weihs.

References

[1] Santos, J., Jacobs, T., and Martinez, E., "Isotherm sensor calibration program for mars science laboratory heat shield flight data analysis," *42nd AIAA Thermophysics Conference*, 2011, pp. 27–30.

[2] Ecke, W., Latka, I., Willsch, R., Reutlinger, A., and Graue, R., "Fibre optic sensor network for spacecraft health monitoring," *Measurement Science and Technology*, Vol. 12, No. 7, 2001, pp. 974.

[3] Finlayson, R. D., Friesel, M., Carlos, M., Cole, P., and Lenain, J., "Health monitoring of aerospace structures with acoustic emission and acousto-ultrasonics," *Insight-Wigston then Northampton*, Vol. 43, No. 3, 2001, pp. 155–158.

[4] McGunigle, R. D. and Jennings, M., "Ultrasonic ablation recession measurement system," *Advances in Test Measurement*, Vol. 1, 1975, pp. 19–24.

[5] Boehrk, H., Leschinski, P., and Reimer, T., "Electrical Resistivity Measurement of Ceramic Matrix Composites under Thermo-Mechanical Load," *Composite Science and Technology*, Vol. 76, June 2013, pp. 1–7.

[6] Schulte, K. and Baron, C., "Load and failure analyses of CFRP laminates by means of electrical resistivity measurements," *Composites science and technology*, Vol. 36, No. 1, 1989, pp. 63–76.

[7] Muto, N., Yanagida, H., Nakatsuji, T., Sugita, M., and Ohtsuka, Y., "Preventing Fatal Fractures in Carbon-

Fiber–Glass-Fiber-Reinforced Plastic Composites by Monitoring Change in Electrical Resistance," *Journal of the American Ceramic Society*, Vol. 76, No. 4, 1993, pp. 875–879.
[8] Wang, S. and Chung, D., "Self-monitoring of strain and damage by a carbon-carbon composite," *Carbon*, Vol. 35, No. 5, 1997, pp. 621–630.
[9] Wang, X., Wang, S., and Chung, D., "Sensing damage in carbon fiber and its polymer-matrix and carbon-matrix composites by electrical resistance measurement," *Journal of materials science*, Vol. 34, No. 11, 1999, pp. 2703–2713.
[10] Smith, C. E., Morscher, G. N., and Xia, Z., "Monitoring damage accumulation in ceramic matrix composites using electrical resistivity," *Scripta Materialia*, Vol. 59, No. 4, 2008, pp. 463–466.
[11] Krenkel, W., *Entwicklung eines kostengünstigen Verfahrens zur Herstellung von Bauteilen aus keramischen Verbundwerkstoffen*, Ph.D. thesis, 2000.
[12] Weihs, H., "Sounding Rockets for Entry Research: SHEFEX Flight Test Program," ESA-Symposium European Rocket and Balloon Programmes and Related Research, 6 2013.
[13] Krenkel, W., Heidenreich, B., and Renz, R., "C/C-SiC Composites for Advanced Friction Systems," *Advanced Engineering Materials*, Vol. 4, No. 7, 2002, pp. 427–436.
[14] Stäbler, T., Böhrk, H., and Voggenreiter, H., "Development of a Resistance Measurement Test Bench for Health Monitoring of Heat Shields," AIAA, 7 2015.
[15] von Weiss, A., *Allgemeine Elektrotechnik*, Friedr. Vieweg und Sohn Verlagsgesellschaft mbH, 1981.

Specificity Bio-identification of CNT-Based Transistor

Sheng-Yu Wang[1] and Hue-Min Wu[2*]

[1] Department of Chemical & Materials Engineering and Graduate Institute of Nanomaterials, Chinese Culture University, Taiwan 11114
[2] Department of Optoelectric Physics, Chinese Culture University, Taiwan 11114

E-mail: hueminwu@faculty.pccu.edu.tw

Abstract. In this research, we report a simple and general approach to π-π stacking functionalization of the sidewalls of CNTs by 1-pyrenebutanoic acid, succinimidyl ester (PSE), and subsequent immobilization of insulin-like growth factor 1 receptor (IGF1R) onto SWNTs with a high degree of control and specificity. The selection of PSE provides visualization and characterization of individual CNTs based on its strong luminescence. In addition, we designed a simple and efficient electrode with a staggered pattern to increase the effect of electrophoresis by using electric field for the macroscopic alignment of CNTs to complete a field-effect device for CNT-based biosensors. Scanning Electron Microscopy (SEM) was used to investigate the morphology of the biosensors. The results of four-point probe method demonstrated high selectivity and sensitivity of detection. The functionalization of SWNTs was investigated by Fourier transform infrared spectroscopy (FTIR). Experimental results imply that specific binding between IGF1R and its specific mAb results in a dramatic change in electrical current of CNT-based devices, and suggest that the devices are very promising biosensor candidates to detect circulating cancer cells.

1. Introduction

Functional carbon nanotubes (CNTs) exhibit a unique combination of excellent mechanical, electrical and electrochemical properties, which has stimulated increasing interest in the application of CNTs as components of biosensors. The use of CNTs for sensing technology has attracted intensive research interest in the last several years because interactions between target molecules and CNTs can significantly change the electronic properties of CNTs.[1,2] Carbon nanotubes are unique among solid-state materials in that every atom is on the surface. Surface chemistry could therefore be critical to the physical properties of CNTs and their applications.[3,4] Therefore, the sidewall functionalization of CNT in noncovalent ways is important to the purposes of soluble nanotubes, self-assembly on surfaces, and chemical sensors to preserve the nanotube structure and thus their electronic characteristics. On the other hand, immobilization of biomolecules on carbon nanotubes has been pursued in the past as well, motivated by the prospects of using nanotubes as new types of biosensor materials.[5,6] Recent reports have shown that nanoscale electronic devices can be used to detect a change in electrical properties when receptor proteins bind to their corresponding antibodies functionalized on the surface of the device.[7] A prerequisite for research in this area is the development of chemical methods to immobilize biological molecules onto carbon nanotubes in a reliable manner. Nowadays, clinical cancer imaging technologies do not possess sufficient spatial resolution for early detection based on the molecular signatures of proteins that are overexpressed in cancer cells.[8] Development of new technologies for reliable early detection of cancer from biological fluids via minimally invasive methods is still a high priority. In this research, we report a carbon nanotube field effect transistor array, with a simple and general approach to noncovalent functionalization of carbon nanotubes, and subsequent immobilization of various biological molecules onto nanotubes with a high degree of control and specificity.

2. Experiment

2.1. Functionalization of CNTs
The noncovalent functionalization involves a bifunctional molecule, 1-Pyrenebutyric Acid N-Hydroxysuccinimide Ester (PSE) (SIGMA-Albrich, Inc purity95%, 114932-60-4), irreversibly adsorbed onto the inherently hydrophobic surfaces of CNTs in an organic solvent, isopropanol. The CNTs 1.5mg were soaked in 10ml IPA, and dispersed by 3 hrs sonification. And then 0.5mg PSE was added to the SWCNTs solution dispersed by 10mins sonification. This leads to the functionalization of CNTs (f-CNTS) with succinimidyl ester groups that are highly reactive to nucleophilic substitution by primary and secondary amines that exist in abundance on the surface of most proteins.

2.2. Immobilization of anti-protein
The nanotube-antibody device was prepared by modified f-CNTS with specific-IGF1R mAb and non-specific IgG mAb respectively. Specific anti-IGF1 receptor mouse monoclonal antibody (Life Span Ltd. LS-C41502) was mixed with phosphate buffered saline (PBS: pH 7.4) with 1:500 dilution and preserved at 4°C environment until use. Non-specific IgG mAb (Jackson ImmunoResearch Ltd. Antibody concentration 2.4mg/ml 115-005-003) was diluted 1:100 with sterile phosphate buffered saline (PBS: pH 7.6) and preserved at 4°C environment. MCF7 breast cancer cells was used in this study due to its high expression of IGF1 receptor.[9] The CNTs functionalized with IGF1R-specific, which exhibits highly sensitive and selective sensing, interact with IGF1R overexpressed by MCF7cell line.

2.3. Processing of electrode device
Carbon nanotube-based field effect transistor devices have been fabricated on a doped P^+-Si that was coated with a layer of silicon dioxide. The staggered titanium metal electrodes, which were designed as metal arrays of source and drain separated with a spacing of 60μm, were patterned using microfabrication techniques such as photolithography, metal evaporation, and lift-off. Following that the nanotubes were integrated between the electrodes by electrophoretic self-assembly technique. This process was carried out using AC electrophoresis with a frequency of 60 Hz and voltage of 10.0V peak-to -peak.

2.4. Measurements and analysis
The electrical measurements were performed using a probe station attached to a semiconductor parameter analyzer after the electrophoretic self-assembly. The silicon substrate was used as back gate. Electrical conductance of each bare-CNTFET device was taken as a reference for each measurement to eliminate any possible variation between the device. The real-time electrical conductance measurements have taken place in three steps: starting with bare-CNTFET, then antibody exposure to the CNTFET device, and finally application of MCF7 breast cancer cells to the surface of the CNTFET device. The surface image of the device was analyzed by scanning electron microscope. The

IOP Conf. Series: Journal of Physics: Conf. Series **939** (2017) 012031 doi:10.1088/1742-6596/939/1/012031

CNTs functionalization and immobilization of antibody were investigated by Fourier transform infrared spectroscopy (FTIR).

3. Results and Discussion

The optical micrograph and scanning electron microscopy images of a fabricated CNTFET device is shown in Fig.1(a). By electrophoretic self-assembly technique the well separated f-CNTs network was obtained in between source and drain Ti-electrodes, which is clearly identified in the Figure 1(b). The measured width of the source drain channel was about 62μm.

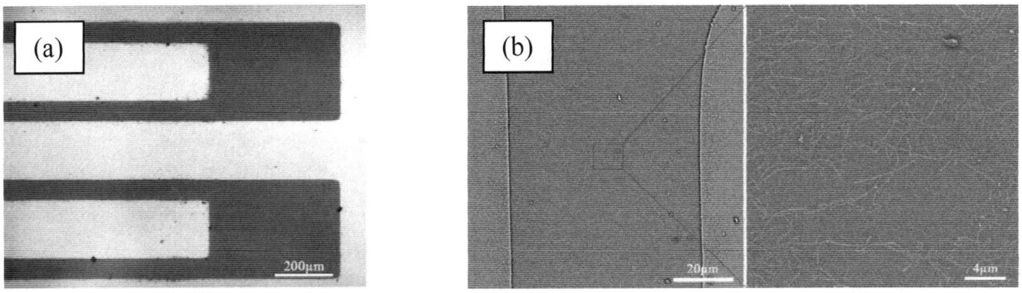

Fig.1.(a) Optical micrograph of the staggered patterned CNTFET, and (b) SEM image of a fabricated CNTFET indicating the electrophoretic self-assembled CNT network.

The electronic transport properties measured, using Si substrate as back gate, before immobilization of anti-protein is shown in figure 2. The device exhibited significant gate effect as determined by I-V_G measurements. The source-drain current of the devices increased with increasing gate voltage indicating that the dominant conduction process is due to electron transport, i.e. n-type semiconducting CNTs. This result indicating the conductance of semiconducting CNTS could be modulated by gate voltages. A CNTFET can act as ambipolar device, both an n-type and a p-type transistor, depending on the FET voltage. The conductance in the n-type region is larger than the p-type region due to the work function of Ti electrodes. The Fermi level of the Ti electrodes lies close to the conduction band of the CNT, making a path for the injection electrons. Figure 3 shows the FTIR spectra of CNTFET before and after functionalization and immobilization of CNTs. Shown in curve (a) is the absorption spectra of bare-CNTs as the control group. Curve (b) shows the absorption peaks of PSE functionalized CNTs in IPA. It can be seen from the figure that the two less intense peaks at 2850 ~ 3000cm^{-1} are C-H stretching vibration modes of functional groups. The 1592 cm^{-1} peak is generated from stretch aromatic hydrocarbon carbon sp^2 orbital vibrations. There is a multi-frequency observed around 1600cm^{-1} absorption peak indicates that the Benzene group of pyrene butyric acid attached on the SWNT sidewall via physical adsorption, resulting π -π stacking interactions and makes sp^2 hybrid orbital of carbon and carbon benzene ring oscillating between the apparent absorption frequencies in the range of the benzene ring. The broadening of IPA

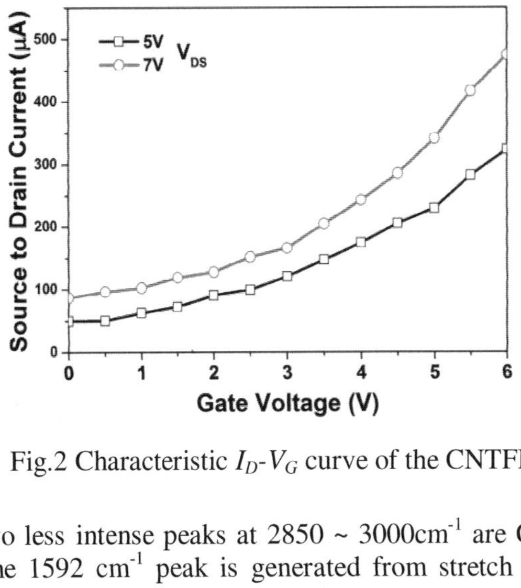

Fig.2 Characteristic I_D-V_G curve of the CNTFET

peak in the range of 700cm^{-1} ~ 800cm^{-1} owing to the C-C bending vibration and C-H stretching vibration of the tar-based structural brain benzene ring suggests a considerable amount of pyrene butyric acid adsorbed on the SWNT sidewall.

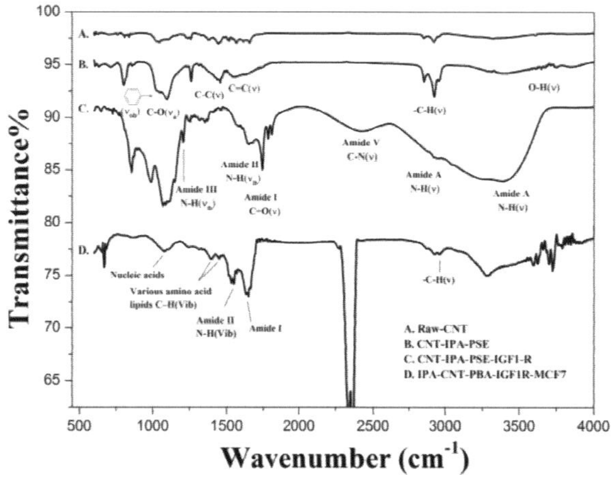

Fig.3 FTIR spectra of (a) bare-CNTs, (b) pyrene butyric acid modified f-CNTs in IPA, (c) specific-IGF1R mAb on the pyrene butyric acid modified f-CNT, and (d) MCF-7 cell with specific-IGF1R mAb immobilized f-CNTs.

Curve (c) shows the immobilization of specific-IGF1R mAb on the pyrene butyric acid modified f-CNT. The biological characteristic of the amino bonds after IGF1R protein molecules reaction is obvious. The stretching vibration modes of the amino functional groups of N-H, C-N and C=O bonds are found at 3450cm^{-1}, 3000cm^{-1}, and 2400cm^{-1} respectively. Three characteristic peaks of amide I, II, and III appeared at frequencies 1700cm^{-1}, 1500cm^{-1} and 1300cm^{-1}. This result confirms the existence of specific-IGF1R anti-protein functional groups which acts as the biological probe on CNTFET device. Curve (d) represents the FTIR spectrum of MCF-7 cell reacted with specific-IGF1R mAb immobilized f-CNTs for 2 hrs. The signals at 2800-3000cm^{-1} are dominated by symmetric and asymmetric stretching vibrations of CH_2 and CH_3 groups, mainly contained in fatty acids of cells [10]. The amide I stretching of carbonyl from peptide bond was observed at 1641cm^{-1}, a band sensitive to secondary structure of the protein. Deformation of protein amide N-H bond (Amide II) results in a signal at 1540cm^{-1}. The signals in the region between 1450-1400cm^{-1} result from various amino acid side chains and some lipids. The absorptions bands at 1238cm^{-1} and 1080cm^{-1} are characteristic of asymmetric and symmetric phosphodiester vibrations of nucleic acids [11].

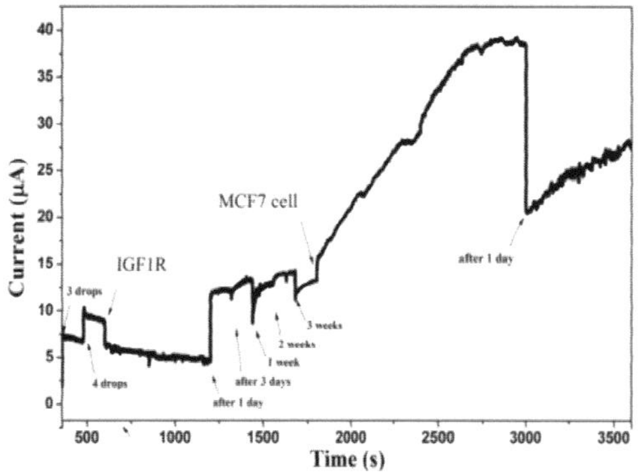

Fig.4 Conductance measurements of specific IGF1R mAb immobilized f-CNTs and adsorbed with MCF-7 breast cancer cells.

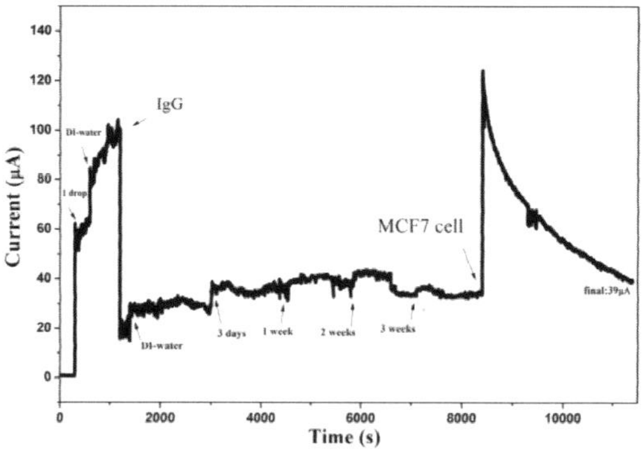

Fig.5 Conductance measurements of non-specific IgG mAb immobilized f-CNTs and adsorbed with MCF-7 breast cancer cells.

The electrical conductance measurements of modified f-CNTs were carried out in real-time at room temperature by four point probe method. To monitor breast cancer cell binding events, a 10 μl of the

24 μg/ml mAb solution was applied on the surface of f-CNT device, while the current was monitored continuously. Once the current stabilized, 5 μl of MCF-7 breast cancer cells (about 10000 cell) were applied to the device and the current was monitored continuously. Figure 4 demonstrates the effects of specific-IGF1R mAb adsorption, followed by MCF-7 cell, which generated about 3-fold increase in the device conductance. Similar result was obtained when the measurement was taken again after one day. Figure 5 shows the conductance of the device during nonspecific-IGg mAb adsorption, followed by the adsorption of MCF-7 breast cancer cells. The conductance rises immediately and dropped to the same level of the conductance before MCF-7adsorption. These results suggest a charge transfer between mAb modified f-CNTs and breast cancer cells. Furthermore, specific binding between IGF1R and its specific mAb results in a dramatic change in electrical conductance of the device.

4. Conclusions

This research has demonstrated the label-free detection of MCF-7 breast cancer cells using specific-IGF1 receptor mAb modified CNTFET device. Experimental results indicated that it is the specific interaction between the IGF1R and its specific antibody that results in a large change in conductance. These observations imply a pathway, with high sensitivity and selectivity, that the mAb-CNTFET approach could be used to sense overexpressed protein biomarkers on breast cancer cells. Furthermore, these findings suggest applications of mAb-CNTFET to detect a wide variety of surface markers on diseased cells

References

[1] R. Villamizar, A. Maroto, F.X. Rius, Improved detection of Candida albicans with carbon nanotube field-effect transistors Sensors and Actuators B, 136 451-457 (2009)
[2] B. Lee, A. Padmakar, D. Kichambare, and A.Star, Carbon Nanotube Field-Effect-Transistor-Based Biosensors. Advanced Materials, 19, pp. 1439–1451 (2007)
[3] R.J.Chen, Y.Zhang, D.Wang, and H.Dai, Noncovalent sidewall functionalization of single-walled carbon nanotubes for protein immobilization. J. Am.Chem. Soc. 123, pp. 3838-9 (2001)
[4] S.H. Yoshimura, S. Khan, H. Maruyama, Y. Nakayama, and K. Takeyasu., Fluorescence Labeling of Carbon Nanotubes and Visualization of a Nanotube–Protein Hybrid under Fluorescence Microscope, Biomacromolecules 12(4), pp. 1200–1204 (2011)
[5] Y.-K. Baek, D.-H. Jung, S.M.Yoo, S.Shin, J.-H. Kim, H.-J. Jeon, Y.-K. Choi, S.Y. Lee, and H.-T. Jung, Label-Free Detection of DNA Hybridization Using Pyrene-Functionalized Single-Walled Carbon Nanotubes: Effect of Chemical Structures of Pyrene Molecules on DNA Sensing Performance. Journal of Nanoscience and Nanotechnology 11, pp. 4210–4216 (2011)
[6] X.Q. Chena, T. Saito, H. Yamada and K. Matsushige., Aligning single-wall carbon nanotubes with an alternating-current electric field. Appl. Phy. Letters Vol 78 No.23, pp. 3714-3716 (2001)
[7] G. Zheng, F. Patolsky, Y. Cui, W. Wang, and C. Leiber, Multiplezed electrical detecton of cancer surface markers with nanowire sensor arrays Nat. Biotechnal. 23 1294-301 (2005)
[8] N. Shao, E. Wickstrom, and B. Panchapakesan, Nanotube-antibody biosensor arrays for the detection of circulating breast cancer cells, Nanotechlogy 19 465101-12 (2008)
[9] N. Shao, S. Lu, E. Wickstrom and B. Panchapakesan, Integrated molecular targeting of IGF1R and Her2 surface receptors and destruction of breast cancer cells using single wall carbon nanotubes, Nanotechnology 18 315101 (2007)
[10] L.Di Giambattista, D. Pozzi, P. Grimaldi, S. Gaudenzi, S. Morrone, AC. Castellano, New marker of tumor cell death revealed by ATR-FTIR spectroscopy, Anal. Bioanal Chem 399 2771-8 (2011)
[11] B.-B. Wu, Y.P. Gong, X.H. Wu, Y.Y. Chen, F.F. Chen, L.T. Jin, B.R. Cheng, F. Hu and B. Xiong, Fourier transform infrared spectroscopy for the distinction of MCF-7 cells treated with different concentrations of 5-fluorouracil. J. Translational Medicine 13, 108-116 (2015)

IOP Conf. Series: Journal of Physics: Conf. Series **939** (2017) 012032 doi:10.1088/1742-6596/939/1/012032

Au and Ag sputter deposition on printer paper

A Kaidatzis[1], V Psycharis[1], M U González Sagardoy[2], J M García Martín[2] and D Niarchos[1]

[1]Institute of Nanoscience and Nanotechnology, NCSR "Demokritos", Athens, Greece
[2]IMM-Instituto de Microelectrónica de Madrid (CNM-CSIC), Tres Cantos, Madrid, Spain

Corresponding author: a.kaidatzis@inn.demokritos.gr

Abstract. We study metal-coated paper for low cost sensor applications. Sputter deposition is used to coat commercially available stationary printer paper with Au or Ag thin films. Two kinds of substrates are used: copy paper and glossy photo paper. The surface morphology, coating structure and electrical resistivity of the samples is studied. We find that after deposition of a nominally 100 nm thick metal layer the templates retain their initial morphology, which is fibre-like at the copy paper and flake-like at the photo paper. The metal-coated paper is conductive in both cases, with resistivity a few times higher than the bulk value. The photo paper samples have lower resistivity than the copy paper ones. Moreover, no observable resistivity variation is observed after 8 weeks of room temperature storage in a desiccator.

1. Introduction

Paper is a low-cost, foldable, lightweight, environmentally-friendly material that can be fabricated at thin sheets, easily stackable, stored and transported; it is currently being investigated as substrate for inexpensive healthcare sensors fabrication [1, 2] and for foldable electronics applications [3, 4]. In most cases, metal-coated paper-sheets are used as electrodes. The main approaches for metalizing paper include physical vapor deposition [2-4], conducting inkjet printing [4], and electrodeposition [5], while various types of paper substrates and metal coatings have been studied [2-5].

In this work we study common stationary paper, sputter-coated with Au or Ag. The advantage of sputter-deposition is that allows for the metallization of paper substrates without using liquid solvents that could degrade paper properties. Additionally, the use of stencil masks with μm-size features during deposition could allow for the creation of patterns without using lithography-based patterning that includes lift-off processes and wet etching [2]. Finally, sputtering is an industry-compatible technique, allowing for the deposition of large area coatings. Thus, a totally dry process may be developed for fabricating micro-patterned sensor arrays on paper.

2. Experimental

Paper substrates were coated at room temperature using an ultra-high vacuum (base pressure 1.5×10^{-9} Torr) magnetron sputtering system (ATC 2200-V, AJA Inc., USA). Significant degassing was observed during load-lock pump-down, due to the porous morphology of the paper substrates which results in high water absorption. The source material was metallic Au or Ag (99.995% purity) and the working gas was high purity Ar at 3 mTorr pressure. Direct Current (DC) power was applied to the target (7.5 W/cm^2), leading to a growth rate of 0.48 nm/s. The substrates used were commercially available copy paper (density: 80 g/m^2, thickness: 100 μm), or glossy photo paper (density: 200 g/m^2, thickness: 200 μm). Test samples were deposited side-by-side on Si(100) substrates for comparison purposes.

The surface morphology of the samples has been studied using optical microscopy and Atomic Force Microscopy (AFM). A Bruker Icon AFM has been employed using commercial AFM probes (Nanosensors PPP-FMR). X-ray diffraction (XRD) structural analysis has been performed using a Siemens D500 diffractometer with Cu-K_α radiation, in steps of 0.03° and counting time 12 s/step. The room-temperature film resistivity has been determined by Van der Pauw (VdP) measurements performed on square samples using a programmable current source (Keithley 220) and a multimeter (Keithley 2000). Electrical contacts were made at the sample edges using silver paint.

3. Results and discussion

Figure 1 shows the surface morphology characterization as obtained by optical microscopy and AFM. In figure 1a Au-coated copy paper is shown; its morphology is dominated by the highly-entangled cellulose fibers of the paper substrate, revealing the high porosity of copy paper. Figure 1b shows an AFM image of the copy paper: the apparent metal coated fibers have diameters of several μm and the surface roughness is approximately 1 μm. Figure 1c shows an AFM image of the photo paper surface before metal deposition. The surface is substantially smoother compared to the copy paper, with sub-μm flake-like features dominating its morphology. No considerable alteration of the surface morphology is observed after deposition of a 100 nm thick Au layer (figure 1d); the surface roughness is approximately 60 nm before and after deposition. It should be noted that similar results are also obtained for Ag-coated paper substrates (data not shown).

Figure 1 (a) Optical micrograph and (b) AFM image of Au-coated copy paper. AFM images of (c) photo paper and (d) Au-coated photo paper.

The fabrication of high sensitivity and accuracy sensors for quantitative analysis requires the patterning of μm or sub- μm features with sharp edges. From the surface morphology characterization of our samples, it becomes immediately apparent that the fibrous morphology of copy paper renders it unsuitable for such applications and it could only be used as a substrate for fabricating mm-sized qualitative indicators. However, the highly porous morphology of copy paper would be advantageous for applications that require high effective metalized surface area, e.g. for the fabrication of paper-based super-capacitors [5]. On the other hand, the much smoother surface morphology of photo paper allows for the patterning of μm-scale features and the fabrication of more complex configurations.

Figure 2 shows XRD graphs of the samples. The XRD graphs of uncoated copy and photo paper show the characteristic calcite peaks (indicated by stars), due to the calcite additives used for paper manufacturing. The XRD graphs of the test coatings on Si show the characteristic peaks of Au or Ag crystallized to *fcc* structure. Furthermore, the Si(100) substrate gives rise to the strong peak at around 68°. All paper samples were mounted onto Si substrates for facilitating handling; this explains the Si peak present at the copy paper samples. However, no Si peak appears at the photo paper samples, as it has twice the thickness of the copy paper. In all cases, the characteristic peaks of *fcc* Au or Ag are apparent at the coated-paper samples, revealing the polycrystalline microstructure of the deposited coatings.

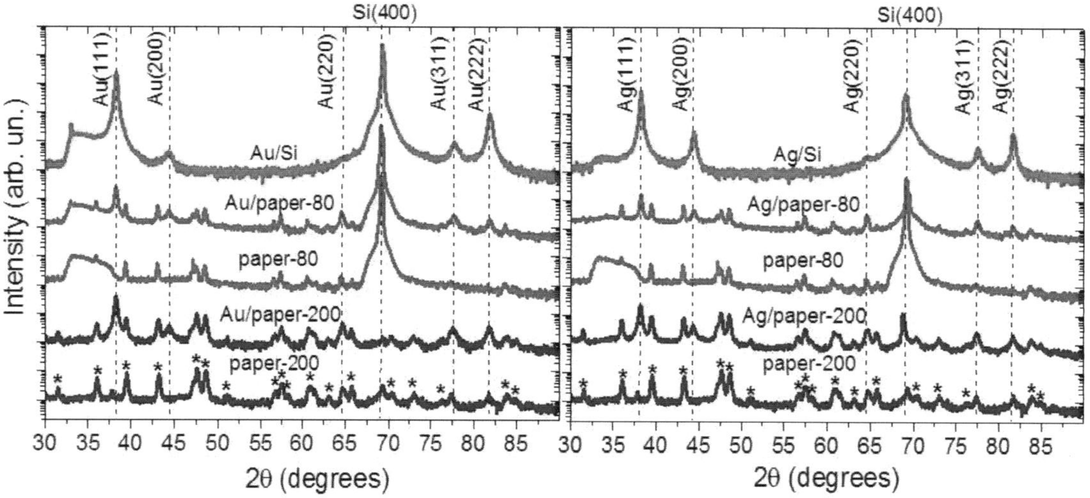

Figure 2 XRD θ-2θ graphs of the studied samples: (left) Au-coating, (right) Ag-coating. The substrates used were Si, copy paper (paper-80), and photo paper (paper-200).

Figure 3 shows the resistivity of the samples as determined by Van der Pauw measurements. For comparison, the bulk Au and Ag resistivity values are also shown. The resistivity increase observed at the 100 nm thick films deposited onto Si is due to the additional electron scattering at the top and bottom film interfaces. Resistivity further increases at the metal-coated photo paper samples: it reaches 10.4 μΩ·cm for Au coating and 4.1 μΩ·cm for Ag coating; this increase is attributed to the flake-like morphology of these samples which results in higher defect density and thus higher electron scattering. Finally, resistivity increases even further at the metal-coated copy paper samples it reaches 17.5 μΩ·cm for Au coating and 12.1 μΩ·cm for Ag coating; in this case, the fibrous morphology of the samples probably results in longer conducting paths, in addition to the increased defects density. It should be noted that the resistivity of the samples has been measured after 8 weeks of storage in a desiccator and no variation is observed, within the experimental error of the measurement method (data not shown). Since metal-coated paper substrates are studied for use as electrodes in sensors and electronic circuits, low resistivity is important for minimizing resistive heating and power dissipation upon device operation. In this context, metal-coated photo paper is more suitable for use as a flexible, low cost, environmentally friendly, and conductive substrate.

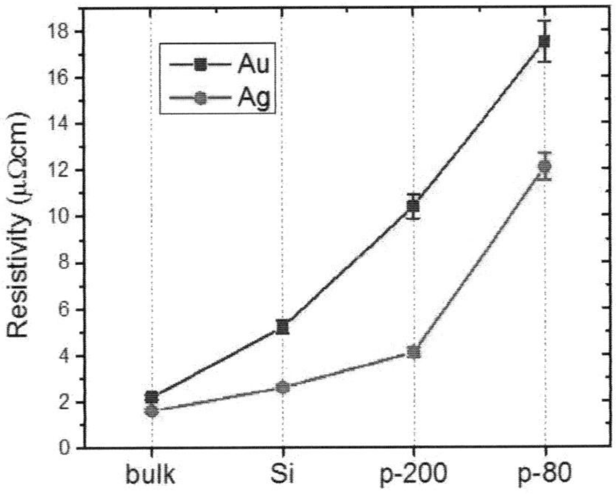

Figure 3 Resistivity values of the coated samples

4. Conclusions

We have studied Au or Ag coated copy and photo paper. It is shown that metal-coated photo paper is more suitable for flexible electronics and inexpensive quantitative sensors applications due to its low roughness, which allows for patterning μm-scale features and also results in lower resistivity. On the other hand, metal-coated copy paper would be more suitable for large-area qualitative indicators, easily detected by the naked eye, or for applications that require high effective surface.

Acknowledgments

Funding from the E.C. through a FP7-ICT project (Grant No. 318144) and MINECO (ref.: MAT2014-59772-C2-1-P) is acknowledged. We would like to thank Dr Christos Tsamis for stimulating discussions.

References

[1] Martinez A W, Phillips S T and Whitesides G M "Three-dimensional microfluidic devices fabricated in layered paper and tape" 2008 *PNAS* **105** 19606–11
[2] Tao H *et al.* "Metamaterials on paper as a sensing platform" 2011 *Adv. Mat.* **23** 3197–201
[3] Siegel A C, Phillips S T, Dickey M D, Lu N, Suo Z and Whitesides G M "Foldable Printed Circuit Boards on Paper Substrates" 2010 *Adv. Funct. Mater.* **20** 28-35
[4] Gomes L, Marques A, Branco A, Araújo J, Simões M, Cardoso S, Silva F, Henriques I, Laia C and Costa C "IZO deposition by RF and DC sputtering on paper and application on flexible electrochromic devices" 2013 *Displays* **34** 326–33
[5] Mostafalu P and Sonkusale S "Paper-based super-capacitor using micro and nanoparticle deposition for paper-based diagnostics" 2013 *IEEE SENSORS Proc.* 1234–37

Sensing properties of gold contact in diode type gas sensors

S. Rahbarpour

Electrical Engineering Department, Shahed University, Tehran 159-18155, Iran.

s.rahbarpour@shahed.ac.ir

Abstract. Gold nanoparticles have been increasingly used as catalytic material since 1997, when their chemical activity was discovered. Here, chemical activity of gold contacts in Au-TiO_2-Ti structures is reported and verified experimentally. Diode type Au-TiO_2-Ti structures was fabricated by Au deposition on thermally oxidized titanium film. The samples demonstrated I-V characteristics of a typical Schottky diode at operated temperatures up to 400 °C, the highest examined temperature. The Schottky barrier height (SBH) of different fabricated Au-TiO_2 junctions was measured. It was shown that the measured SBHs depended strongly on the temperature and particle size of utilized gold contacts and could vary in the range of 1 to 1.8 eV. The hydrogen sensing properties of the samples were also investigated. The samples demonstrated high sensitivity to hydrogen; 1 Vol% of hydrogen in air could increase the reverse current of the diodes ~1000 times. The results are described based on the reactions taking place at the Au/TiO_2 interface.
Keywords: gold, particle size, diode type gas sensors, oxygen adsorption, hydrogen adsorption, titanium oxide.

1. Introduction

Gold has long been recognized as the noblest metal in the world. It is widely used in Electronic devices as metal contact in the form of continues thin film. Considering high price of gold, in some applications it is replaced with other metals. In some applications using gold or other noble metal contacts is inevitable. For example in the field of gas sensors, the corrosive environment and usually high operating temperature of the sensors necessitate the utilized metal contact to be a noble one. Noble metals have high electrical work function and, hence, make Schottky junction with conventional metal oxides utilized in gas sensor fabrication, such as SnO_2, TiO_2 and ZnO. Gas sensing properties of these junctions is mainly attributed to adsorption/desorption reactions take place at the metal surface. Many studies have been done to investigate the chemistry of these reactions occur on the surface of Pt [1,2], Pd [3,4] and Ag[5]. Although high gas sensitivity reported in various Au-metal oxide junctions [6], reactions happen on gold-metal oxide junctions are yet remained unstudied. Considering high nobility of gold, such study was assumed to be unreasonable.

At 1997, Haruta showed that gold nanoparticles have completely different characteristic than their bulk structures [7]. While bulk gold has no chemical reaction at room temperature, gold

IOP Conf. Series: Journal of Physics: Conf. Series **939** (2017) 012033 doi:10.1088/1742-6596/939/1/012033

nanoparticles are chemically active and have unique catalytic properties such as CO oxidation, propene oxidation], NO reduction and many other catalytic activities reported in the background literature [8]. Catalytic properties of gold nanoparticles become more evident by decreasing gold particle size to bellow 5nm. Several nano-meter gold particles are usually supported by metal oxides [9] to be used as catalysts. Mills et. al. proved that although the surface of bulk gold is inert, it can be chemically activated by placing gold nanoparticles on it [10]. These gold nanoparticles maintain their chemical activity. He theoretically confirmed that the main reason of rarely reported chemical activity of fabricated gold layers is the surface roughness of gold layers which is usually vanished by increasing the thickness of layer formed. The effect of gold surface roughness on its chemical activity is disregarded by researchers in the field of gas sensors and mechanism of gas detection in Au-metal oxide gas sensors is not completely understood yet.

Here, we have reported the fabrication and characterization of various Au-TiO_2-Ti structures with different Au grain sizes. Fabricated samples performed high sensitivity among very low concentration hydrogen in air; 1 Vol% of hydrogen could increase the reverse current of the diodes up to 1000 times. We experimentally proved that gas response of the fabricated samples is related to the shape and structure of gold contact utilized in the device. Our results have important implications on previously reported gas sensing of Au-metal oxide junctions and clarify reactions take place on gold surface.

2- Experimental

To investigate electrical properties of Au-TiO_2 junction, Au contacts were deposited on TiO_2 layer by sputtering technique. TiO_2 layers were formed by partially oxidation of titanium layers deposited by electron beam evaporation on fused silica substrates. The thickness of the oxide layer and, hence, oxidation conditions were selected based on TiO_2 resistance considerations and several oxidation runs at different temperatures. Selected oxidation procedure was 30 minutes in an air filled electrical furnace operated at 600 °C. The whole structure was positioned on a micro-heater which provided the elevated operating temperature up to 400 °C. The schematic diagram of fabricated Au-TiO_2-Ti structure and SEM micrograph of Ti/TiO_2 interface are shown in Fig. 1a&b respectively.

To achieve different gold particle size distribution and comparing its effect on the SBH of the junction, different thickness of gold were deposited. Growth mechanism of gold on TiO_2 is reported [11]. It is shown that, at the beginning of gold deposition, gold particles nucleate on the surface and grow to larger size if deposition is continued. Hence, deposition of gold layers with different thicknesses leads to layers with different gold particle size. In Fig.2, sample SEM micrographs of the surface of different gold contacts are shown. Comparing Fig.2a and Fig.2b, growth of gold particles by increasing the film thickness is obvious.

The current-voltage diagrams of fabricated devices were obtained at different conditions using a Keithley 238 source-measure unit (SMU) to estimate the height of Schottky barrier potential established at Au-TiO_2 junctions. Gas response of the samples is defined as:

$$SR = {I_g}/{I_a}, \quad (1)$$

in which I_a and I_g are the junction reverse currents measured in clean and contaminated air, respectively. Detail of measurement method is explained in our previous works [12-14].

3- Results and discussion

To study the effect of gold particle size on electronic properties of Au-TiO_2 junctions, different thickness of gold were deposited on TiO_2 as active contact. Resulted Au-TiO_2 junctions were studied by measuring I-V diagram and SBH calculation. Examined thicknesses were in the range of 20 to 50 nm. The I-V characteristic diagrams of a 35 and 20 nm thick Au-TiO_2-Ti junctions measured at different temperatures are shown in Fig.3a&b respectively.

The Schottky barrier height (SBH) established at the junction can be calculated from the following well known relationship:

$$\varphi = KTLn\left({AA^*T^2}/{I_0}\right) \quad (2)$$

in which A is the junction area, A* is the Richardson constant for TiO_2, T is operating temperature and I_0 is the saturation current of the junction. Estimated SBH values (See Table 1) show that increasing the operating temperature of the Au-TiO_2 junction cause the SBH to increase. However, estimated SBHs at temperatures below 100 °C are comparable with reported SBH at Au-TiO_2 junction by other researchers [15]. To describe the increase of SBH at higher temperatures, it should be noted that the SBH is also related to the work function of the gold (W_{Au}) and the electron affinity of the TiO_2 (χ_{TiO2}) by Eq.3:

$$\varphi = W_{Au} - \chi_{TiO2} \quad (3)$$

W_{Au} is reported to be 5.1 eV and $\chi_{TiO2} = 3.9$ eV which results in 1.2 eV SBH. The actual SBH is usually lower due to non-ideal conditions at the interface [16]. Reported SBH is ~1.1 eV. χ_{TiO2} is hardly affected by changing atmospheric and operating temperature [17]. Hence, the gold work function variations is responsible to the SBH incrase shown in Table 1. Gottfried et. Al, showd that 1 ML oxygen coverage on gold leads to 1 eV increase in its workfunction [18]. Alghough 1ML oxygen coverage never occures spontanously in our sitution, we attributed Increase of W_{Au} to lower oxygen adsorbtion on gold surface. According to high energy of oxygen adsorption on gold surface, gold has low tendency to adsorb oxygen. This energy can be provided by thermal energy. Hence, increasing the

IOP Conf. Series: Journal of Physics: Conf. Series **939** (2017) 012033 doi:10.1088/1742-6596/939/1/012033

surface temperature increases the adsorption rate. The effect of temperature on oxygen adsorption to gold is reported by some researchers [19]. However, they didn't define its consequence effect on W_{Au}. Considering the data given in table one, increase of W_{Au} is concluded. Hence, observation of any increase in expected SBH value at Au-TiO_2 junction (that is 1.1 eV) implicates that oxygen species are adsorbed on the the gold surface from the ambient. To evaluate this hypothesis partial pressure of oxygen was lowered and the SBH of the junction was monitored continuously. Lowering the pressure to 40 mbar caused the SBH to decrease to ~1 eV. At this pressure oxygen species are completely removed from the gold surface. This result expresses that removing adsorbed oxygen from the gold surface, decreases W_{Au} to approximately its theoretically predicted value. Comparing I-V's in Fig.3a and Fig.3b, obviously shows reduction of I_0 and increasing the SBH with increasing the thickness of Au. Calculated SBHs for samples with 20 to 50 nm thick Au, were in the range of 1.8 to 1.4 eV at 300 °C. Results showed that increasing the gold thickness, equivalent to larger gold particle size, decreases the SBH appear at the junction. The SEM micrographs of gold surface related to the margin lengths were shown in previous section and proved the gold particle growth during the deposition process (See Fig.2).

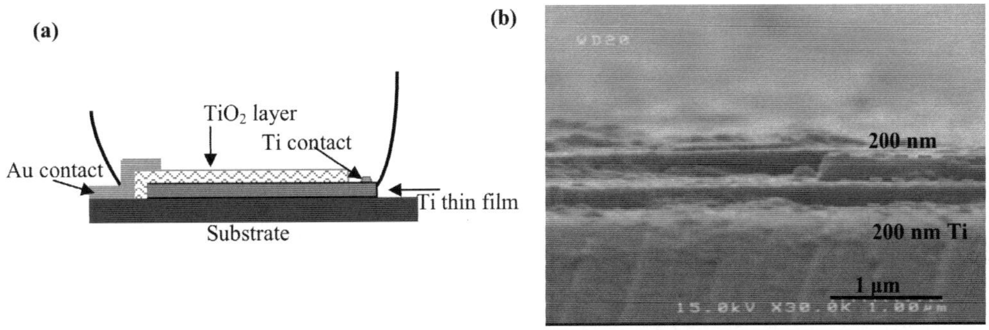

Fig. 1: The schematic diagram of fabricated device (a) and the SEM micrograph taken from interface of thin titanium film and titanium oxide layer (b).

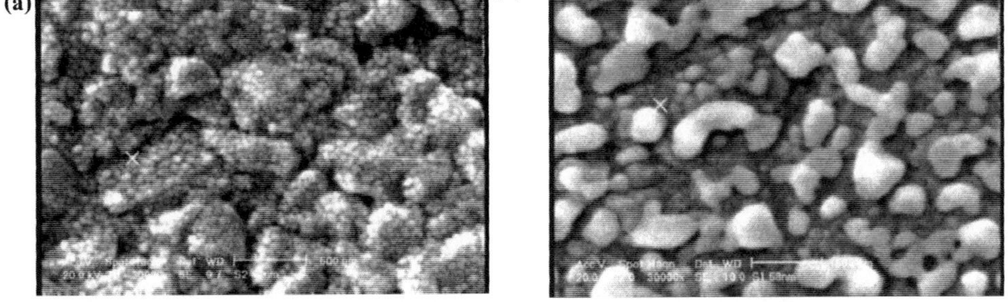

Fig. 2: SEM micrographs of the surface of a 20 nm (a) and 50 nm (b) gold layers sputtered on TiO_2 surface. The growth of gold particles by increasing the film thickness is obvious. Both scale bars are 500 nm.

Hydrogen responses of a 1.5 eV SBH Au-TiO_2 diode to various levels of hydrogen contamination were also obtained and are given in Fig. 4a. The device showed high sensitivity to low levels of hydrogen. Increasing the hydrogen level caused the I-V diagram of the device to deviate from the diode type profile and turned to ohmic diagram at 1% hydrogen which is equal to 3 orders of magnitude change in the device current (See Fig. 4b). Although the excess barrier of Au-TiO_2 (the amount of SBH greater than 1.1 ev) can be described by oxygen adsorption on the gold surface, the observed ohmicity in hydrogen atmosphere cannot be described by oxygen reduction from gold surface caused by hydrogen atoms [20]. Resistive diagram is described by the assumption of hydrogen adsorption on gold surface. In this case, hydrogen detection mechanism of Au-TiO_2 is due to both oxygen removals from the gold surface by hydrogen and adsorption of hydrogen species to gold surface which in turn reduces the SBH to zero.

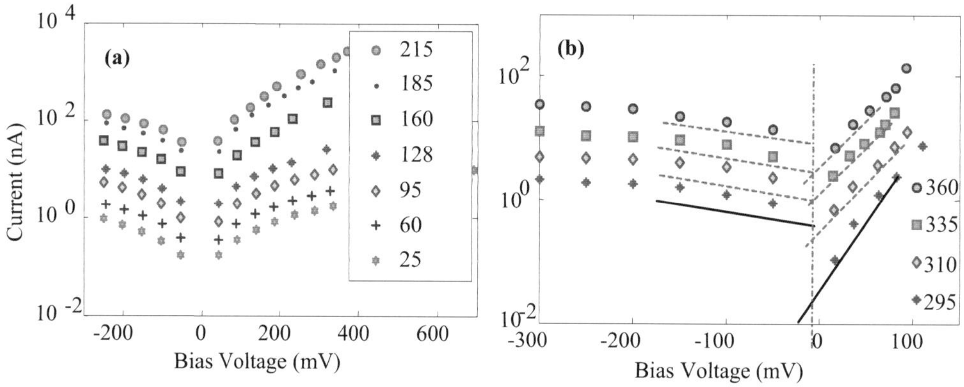

Fig. 3: The I-V diagrams of a (a) 35 and (b) 20 nm thick Au-TiO2 junction measured at different operating temperature.

Table 1: Estimated SBHs derived from measured I-Vs in Fig. 3a (however some I-Vs are not shown in Fig. 3a because of redundancy).

Temperature (°C)	335	315	295	260	215	185	160	128	95	60	25
Saturation current (nA)	220	70	60	52	40	30	8	2	1	0.5	0.2
SBH (eV)	1.62	1.61	1.55	1.45	1.31	1.25	1.22	1.18	1.10	1.01	0.95

4- Conclusion

Our fabricated Au-TiO_2 samples with different gold thicknesses were studied. It is shown that changing the thickness of Au layers leads to different gold particle size and affect its chemical activity. By this method, chemical activity of gold can be investigated easily not only by estimation of their

SBH but also by investigating their gas sensing properties. Other studies are based on complicated and time consuming methods such as LEED, TDS, XPS and miscellaneous methods.

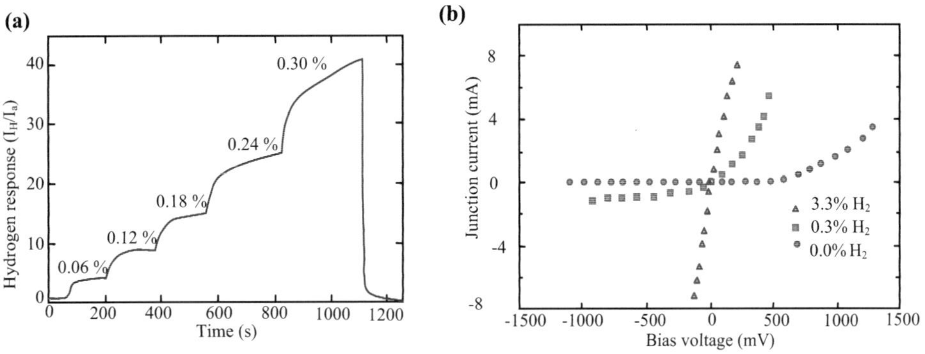

Fig. 4: Transient responses (a) and I-V diagrams (b) of Au-TiO_2-Ti sensor at different levels of hydrogen in air measured at 300 ˚C.

5- References

[1] L. Wang, Y. Wang, K. Yu, S. Wang, Y. Zhang and C. Wei: Sens. Actuators B Chem. Vol. 232 (2016), P. 91.
[2] V.B. Kamble and A.M. Umarji: Sens. Actuators B Chem. Vol. 236 (2016), P. 208.
[3] E. Şennik, O. Alev and Z.Z. Öztürk*: Sens. Actuators B Chem. Vol. 229 (2016), P. 692.*
[4] F. Hooriabad-Saboor, T.Ueda, K. Kamada, T. Hyodo, Y. Mortazavi, A. Khodadadi and Y. Shimizu: Sens. Actuators B Chem. Vol. 223 (2016), P. 429.
[5] M. Li, H. Zhu, B. Wang, J. Cheng, W. Yan, S. Xia and Z. Tang: Sens. Actuators B Chem. Vol. 232 (2016), P. 545.
[6] H. Miyazakia, T. Hyodob, Y. Shimizua and M. Egashira: Sens. Actuators B Chem. Vol. 108 (2005), p. 467.
[7] M. Haruta: Gold Bulletin Vol. 37 (2004), p. 27.
[8] R. Meyer, C. Lemire, Sh.K. Shaikhutdinov and H. J. Freund: Gold Bulletin B Vol. 108 (2004), p. 14609.
[9] M. Valden, X. Lai and D.W. Goodman: Science Vol. 281 (1998), p.1647.
[10] G. Mills, M . S. Gordon and H. Metiu: Journal of Chemical Physics Vol. 118 (2003), p. 4198.
[11] G.Korotcenkov, V.Brinzari and B.K.Cho: Materials Letters. Vol. 147 (2015), p. 101.
[12] F. Hossein-Babae and S. Rahbarpour: Solid-State Electronics Vol. 56 (2011), P. 185.
[13] F. Hossein-Babae and S. Rahbarpour: Sensors and Actuators B: Chemical Vol. 160 (2011) P. 174.
[14] F. Hossein-Babaei: ELECTRON LETT Vol. 44 (2008), p. 161.
[15] E.W. McFarland and J. Tang: letters to nature Vol. 421 (2003), p. 616.
[16] F. Hossein-Babaei, S. Abbaszadeh and M.S. Esfahani: IEEE SENS J Vol. 9 (2009), P. 237.
[17] N. Yamamoto, S. Tonomura, T. Matsuoka and H. Tsubomura: Surf. Sci. Vol. 92 (1980), p. 400.
[18] J. M. Gottfried, K. J. Schmidt, S. L. M. Schroeder and K. Christmann: Surface Science Vol. 511 (2002), p. 65.
[19] M.A. Chesters and G.A. Somorjai: Surface Science Vol. 52 (1975), p. 21.
[20] S.Rahbarpour and F.Hossein-Babaei: Key Engineering Materials Vol. 495 (2012), p 289.

IOP Conf. Series: Journal of Physics: Conf. Series **939** (2017) 012034 doi:10.1088/1742-6596/939/1/012034

Electric field feedback for Magneto(elasto)Electric magnetometer development

M-T. Yang, X. Zhuang, M. Lam Chok Sing, C. Dolabdjian, P. Finkel, J. Li and D. Viehland
Greyc-Ensicaen, France
May-tia.yang@ensicaen.fr

Abstract. Magneto(elasto)Electric (ME) sensors based on magnetostrictive-piezoelectric composites have been investigated to evaluate their performances to sense a magnetic signal. Previous results have shown that the dielectric loss noise in the piezoelectric layer exhibits as the dominant intrinsic noise at low frequencies, which limits the sensor performances. Also, it has intrinsically no DC capability. To avoid a part of this limitation, modulation detection methods are evaluated through a frequency up-conversion technique [1-4]. Moreover, classical magnetic field feedback techniques can be used to increase the dynamic range, the sensing stability and the system linearity, too. In this paper, we propose a new method to feedback the system by using both the magneto-capacitance modulation and an electric field feedback technique. Our development shows the feasibility of the method and the results match with the theoretical description and material properties. Even if the present results are not totally satisfactory, they give the proof of concept and yield a way for the development of very low power magnetometers.

1. Introduction

A classical *ME_EE* modulation system [3] consists of a magnetoelectric composite as a sensitive element for picking up magnetic fields, a low noise operational amplifier to stress the sensor at the operating carrier frequency and to amplify the signal as well as a demodulation stage whose output is connected to a simple feedback loop. A ME laminate with multi-push-pull mode configuration [5] was chosen as the sensor. It provides a high sensitivity for magnetic-to-electric conversion due to a well-designed structure and the colossal piezoelectric and efficient magnetic coefficients of the ferroelectric and ferromagnetic materials. A rectangular piezoelectric layer of dimensions 40×10×0.2 mm^3, made of ferroelectric single crystal PMN-PT, is sandwiched between two ferromagnetic layers. Each ferromagnetic layer is made of three Metglas sheets with dimensions 80×10×0.15 mm^3. Two pairs of interdigital electrodes are stacked on both the top and bottom surfaces of the piezoelectric layer with a center-to-center space of 1.5 mm.

2. Modeling

2.1.1 Transfer function

In order to investigate the relation between the transfer functions in open loop and for a feedback loop, we need to specify the transfer function for an applied magnetic signal, first. Referring to Fig. 1, the open loop transfer function is obtained by taking into account the sensor sensitivity, the charge amplifier and the demodulation circuit gains. It yields

$$A = \frac{\alpha^{NL}_{ME\_EE}}{C_1} G_1 G_2 G_3 \tag{1}$$

where $\alpha^{NL}_{ME\_EE}$ and C_1 are the nonlinear ME charge coefficient and the capacitance of the charge amplifier. G_1, G_2 and G_3 are, respectively, the gains of the different succeeding stages. A quantitative evaluation of the deformation induced by a magnetic or electric field provides the key information for the feedback system. In a ME layered sensor, the mechanical capacitance, C_{mech}, is given by the constitutive equations. It yields a relationship between the mechanical force, F, and the micro displacement, Δu, both along the length direction of the sensor, as $\Delta u = F\, C_{mech}$. This force, F, can be generated by either a magnetic or electric field, which depends on the following expressions $F_m = \varphi_m H$ or $F_p = \varphi_p E$ [6-7]. The mechanical capacitance can be deduced from the mean flexibility, s_{33}, the sensor length, l, and the cross section area, A. It yields: $C_{mech} = s_{33}\, l/A$. The sensor mean flexibility can be obtained by taking into account those of the magnetostrictive layer and the piezoelectric layer and the thickness ratio n, between two layers [8]. Thus, an equivalent displacement due to a magnetic and an electric field results

$$\varphi_m H\, C_{mech} = \Delta u = \varphi_p E\, C_{mech}. \tag{2}$$

Provided that the magnetic and piezoelectric phases match perfectly, the ratio between a magnetic field and an electric field is given by

$$\frac{H}{E} = \frac{\varphi_p}{\varphi_m}. \tag{3}$$

Taking account of the magnetic susceptibility, χ, the feedback factor, β, representing the flux-to-voltage ratio in (T/V), is

$$\beta = \frac{B}{V} = \frac{\mu_0 H}{E\, l_p} = \frac{\varphi_p \mu_0}{\chi \varphi_m l_p} \tag{4}$$

where $\chi\varphi_m = 15$ N/(A.m^{-1}), $\varphi_p = 3\times10^{-5}$ N/(V.m^{-1}) and $l_p = 1.5$ mm, the center-to-center distance between interdigital electrodes [3].The value of the flux-to-voltage ratio is $\beta = 1.7\times10^{-9}$ T/V. The term, $A/(1+A\beta)$, represents the feedback transfer function gain. This gain cannot be measured directly since the transfer function A combines the magnetoelectric coefficient $\alpha_{ME/EE}$ which is related to the sensed magnetic field. However, we can apply an input voltage, V_{inFB}, at the positive input of the charge amplifier (*cf.* Fig. 1), to evaluate this term, $A\beta$. Therefore, we have

$$A\beta \approx \frac{1}{3} \alpha^{NL}_{ME\_EE} G_1 G_2 G_3 \tag{5}$$

with the nonlinear voltage coefficient $\alpha^{NL}_{ME\_EE}$. The factor of 1/3 is due to voltage divider induced by the resistances. The two equations (1) and (3) yield the predicted transfer function with a feedback loop as

$$\frac{A}{1+A\beta} \approx 3\frac{\alpha^{NL}_{ME\_EE}}{\alpha^{NL}_{ME\_ME}} \approx \frac{1}{\beta}. \tag{6}$$

Up to certain point, the feedback loop permits to increase the bandwidth. Admitting that the cutoff frequency of the open loop transfer function is f_{c_OL}, we can deduce the closed loop bandwidth f_{c_FB}, with a feedback transfer function

$$T_{rFB}(f) = \frac{T_{rOL}(f)}{1+T_{rOL}(f)\beta} = \left(\frac{A}{1+A\beta}\right)\frac{1}{1+j\dfrac{2\pi f}{2\pi f_{c\_FB}}} \quad \text{with } f_{c\_FB} = f_{c\_OL}(1+A\beta). \tag{7}$$

2.1.2 Equivalent magnetic noise (EMN)

By using the *ME/EE* modulation technique on a ME sensor in longitudinal vibration mode, the noise from the excitation source and the amplification electronics dominate over the noise performances of the sensor [9]. In the feedback loop, the voltage divider consisting of the three resistances, *R*, has little noise contribution since the associated voltage noise is small compared to other voltage sources. Presently, we believe that the dominant noise sources are originating from the excitation and amplification electronics. Thus, the total output voltage noise can be expressed as [2],

$$e^2_{nOL}(f) = G_I^2 G_2^2 G_3^2\left(\left|\frac{R_1}{1+j2\pi R_I C_I f}\right|^2\left(i_{nRI}(f)^2 + i_{namp}(f)^2\right) + \left|\frac{C_I + C}{C_I}\right|^2\left(e_{namp}(f)^2 + e_{nexc}(f)^2\right)\right) \tag{8}$$

where $i_{nRI}(f)$ is the current noise source of the resistance R_I, $i_{namp}(f)$ and $e_{namp}(f)$ are, respectively, the equivalent input current and the voltage noise sources of the charge amplifier. $e_{nexc}(f)$ is the noise source associated to the driving carrier voltage, V_{exc}. Thus, the output voltage noise spectral density and the transfer function with a feedback loop is defined by

$$e_{nsFB}(f) = \frac{e_{nOL}(f)}{1+A\beta}. \tag{9}$$

The EMN of this system is defined by the ratio of the output voltage noise, $e_{nOL}(f)$, and the magnetic signal transfer function, $T_{rOL}(f)$. In open loop mode, EMN can be directly evaluated from the output electric noise and the transfer function. Also, the EMN in feedback loop mode is expected to be equal to

$$b_{nFB}(f) = \frac{e_{nFB}(f)}{T_{rFB}(f)} = \frac{\dfrac{e_{nOL}(f)}{1+A\beta}}{\dfrac{T_{rOL}(f)}{1+A\beta}} = \frac{e_{nOL}(f)}{T_{rOL}(f)} = b_{nOL}(f). \tag{10}$$

3. Experimental results

Several experiments were performed by applying a low frequency magnetic reference signal of two hertz. The latter was applied by a pair of Helmholtz coils in series with a resistance of 1 kΩ and having a transfer function of 6.8×10^{-4} T/A. The ME composite senses this magnetic field and induces a modulated signal which appears at the charge amplifier output. The voltage excitation carrier has a frequency of 25.3 kHz corresponding to the longitudinal sensor resonant frequency. The output signal from the charge amplifier is applied to the positive input of a differential amplifier presenting a high pass filter behavior having a voltage gain of 10 and a cut off frequency of 160 Hz. A sinusoidal voltage, which is synchronized in phase and in frequency with that of the excitation carrier, is connected to the negative input of this subtractor to reduce the output carrier amplitude present in U_{ex}, as shown in Fig. 1. This is a critical step before the demodulation process since a too large carrier level prevents the noise measurements when using a spectrum analyzer. After demodulation and a further amplification through an amplifier G_3, the signal is feedback towards the positive output of the charge amplifier. The schematic of the setup is given in figure 1.

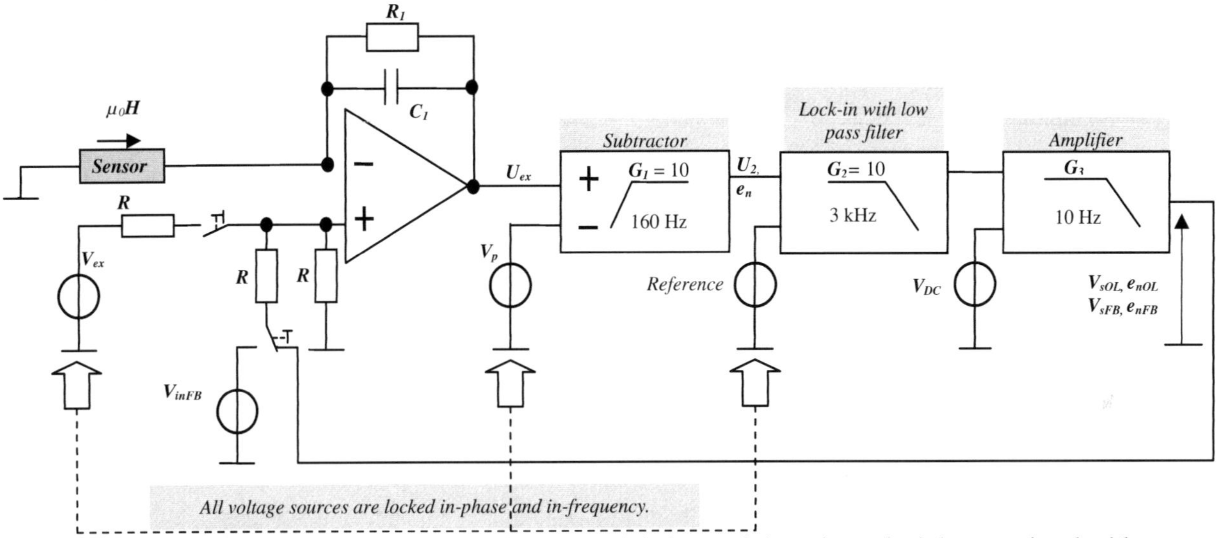

Fig. 1: Schematic of the magneto-capacitance modulation and detection principle, associated with an electric field feedback (C_l = 100 pF, R_l = 10 GΩ).

In order to characterize the sensing performance with the feedback loop, we measure each of the following transfer functions: the open forward gain *A*, the gain *Aβ* of the feedback channel and the closed gain with the feedback loop *A* / (1+*Aβ*). The pair of Helmholtz coils is used to generate the sinusoidal-sweeping magnetic signal along the length direction of the sensor.

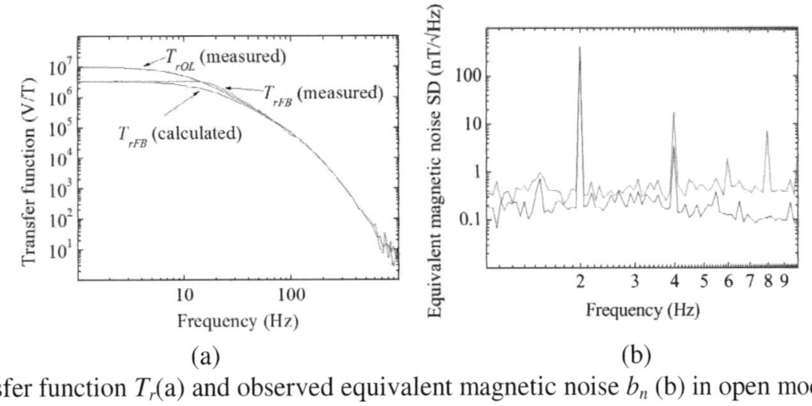

(a) (b)

Fig. 2: Transfer function T_r(a) and observed equivalent magnetic noise b_n (b) in open mode (black) and feedback loop (red). In (a), the blue curve is the evaluated response.

By measuring the ratio between the output voltage and the input reference, one can separately measure the values of *A* and *A* / (1+*Aβ*) for a gain of G_3 = 50, as shown in Fig. 1.The transfer function of the feedback channel, *Aβ* can be deduced from *A* and *A* / (1+*Aβ*) by taking into account the equality $T_{rFB\_mes} = \left(\frac{A}{1+A\beta}\right)$. It yields

$$A\beta = \frac{1}{T_{rFB\_mes}} - 1 \tag{11}$$

where the index *mes* refers to measured values. The experimental and expected performances are compared in Fig. 2. According to our previous studies, the nonlinear charge coefficient $\alpha_{ME\_EE}^{NL}$ for this given sample has a value around 250 nC/T at low frequencies for a voltage excitation of V_{exc} = 400 mV_{pk}. Both the calculated and measured transfer functions in the feedback mode show a strong coherence with our previous measurements.

Conclusions

We have investigated the magnetic transfer function, the noise performance of a ME sensor working under modulation techniques, relying on an electric excitation and an electric feedback in the detection system in order to improve the linearity and the dynamic range. The performances between the measured values and the predicted ones have been compared. The gains and bandwidths have been found similar. The results give the proof of concept of the method and yield a way for the development of very low power magnetometers. To date, the performances of the feedback loop are limited by the nonlinear ME charge sensor coefficient, $\alpha_{ME\_EE}^{NL}$, mainly.

Acknowledgment

This work is supported by the ONRG (Office Naval Research Global) and the region of Normandy.

References

[1] Xing Z, Zhai J, S Dong, Li J, Viehland D and Odendaal G 2008 *Meas. Sci. Technol.* **103** 033903
[2] Zhuang X, Lam Chok Sing M, Cordier C, Dolabdjian C, Das J, Gao J, Li J, Mclaughlin K and Viehland D 2011 *IEEE Sensors J.* **11** 2183-88
[3] Zhuang X, Lam Chok Sing M, Cordier C, Dolabdjian C, Wang Y, Li J and Viehland D 2015 *IEEE Trans. Magn.* **51** 10.1109
[4] Reerman J, Zabel S, Kirchhof C, Quandt E, Faupel F and Schmidt G 2016 *IEEE Sensors J.* **16** 4891-4900
[5] Wang Y, Gray D, Berry D, Gao J, Li M, Li J, Dwight V 2011 *Adv. Mater.* **23** 4111-14
[6] Dong S, Zhai J, J. F. Li J F, and Viehland D 2006 *App. Phys. Lett.* **89** 243512
[7] Bichurin M I, Petrov V M, and Srinivasan G 2002 *J. Appl. Phys.* **92** 9681
[8] Zhuang X, Lam Chok Sing M, Dolabdjian C, Finkel P, Li J et Viehland D 2014 *Key. Eng. Mat.* **605** 344
[9] Zhuang X, Lam Chok Sing M and Dolabdjian 2012 *IEEE. Trans. Magn.* **49** 120-3
[10] Zhuang X, Lam Chok Sing M, Dolabdjian C, Finkel P, Li J et Viehland D 2014 *IEEE. Sensors J.* **14** 150-8

Design and development of a magnetometer calibration device

S Angelopoulos[1]
School of Electrical and Computer Engineering,
Laboratory of Electronic Sensors,
National Technical University of Athens, Zografou Campus, Athens, 15780, Greece

E-mail: spyrosag@central.ntua.gr

Abstract. This paper describes the development of a new magnetometer calibration device, which is able to provide accuracy of 1fT/√Hz. The mentioned device is able to eliminate the ambient magnetic field, using an active shielding technique. This can be achieved by the use of a solenoid or a pair of Helmholtz coils. In order to measure the existing magnetic field, it is necessary to develop and use accurate magnetometers with amorphous ribbons as core magnetic materials. The whole system works as a closed-loop system, which is able to control and adjust the produced counter magnetic field.

1. Introduction

Magnetometers are used in a variety of applications, in a wide range of fields, which include sciences, transportation and industry [1-3]. As a result, it is mandatory to be able to maintain their accuracy at the required level. This is achievable by the use of a magnetometer calibration device. Such a device, is able to eliminate the ambient magnetic field, in order to produce a space where controllable magnetic measurements can be conducted.

The elimination of the ambient magnetic field is called magnetic shielding and is divided into two different methods; the passive magnetic shielding and the active magnetic shielding. The main parameter that differentiates the two methods is the frequency of the magnetic field that they are able to eliminate. The high frequency magnetic field can be eliminated more easily using passive magnetic shielding methods. These techniques are based on the use of materials with high magnetic permeability, such as aluminium or mumetals. Those materials are used as a cover, shielding a certain space. As a result, the external magnetic field "prefers" the easiest way of transmission, which is through the materials with the highest magnetic permeability, ignoring the internal space which they cover. The disadvantages of passive magnetic shielding, except for the frequency range, is the requirement of a thorough covering of the space where the measurements will be conducted and the cost of the shielding [4-7].

In order to avoid passive shielding techniques, active shielding can be used. These techniques are based on the generation of a counter-acting magnetic field, which will be able to oppose to the ambient magnetic field. The main advantage of the active magnetic shielding is its ability to be adapted via a closed-loop system to the variations of the external magnetic field. A sensing element, e.g. a magnetometer, can provide the input of the system, in order to precisely adapt the required opposed

[1] Tel: +30 2107723737, E-mail: spyrosag@central.ntua.gr

magnetic field. As a result, the active shielding techniques can be used even in the case of either DC magnetic fields, or low frequency magnetic fields.

2. Operating principles

The main purpose of the magnetometer calibration device is the zeroing of any offset of the magnetometer's reading, in order to correlate its output voltage value to the magnetic field's strength value. This is possible to be applied on more than one axes. In this case, it is required to use three perpendicularly placed magnetometers, in order to measure the strength of the magnetic field in each direction.

The accuracy of the measurement of the existing magnetic field depends on the choice of the type and the design of the magnetometer, which will be used as the sensing element of the device. More precisely, the choice of the core material of the magnetometer is a matter of great significance. The best choice is to use amorphous magnetic ribbons or wires as core materials, such as FeSiB or CoFeSiB, which are able to provide rapid output response to the changes of the magnetic field [8-10].

The calculation of the offset value of the magnetometer is based on the measurement of a time difference. Specifically, Figure 1 resembles the output signal of the receiving coil of the magnetometer, in the case of absence of any external magnetic field. The excitation signal of the magnetometer is sinusoidal, however, there are positive and negative peaks at the received signal. These peaks indicate that the core material of the magnetometer has reached its positive or negative magnetic saturation value, because of the existence of a magnetic field, which is produced by the excitation coil. It is obvious that the time difference between a positive and the following negative peak is the same, which can be explained by the use of sinusoidal excitation signal and the hysteresis loop of the soft amorphous magnetic material.

Figure 2 represents the received signal in the case of presence of an external magnetic field. It is easily observed that the minima and the maxima of the output signal have been changed, despite the fact that the excitation signal is still the same. This is due to the reduction of time which is required for the core material to reach its magnetic saturation value, because of the additional magnetic field strength, which affects it.

As a result, a device which is able to accurately measure and eliminate the time difference between the positive and the negative peaks of the output signal, can lead to the elimination of the undesirable magnetic field.

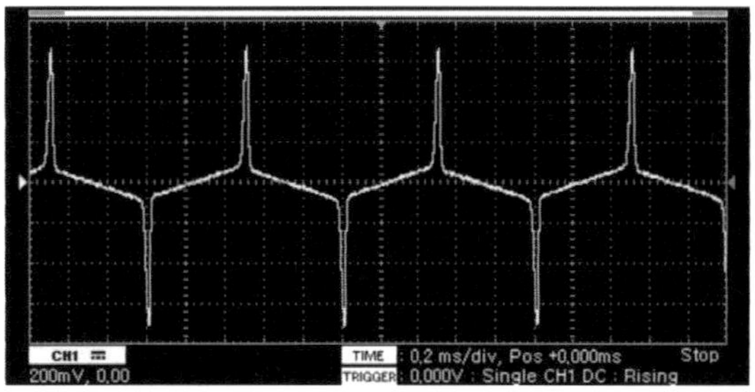

Figure 1. Fluxgate sensor's signal for zero external magnetic field.

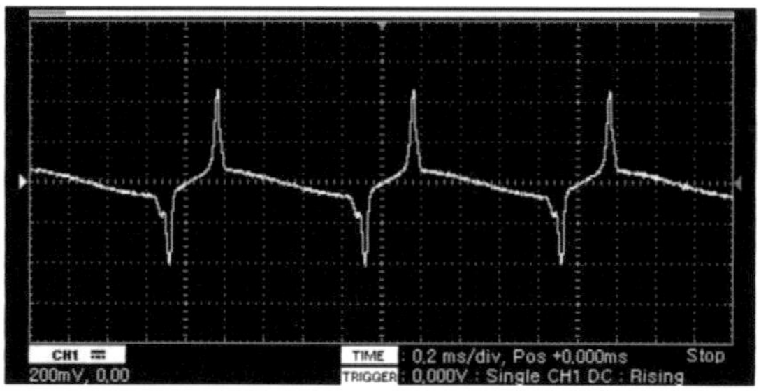

Figure 2. Fluxgate sensor's signal during the presence of an external magnetic field.

3. Development of the magnetometer calibration device

3.1. The arrangement

The chosen arrangement is based on the active shielding method of calibration. It includes a signal generator, which produces the excitation signal, a magnetometer as a sensing element, a solenoid as a magnetic shielding device and an oscilloscope, in order to view the final signal. The above-mentioned parts will be described and analysed below.

3.2. Development of the magnetic shielding device

As it was mentioned before, the calibration device is based on the active shielding method. This means that the magnetic shielding is created by the generation of a magnetic field opposed to the externally forced magnetic field. The generation of this magnetic field is achieved by the use of a solenoid. The developed solenoid, which was used during the experiments, was made by two layers of 290 windings each, made of Ø1mm enamelled copper wire. The overall length of the solenoid is 35cm and its radius is 10cm. The space surrounded by the solenoid acts as the magnetometer calibration space, where the magnetic field can be precisely controlled, through the adjustment of the supplied DC current, produced by a controllable power supply unit.

3.3. Development of the magnetometer

The developed magnetometer was designed as a fluxgate sensor. The chosen design was based on the literature [11-17]. Its excitation and receiving coils were made by 2 layers of 550 windings each, using Ø0.1mm enamelled copper wire. The chosen core magnetic material was a wire of FeSiB (Ø111μm) [18-21]. The total length of the sensor is 7cm and its diameter is 3.2mm. The final sensor was placed on a perfboard.

4. Experimental results & Conclusions

In order to conduct the experimental procedure, the sensor was placed inside the solenoid. The solenoid was connected to a power supply unit. Furthermore, the excitation coil of the sensor was connected to a function generator and its receiving coil to a digital oscilloscope.

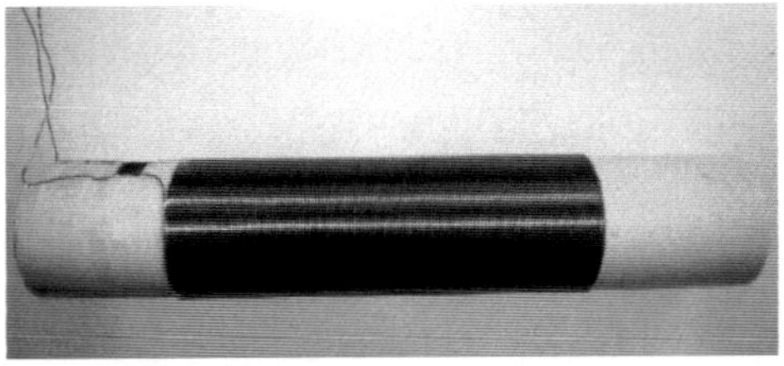

Figure 3. The developed solenoid.

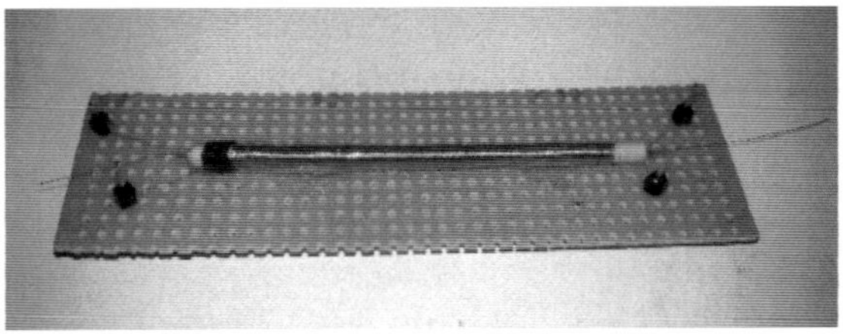

Figure 4. The fluxgate sensor.

Using a sinusoidal $5V_{pp}$, 1kHz excitation signal, the received signal was the one shown in Figure 1. The signal was changed when we supplied the solenoid with DC current, through the power supply unit. As a result, it was possible to capture different output signals, based on the different supplied current values. Figure 5 expresses the measured time difference between the sequential maxima and minima of the received signal, versus the supplied current to the solenoid. The results were similar when the polarity of the supplied DC current was reversed. As it was expected, the relation between the mentioned time difference and the supplied current values, is linear.

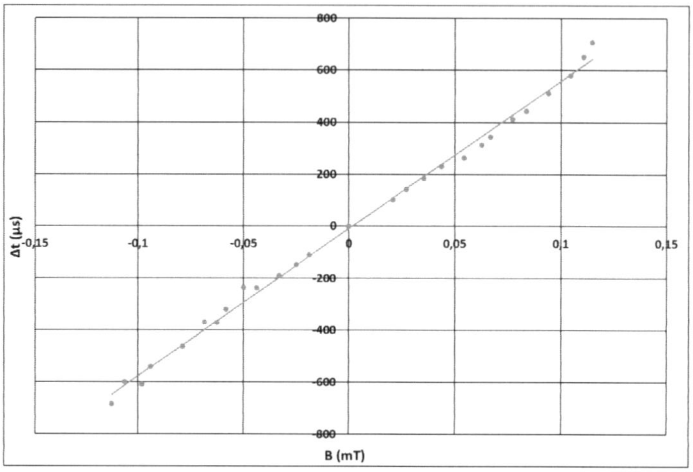

Figure 5. Correlation between the magnetic field and the time difference of the peaks.

As a result, we conclude that it is possible to measure the time difference between the peaks of the received signal, based on the operating principles which were explained above, in order to calibrate a magnetometer. The main parameters of the developed device, which can be further improved, are the magnetic shielding space, which can be developed as a 3D arrangement of Helmholtz coils and the time measurement and current supply system, which can be constructed as a highly accurate closed-loop system.

5. References

[1] Hristoforou E 2002 *J. Opt. Adv. Mat.* **4** 245-260
[2] Hristoforou E, Hauser H and Dimitropoulos PD 2006 *IEEE Sensors* **6** 372-379
[3] Hristoforou E and Ktena A 2007 *J. Magn. Magn. Mater.* **316** 372-378
[4] Kelha V, Peltonen R, Penttinen A, Ilmoniemi R and Heino J 1982 *IEEE Trans. On Magn.* **18** 260-269
[5] Harakawa K, Kajiwara G, Kazami K, Ogata H and Kado H 1996 *IEEE Trans. On Magn.* **32** 5226-5229
[6] Tashiro K and Sasada I 2005 *IEEE Trans. On Magn.* **41** 4081-4083
[7] Knappe-Grüneberg S, Schnabel A, Wuebbeler G and Burghoff M 2008 *J. of Appl. Physics* **103** 925-927
[8] Ripka P 2000 *J. of Magn. and Magn. Mat.* **215/216** 735-739
[9] Benyosef L C, Stael G and Bochner M 2008 *Mat. Res.* **11** 145-149
[10] Razmkhah S, Eshraghi M J, Forooghi F, Sarreshtedari F and Fardmanesh M 2011 *El. Eng. ICEE*
[11] Hristoforou E, Chiriac H and Neagu M 1997 *IEEE Trans. Instr. & Meas.* **46** 632-635
[12] Dimitropoulos P, Avaritsiotis J N and Hristoforou E 2001 *Sens. and Act. A* **90** 56-72
[13] Dimitropoulos P D, Avaritsiotis J N and Hristoforou E 2003 *Sens. and Act. A* **107** 238-247
[14] Petridis C, Petrou I, Dimitropoulos P D and Hristoforou E 2007 *Sens. Let.* **5** 98-101
[15] Petridis C, Ktena A, Laskaris E, Dimitropoulos PD and Hristoforou E 2007 *Sens. Let.* **5** 93-97
[16] Petridis C, Dimitropoulos PD and Hristoforou E 2009 *IEEE Sens. J.* **9** 128-134
[17] Petrou J, Skafidas J and Hristoforou E 2013 *Sens. Let.* **11** 91-95
[18] Hristoforou E and Reilly R E 1991 *J. Appl. Phys.* **69** 5008-5010
[19] Hristoforou E and Niarchos D 1993 *IEEE Trans. Magn.* **29** 3147-3149
[20] Hristoforou E, Chiriac H, Neagu M, Darie I 1994 *J. Phys. D: Ap. Ph.* **27** 1595-1600
[21] Hristoforou E, Chiriac H and Nagacevschi V 1999 *Sens. & Act. A* **76** 442 – 447

IOP Conf. Series: Journal of Physics: Conf. Series **939** (2017) 012036 doi:10.1088/1742-6596/939/1/012036

Design and development of a magnetic device for mesenchymal stem cell retaining in deep targets

G. C. Banis
Dpt. of Electrical & Computer Engineering,
National Technical University of Athens, Zografou Campus, Athens, 15780, Greece
E-mail: gbanis@central.ntua.gr

Abstract. This paper focuses on the retaining of mesenchymal stem cells in blood flow conditions using the appropriate magnetic field. Mesenchymal stem cells can be tagged with magnetic nanoparticles and thus, they can be manipulated from distance, through the application of an external magnetic field. In this paper the case of kidney as target of the therapy is being studied.

1. Introduction

Super paramagnetic iron nanoparticles (SPIONs) and magnetic techniques are being used for various applications and play a significant role in the field of drug delivery [1-6]. They can be used as a means of carrying a drug or a molecule to a specific site due to their magnetic susceptibility, via the force applied by an external magnetic field, while monitoring their position by magnetic field sensors [7-11].

Mesenchymal stem cells' positive effects have been proved and their potential uses can be countless due to their regenerative character. However, it is crucial to guide them precisely at the target otherwise, other vital organs and tissues are susceptible to teratogenesis, cancers and other unwanted effects [12-18].

2. Theoretical background

The force applied by a magnetic field with magnetic field intensity H at a magnetic nanoparticle of radius α(m) and magnetic susceptibility χ is:

$$\overrightarrow{F_M} = \frac{4\pi\alpha^3}{3}\frac{\mu_o\chi}{(1+{}^{\chi}/_{3})}\vec{H}\frac{d\vec{H}}{d\vec{x}} = \frac{2\pi\alpha^3}{3}\frac{\mu_o\chi}{(1+{}^{\chi}/_{3})}\nabla\left(\left|\vec{H}\right|^2\right) \quad \text{(eq. 1)}$$

The magnetic susceptibility of the SPIONs is $\chi\sim20$ and that of the human body is $\chi\sim10^{-6}$-10^{-4} hence, the efficiency of the magnetic field on the SPIONs is not being remarkably affected by the human tissues [19-20]. The small radius of the magnetite assures a super paramagnetic behavior under the influence of the magnetic field. Therefore, in the absence of magnetic field, the nanoparticles are not magnetized and vice versa. The magnetic force depends strongly on the size of the nanoparticles. Furthermore, the magnetic force is being increased accordingly to the increase of the magnetic field gradient; the steeper is the gradient, the bigger the force applied on a nanoparticle [21-23]. It is worth notifying that in stem cells, the SPIONs are aggregated into lyssosomes and create clusters of a few μm and thus, the force applied on them is bigger than that applied on one nanoparticle [24-30].

3. Experimental components

The experimental setup consisted of two main components. First, the electromagnet, which had the characteristics shown in figure 1. A KEPCO BOP 72-6M was used to apply 32V at the coils. Secondly, regarding, the circulatory system, a mix of ethanol and oleic acid was used for the dispersion of the SPIONs. The pump used, provided 833ml/min which is approximately 200ml/min more than the blood flow in each renal artery. The tubes were 2400mm long, with inner diameter 5 mm, approaching the diameter of the blood vessels.

Gap	100*100*80mm^3
Magnetic circuit length	870mm
Core surface	10.000mm^2
H(A) from BH curve for B=1T	200A/m
Current Intensity J	2,5(A/mm^2)
Copper wire diameter	2mm
.Number of turns/coil N	1250
Resistance/coil R	4,12Ω
I*N at 6^A	15000
Max magnetic field at the gap B	0,23T

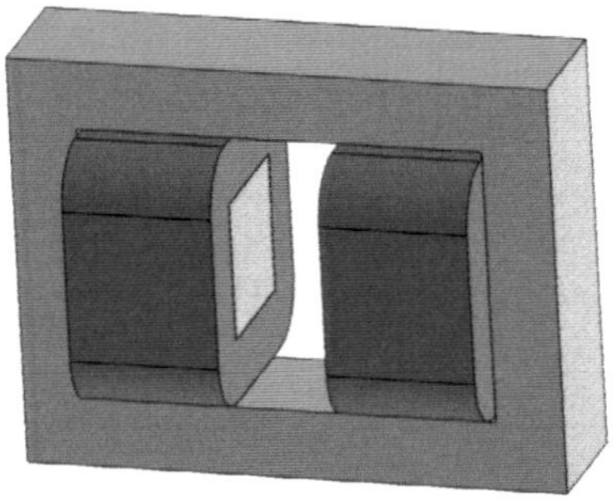

Figure 1: Characteristics and shape of the electromagnet

4. Experimental setup

The purpose of the experiment was to locate the areas and define the conditions under which the SPIONs could be retained in the gap of the electromagnet. The tubes were placed at specific positions in the gap between the poles of the electromagnet and after a few circles of the circulatory system, the flow was being shut off so that measures could be obtained. The tube was shaped as a loop and four sets of measurements were conducted. At each location four flows were tested: Upwards (U), downwards (D), clockwise (C) and counterclockwise (CC).
During the first experiment the loop was in touch with the corner of the right core of the electromagnet and the concentration of the SPIONs was rich, as expected. The second experiment was conducted 3mm away from the same core. The concentration was still rich, while upward and clockwise flows provided the best results. The third experiment was conducted 6mm away from the core and the concentration was lower in all the cases although, a higher concentration could be observed at the same flow cases (upwards and clockwise). Similar results were provided at 9mm distance while, at 12mm, the SPIONs retained were significantly decreased. The direction of the flow played a major role on the effectiveness of each trial. More specifically, at the upward flow, the fluid pressure drop was gradually increasing at the direction of the flow, due to the vertical elevation ($\Delta p = \rho \times g \times \Delta H$) (eq. 2) and thus, the force from the flow was decreasing as well, providing a more efficient retaining of the SPIONs, comparing with the downwards flow.
At the clockwise direction the SPIONs, after passing through the curve, which created a pressure drop of the flow, they were approaching a higher magnetic field. Due to their decreased velocity and thus, decreased force from the fluid flow, the magnetic force provided more effective results, in contrary with the counter clockwise flow.

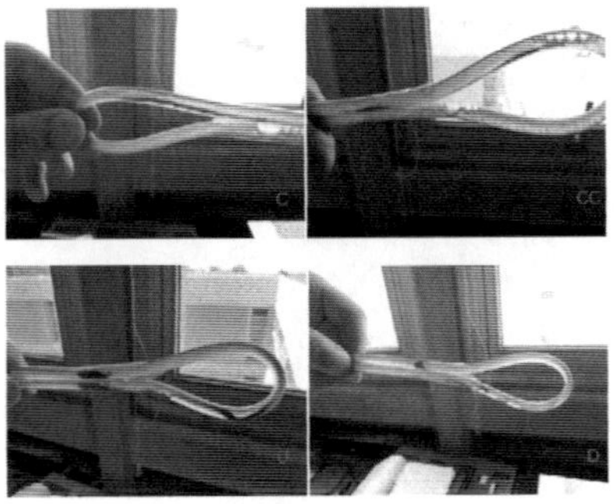

Figure 2: Results of magnetic field for c) clockwise cc) counter clockwise u) upwards d) downwards

The next step was to add magnetic tips at the poles of the electromagnet in order to provide a steep gradient and higher magnetic field intensity in specific locations of the gap.

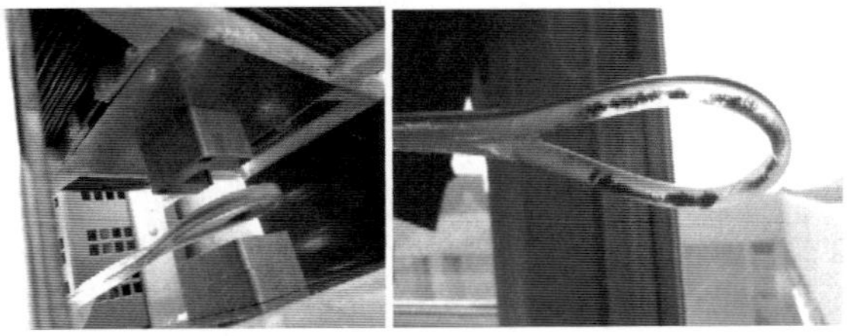

Figure 3: Magnetic tips on the core and results of the generated magnetic field

The magnetic field provided by the new geometry enabled SPIONs to be retained at 30mm away from the cores in contrast to 9mm which was achieved previously using the homogeneous magnetic field.

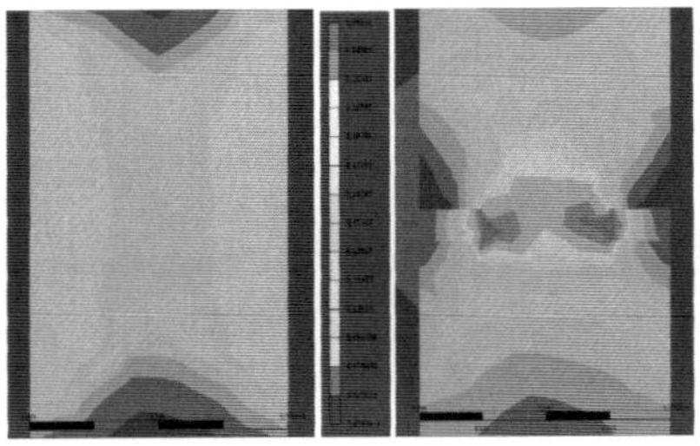

Figure 4: Magnetic field intensity a) without magnetic tips and b) with magnetic tips

5. Conclusions

The effectiveness of the magnetic field regarding the retaining of the SPIONs, was quantified optically. In order to retain SPIONs, the magnetic force on a SPION (eq. 1) should overcome the force of the fluid flow derived from Stoke's Law [31-34]:

$$F = 6\pi v R\vec{v} \qquad \text{(eq. 3).}$$

The gradient of the magnetic field plays a significant role in the case, as seen from eq. 1. However, gradient can be provided either by the magnetic field heterogeneity or by the fluid flow stream which add up.

By defining the centerline radius of bend and by knowing the internal diameter of the tube, it is possible to calculate the pressure drop at each point.

Applying this conclusion to a human kidney model, the curvy and multidirectional character of the blood vessels would provide major retaining efficiency. Hence, the locations that the mesenchymal stem cells would be retained and aggregated can be predicted and therefore the therapy could be customized, depending on the needs of each patient, by focusing the magnetic field at the area of the kidney where localized therapy is demanded [35-38].

6. References

[1] E. Hristoforou, 2002, J. Opt. Adv. Mat. 4, 2, p. 245-260.
[2] E. Hristoforou, H. Hauser, PD Dimitropoulos, 2006 IEEE Sensors, 6, p. 372-379,.
[3] E. Hristoforou, A. Ktena, 2007, J. Magn. Magn. Mater., 316, p. 372-378,.
[4] Q. A. Pankhurst, J Connolly, S K Jones and J Dobson 2003, *J. of Phys.D:.,36,13*
[5] B. Shapiro, S. Kulkarni, A. Nacev, S. Muro, P. Stepanov, I. Weinberg 2015, *Wiley Interdiscip Rev Nanomed Nanobiotechnol.* 7(3), 446–457
[6] G.S.E. Antipas, E. Statharas, P. Tserotas, N. Papadopoulos and E. Hristoforou, 2013, ChemPhysChem, 14 (9) , pp. 1934-1942,
[7] P. Dimitropoulos, J.N. Avaritsiotis, E. Hristoforou, 2001 Sens. and Act. A, 90, p. 56-72,
[8] P.Dimitropoulos, J.N.Avaritsiotis, E.Hristoforou, 2003 Sens. and Act.A, 107, p.238-247
[9] C. Petridis, I Petrou, PD Dimitropoulos, E Hristoforou, 2007, Sens.Let., 5, p. 98-101.
[10] C.Petridis, A.Ktena, E.Laskaris, PD Dimitropoulos, E. Hristoforou,2007, Sens L.,5, p. 93-97
[11] C. Petridis, PD Dimitropoulos, E. Hristoforou, 2009 IEEE Sens. J., 9, p. 128-134,.
[12] L. Yi, G. Yu, X. Chenjie, 2013, *Chin. Phys. B Vol. 22,9* 097503
[13] S. Kakar, D. Batra, R. Singh, U. Nautiyal, 2013, *Vol 2,1*
[14] E.P. Furlani, 2006, *J. Appl. Phys. 99, 024912*
[15] M. Kummer, J. Abbott, B. Kratochvil, R. Borer, A. Sengul, B. Nelson, 2010, *IEEE Trans.on Rob. 26(6):1006 - 1017*
[16] A. Lubbe, C. Bergemann, J Brock, D. McClure, 1999, *Phys. and Biophys, Springer*
[17] J. Ritter, A. Ebner, K. Daniel, K. Stewart, 2004, , *J. of Mag. and Mag. Mat. 280, Is.2–3*
[18] V.R Muzykantov, V.P. Torchilin, 2000, *Springer Science + Business Media, LCC*
[19] A. Sarwar, R. Lee, D. Depireux, B. Shapiro, 2013, *IEEE trans. mag.,49,1*
[20] B. Shapiro, Kulkarni, A. Nacev, A. Sarwar, D. Preciado, D. Depiraux, 2014, *Ann. Rev Biomed Eng.* 11;16:455-81
[21] B. Shapiro, K. Dormer, I. Rutel, 2010, *AIP Conference Proceedings* 1311(1):77-88
[22] F. Ullrich, S. Schurrle, R. Pieters, A. Dishy, S. Michels, B. Nelson, 2014, *IEEE (ICRA)*
[23] I. Hoke , C. Dahmani , T. Weyh, 2008, *Heinz-Nixdorf Lehrstuhl für Med El, T.U. München*
[24] J. Kolosnjaj-Tabi, C. Wilhelm, O. Clément, F. Gazeau, 2013, *J. of Nanobiotech. 11*(Suppl 1):S7
[25] Y.X. Wang, S. Xuan, M. Port, J.M. Idee, 2013, *Curr Pharm Des.* 19(37): 6575–6593
[26] Jasmin, A. L. Machado Torres, L. Jelicks, A.C.C. de Carvalho, D. Spray, R. Mendez-Otero, 2012, *Methods Mol Biol.* 906:239-52
[27] N. Landazuri, S. Tong, J. Suo, G. Joseph, D. Weiss, D. Sutcliffe, D. Giddens, G. Bao, W.R. Taylor, 2013, *Small.* 9;9(23):4017-26
[28] J. Cores, T. Caranasos, K. Cheng, 2015, *J. Funct. Biomater.* 6, 526-546;
[29] R. Engel-Herbert T. Hesjedal, 2005, *J. Appl. Phys.* 97, 074504
[30] A. Grief, G. Richardson, 2005, , *J. of Magn. and Magn. Mat.* 293 455–463
[31] N. Pamme, 2006, *Lab Chip.* Jan;6(1):24-38.
[32] P.A. Voltairas, D.I. Fotiadis, L.K. Michalis, 2002, *J. of Biomech.* 35(6):813-21
[33] L. Agiotis, I. Theodorakos, S. Samothrakitis, S. Papazoglou, I. Zergioti, Y. Raptis, 2015, *J.of Magn. and Magn. Mat.* 40 1 956–964
[34] E.K. Ruuge and A.N. Rusetski, 1993, J.*of Mag. and Mag. Mat.* 122 335-339
[35] A Nacev, C Beni, O Bruno & B Shapiro, 2010, Nanomed. (9):1459-66
[36] T. Wei, T. Jianping, L.Yunlong, T. Zhuo, 2012, *Magnetohydrodynamics 48*, No. 3, 543-556
[37] J. Riegler, A. Liew, S, Hynes, D, Ortega, T. O'Brien, R. Day, T. Richards, F. Sharif, Q. Pankhurst, M. Lythgoe, 2013, *Biomaterials* 34 1987e1994
[38] Angelo Ferraro, 2016, IOP Conf. Ser.: Mater. Sci. Eng. 108 012021

IOP Conf. Series: Journal of Physics: Conf. Series **939** (2017) 012037 doi:10.1088/1742-6596/939/1/012037

Electrodeposited iron-based nanofoams as precursors for transducers

Kata Berkesi[1]

School of Electrical and Computer Engineering, Group of Electronic and Magnetic Materials, National Technical University of Athens, 9 Herron Polytechniou Street, Zografou Campus, Athens, 157 80, Greece

kberkesi@metal.ntua.gr, kberkesi@gmail.com

Abstract. The aim of this work is the electrochemical preparation of iron-based oxide (Fe_2O_3, Fe_3O_4) nanofoams on metal substrate by accompanying hydrogen-evolution as promising precursors for transducers. Various metal substrates have been tested and the most satisfying electrolyte composition have been determined. However, various deposition parameters determine various nanofoam properties, like pore-size, density, pore-type, thickness of interconnections etc. The crystallographic composition and morphology of the electrodeposited samples have been studied by X-ray diffraction (XRD) and scanning electron microscopy (SEM), respectively.

1. Introduction

Semiconductor nano- and microstructures may have great utility in transducers industry. For instance, various sensing applicability can be considered of magnetite regarding to their magnetic or superparamagnetic response of external electrical [1,2] and magnetic field [3]. The fast and well described abreaction in change of resistance by changing the temperature make iron-oxides great candidates as thermistors [4,5]. The usability of iron-oxide semiconductors as precursors of transducers are not well established [6 and references wherein]. Even we talk about sensors, thermistors, actuators [7-14] etc., there are several physical parameters [15] concerning the structural and morphological properties that should be satisfied. The main physical parameters regarding the structure of the semiconducting materials include but not are limited to high porosity, high specific surface area, open porous for fast transport of gas/liquid and piezoelectric potential. Different physical and chemical preparation techniques are available for the fabrication of nano- or microscale structures considering layers, wires, tubes and foams. Widespread and most popular preparation methods are sputtering [16,17], evaporation [18,19], chemical vapor deposition [20-25] and electrochemical deposition [26-31]. To bring together the requirements with the given facilities a three dimensional (3D) foam-like nanostructure would mean a distinguished solution. The nanofoams exhibit many excellent properties to the respect of the above listed requisites. The development of the state of the art preparation procedures includes a novel preparation technique of the iron-based nanostructures according to the requirements of increase in efficiency [6]. The electrochemical preparation of nano- and microscale metalfoams, metal alloy foams or metal oxide foams has been started in the last decade, namely it can be considered as novel method. Also, it can emulate with the other techniques

[1] Tel: +30 210 772 1534, E-mail: kberkesi@gmail.com

IOP Conf. Series: Journal of Physics: Conf. Series **939** (2017) 012037 doi:10.1088/1742-6596/939/1/012037

well due to its low-costs, good versatility of parameters in terms of desired properties and simple handling [26].

2. Theoretical Background of electrodeposition using bubbles

The Dynamic Hydrogen Bubble Template (DHBT) is an excellent method for the fabrication of macrostructures with nanoporous side walls [32,33]. The method has been named from the phenomena when a series of hydrogen-bubbles generated on the surface of the substrae act like a dynamic negative template. Figure 1 demonstrates the theoretical approach of the technique. Fig 1(a) Where the hydrogen-evolution takes place on the surface there is no metal/metal-oxide deposition. The size of the bubbles increasing along the distance from the substrate surface. [34] Fig 1(b) shows the SEM-image of copper nanofoams prepared by DHBT [35].

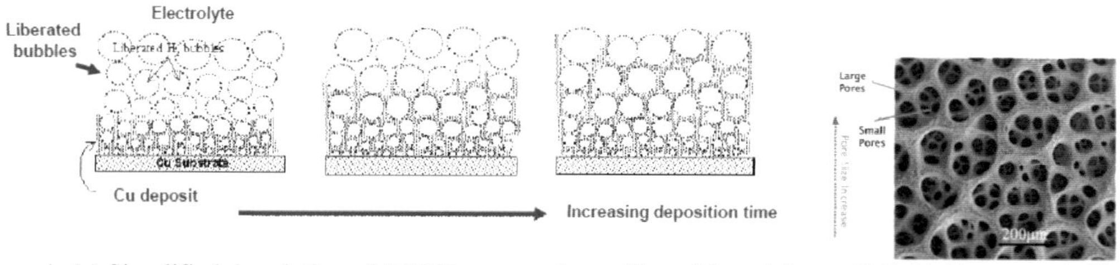

Figure 1. (a) Simplified description of DHBT generated metal/metal-based foams [34] (b) Honeycomb-like Cu-deposit structure made by DHBT [35].

2.1. *Materials and Methods*

The nanostructures were electrodeposited onto copper substrate. The substrates were mechanically polished with 800 and 2000 emery papers, respectively. Henceforward, the polished substrates were ultrasonically cleaned in deionized water, ethanol and acetone, respectively [28]. The 2 cm^2 active surface area of the substrates were ensured by commercial nail-polish and insulating tape. The galvanostatic deposition was carried out in a two electrode cell using a DC Power Supply. Platinum-mesh was used as counter electrode. The electrolytic bath was a strongly alkaline solution with pH 12.5 [27] made of 0.037 M $Fe_2(SO_4)_3$ 0.1 M triethanolamine (TEA) and 2 M NaOH. After mixing the components following the described protocol [29] at T = 80 °C the solution had greyish-green color and no turbidity was observed. The parameters of the deposition process were varied in the terms of current density, bath temperature and deposition time. The range of cathodic current density was chosen j = 5-15 mA.cm^{-2}, the depostion time was given from 5 min to 6 hours and the bath temperature was adjusted T = 60-90 °C. The nominal thicknesses of the prepared structures were calculated by using the Faraday-law [36].

The surface morphology of the deposited layers were studied by scanning electron microscope (Nova NanoSEM 230) and the crystallographic characterization was determined by X-ray diffraction (Bruker D8 Advance, Cu radiation-40kV, 35mA).

2.2. *Results and Discussion*

Several samples were deposited with different conditions. The preferred deposits checked by optical microscope (ZEISS) were sent for structural and morphological characterization.

2.2.1. *Scanning electron microscopy*

Figure 2 demonstrates 4 different SEM-images of the prepared samples. Fig 2(a) was taken on a sample prepared in 300 s deposition period in 60°C bath with -7.5 mA cm^{-2} current density, and the nominal thickness was 2.24 μm. In Fig 2(b) can be seen the morphology of the sample prepared in 300 s at 70°C with-5 mA cm^{-2} current density with nominal thickness of 23.55 μm. Fig 2(c) shows the SEM-result of the sample deposited with the same bath temperature and current density as 2(b) but extended deposition time: Δt = 900 s (d_n = 30.28 μm). Fig 2(d) represents the SEM-image of the sample prepared in the highest bath temperature 90°C with 5 min deposition time, -7.5 mA cm^{-2} current density (d_n = 4.42 μm).

The SEM-images clearly show the structural changes by increasing the bath-temperature. Filamentous deposit structure has been observed at 90 °C. This flake-like structure was observed in the corresponding literatures [30,31], where the authors conclude that the generation of the filaments is in function of more negative deposition potentials but not of the bath temperature. Due to the galvanostatic deposition mode the potential values have been not recorded. For better understanding of the structural formations potentiostatic deposition mode should be done.

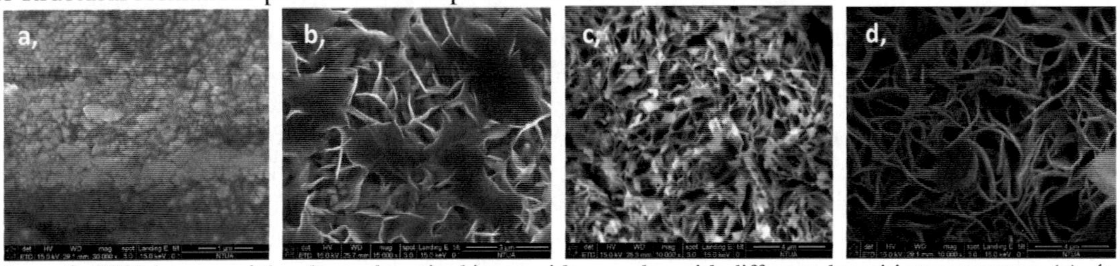

Figure 2. SEM-images of the electrodeposited iron-oxide samples with different deposition parameters (a) Δt = 300 s, T = 60°C, j = -7.5 mA cm^{-2} at 30000X magnification (b) Δt = 300 s, T = 70°C, j = -5 mA cm^{-2} at 15 000X magnification(c) Δt = 900 s, T = 70°C, j = -5 mA cm^{-2} at 10000X magnification (d) Δt = 300 s, T = 90°C, j = -7.5 mA cm^{-2} at 10000X magnification.

2.2.2. X-ray diffraction

Crystallographic study of th e iron-oxide samples were done. Figure 3 represents the identification of the observed peaks. Fig 3(a) shows the results of sample prepared in 3300 s in 70 °C bath with -5 mA.cm^{-2} current density. Grazing incidence XRD patterns of this sample indicate the presence of magnetite (blue lines) and copper substrate (green lines). Grazing angle was 2 (red curve) and 6 degrees (blue curve), respectively. In Fig 3(b) the results of sample with same conditions as in Fig 3(a) can be seen, but in this case the current density was -9 mA.cm^{-2}. Here, grazing incidence XRD patterns indicate iron (grey lines) and copper (green lines). Grazing angle was 2 (red curve) and 6 degrees (blue curve), respectively. Fig 3(c/1) and (c/2) show the results of XRD taken from the same sample with deposition conditions of 2 hours deposition time in 85 °C bath and -12.5 mA.cm^{-2} current density.

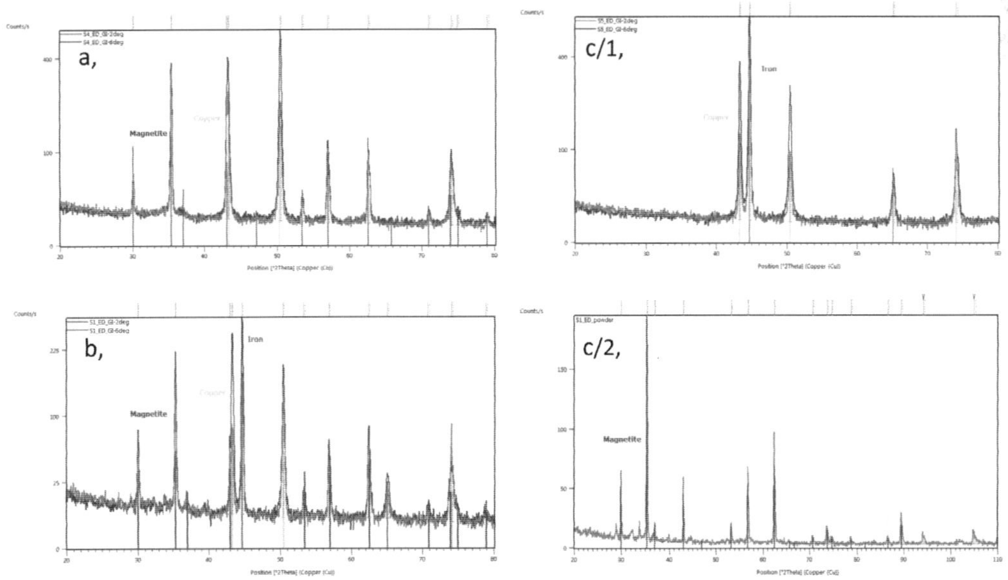

Figure 3. X-ray diffractograms of the iron-oxide electrodeposited samples prepared under different conditions. a, Δt = 3300 s, T = 70 °C, j = -5 mA cm^{-2}; b, Δt = 3300 s, T = 70°C, j = -9 mA cm^{-2}; c/1, Δt = 7200 s, T = 85°C, j = -12.5 mA cm^{-2}; c/2, powder scratched from the surface of sample c/1

In Fig 3(c/1) the grazing incidence XRD patterns were done on the sample with the substrate. Clear evidence of magnetite (blue lines), iron (grey lines) and copper as substrate (green lines). Grazing angle was 2 (red curve) and 6 degrees (blue curve), respectively. Fig 3(c/2) demonstrates grazing incidence XRD pattern of loose powder from the same sample as (c/1). Only magnetite phase (blue lines) is detected. Grazing angle was 2 degrees.

The XRD results show clear evidence of magnetite if the current density does not exceed 8 mA.cm^{-2} cathodic current density (see Fig 3(a) and (b).

3. Conclusion

Attempts have been carried out for the electrochemical preparation of iron-based nanofoams by galvanostatic electrodeposition mode. Different conditions were set in. According to the SEM-results, no foam-like nanostructures were observed on the surface, but filamentous structures can be achieved by increasing the bath temperature. However, this conclusion have been not declared by other authors. Potentiostatic study, cyclic voltammetric studies and potentiostatic deposition mode could help for better understanding the substrate/electrolyte interface reactions and the structural formation of the deposits. Base on the XRD results it can clearly concluded that the electrochemical deposition of iron or iron-oxide structures can be achieved when the current density does not exceed the value of 8 mA.cm^{-2}. More negative cathodic current leads to higher Fe-content in the deposit.

3.1. Future plan

Based on the preliminary experiments it can be concluded that the next steps for iron-based nanofoam preparation might be the application of more negative cathodic current or, in case of potentiostatic mode, more negative potential values (vs. SCE). The problem of the pure iron deposition by increasing the applied current/potential can be prevented by testing different additives and chelating agents in the electrolytic bath. Other possible way for the acceleration of H_2-liberation on the substrate surface could be achieved with higher H^+-content in the bath. Also, the cell construction can be improved by using close cell-structure and external over-pressure to slower and control the generation of H_2-bubbles on the surface substrate for the time of deposition.

Acknowledgment

This work was supported by European Commission HORIZON 2020 HELENIC-REF (Hybrid Electric Energy Integrated Cluster concerning Renewable Fuels) with Project ID: 665318.

4. References

[1] J. M. D. Coey, Magnetism and Magnetic Materials, Cambridge University Press., New York (2009).

[2] E. Hristoforou and A. Ktena, Magnetostriction and magnetostrictive materials for sensing applications, *J. Magnetism and Magnetic Materials* 316 (2007) 372.

[3] K. Larsson et al., Magnetic transducers in biosensors and bioassays, *Analusis* 27 (1999) 617.

[4] N. M. B. El Badramany et al., Electrical resistivity of magnetite and nickel ferrous ferrites above 300 °K, *J. American Ceramic Soc.* 62 (3-4) (1979) 113.

[5] J. Certo et al., α-Fe_2O_3 ceramics as negative coefficient thermistors *J. European Ceramic Society* 11 (1993) 401.

[6] A. Tawfik and A. El-Sharif, Electromechanical properties of magnetite transducer, *J. Magnetism and Magnetic Materials* 278 (2004) 195.

[7] E. Hristoforou and R.E. Reilly, Nonuniformity in Amorphous Ribbon Delay Lines After Stress and Current Annealing, *J. Appl. Phys.*, 69 (1991) 5008.

[8] E. Hristoforou and D. Niarchos, Fast Characterisation of Magnetostrictive Delay Lines, *IEEE Trans. Magn.*, 29 (1993) 3147.

[9] E. Hristoforou et al., Sound Velocity in Magnetostrictive Amorphous Ribbons and Wires, *J. Phys. D: Applied Physics*, 27 (1994) 1595.

[10] H.Chiriac et al., Design and Fabrication of Microminiature Delay Line Using Thin Film Technology, *Sensors and Actuators A*, 59 (1997) 280.

[11] H. Chiriac et al., Magnetoelastic Characterization of Thin Films Dedicated to Magnetomechanical Microsensor Applications, *Sensors and Actuators A*, 68 (1998) 414.

[12] H. Chiriac et al., Magneto-Surface-Acoustic-Waves Microdevice Using Thin Film Technology: Design

and Fabrication Process, *Sensors & Actuators A*, 91 (2001) 107.
[13] M. Pletea et al., Miniaturized Magnetostrictive Delay Line Arrangement Using Multilayer-Like Structure, *Sensors & Actuators A*, 92 (2001) 115.
[14] Youroudi, C. Orfanidou and E. Hristoforou, Circumferentially oriented Ni cylindrical thin films for torque sensor applications, Sensors and Actuators A, 106 (2003) 179
[15] .E. Hristoforou, Magnetic Effects in Physical Sensor Design, *J. Opt. Adv. Mat.*, 4 (2002) 245.
[16] D. Depla et al., Handbook of deposition technologies for films and coatings, 3rd Edition, Chapter 5. Sputter deposition processes, Elsevier Inc. (2016).
[17] M. Santamaria et al., Physicochemical characterization and photoelectrochemical analysis of iron oxide films, *J. Solid State Electrochem.* 17 (2013) 3005.
[18] I. Shah et al., Handbook of deposition technologies for films and coatings, 3rd Edition, Chapter 4. Evaporation:Processes, Bulk microstructures and mechanical properties, Elsevier Inc. (2016).
[19] T. Furubayashi, Magnetite films prepared by reactive evaporation, *J. of Magnetism and Magnetic Materials* 272–276 (2004) e781.
[20] J. O. Carlsson et al., Handbook of deposition technologies for films and coatings, 3rd Edition, Chapter 7. Chemical Vapor Deposition, Elsevier Inc. (2016).
[21] R. Mantovan et al., CVD synthesis of polycrystalline magnetite thin films: structural, magnetic and magnetotransport properties *J. Phys. D: Appl. Phys.* 43 (2010) 065002.
[22] N. Papadopoulos et al., Synthesis and characterization of cobalt precursors for the growth of magnetic thin films by the MOCVD method, *J. Optoelectronics and Advanced Materials* 10 (2008) 1098.
[23] ND Papadopoulos et al., Cyclodextrin inclusion complexes as novel MOCVD precursors for potential cobalt oxide deposition, *Applied Organometallic Chemistry* 24 (2010) 112.
[24] ND Papadopoulos et al., MOCVD Cobalt Oxide Deposition from Inclusion Complexes: Decomposition Mechanism, Structure, and Properties, *Journal of the Electrochemical Society* 158 (2011) 5.
[25] N. Papadopoulos et al., Effects of MOCVD thin cobalt films' structure and surface characteristics on their magnetic behavior, *Chemical Vapor Deposition*, 17 (2011) 211.
[26] M. Paunovic and M. Schlesinger, Fundamentals of Electrochemical Deposition, 2nd Edition, A John Wiley & Sons, Inc. Publication, Hoboken, New Yersey (2006).
[27] H. M. Kothari et al., Electrochemical deposition and characterization of Fe_3O_4 films produced by the reduction of Fe(III)-triethanolamine, *J. Mater. Res.* 21 (2006) 293.
[28] S. Mitra et al., Growth and electrochemical characterization versus lithium of Fe_3O_4 electrodes made via electrodeposition, *Adv. Funct. Mater.* 16 (2006) 2281.
[29] C. Goujon et al., Electrochemical deposition of thick iron oxide films on nickel based superalloy substrates, *Electrochim. Acta* 176 (2015) 230.
[30] D. Li et al, Electrodeposition of micro-nanosize Fe_3O_4 crystals anchored on flexible buckypaper. *J Solid State Electrochem.* 19 (2015) 3053.
[31] S. H: Jeon et al., Effects of deposition potentials on the morphology and structure of iron-based films on carbon steel substrate in an alkaline solution, *Advances in Materials Science and Engineering* 2016 (2016).
[32] B. J. Plowman et al., Building with bubbles: The formation of high surface area honeycomb-like films via hydrogen bubble templated electrodeposition, *Chem. Commun.* 51 (2015) 4331.
[33] T. Nguyen et al., Hydrogen bubbling-induced micro/nano porous MnO_2 films prepared by electrodeposition for pseudocapacitor electrodes, *Electrochim. Acta* 202 (2016) 166.
[34] H. C. Shin et al., Nanoporous structures prepared by an electrochemical deposition process *Adv. Mater.* 15 (2003) 1610.
[35] H. Zhang et al. Effect of bubble behavior on the morphology of foamed porous copper prepared via electrodeposition *J. Electrochem. Soc.* 160 (2013) D441.
[36] M. Faraday, Experimental researches in electricity *Phil. Trans. R. Soc. London* 124 (1834) 77.

IOP Conf. Series: Journal of Physics: Conf. Series **939** (2017) 012038 doi:10.1088/1742-6596/939/1/012038

Microalgae as source of biofuel: technology and prospective

Angelo Ferraro

Institute of Physics, Laboratory of Functional programmable materials of quantum electronics for biomedicine and infocommunication applications. Kazan Federal University. 18 Kremlyovskaya Street, Kazan, 420008, Russian Federation

an.ferraro@libero.it

Abstract. Microalgae are autotrophic organisms found in solitary cells or in groups of single cells connected together. Their natural environment are typically freshwater and marine systems. Microalgae produce, via photosynthesis, approximately one-half of oxygen generated on earth while simultaneously consume carbon dioxide (CO_2). Among the technologies being examined to produce green fuels (e.g. biodiesel, bioethanol and syngas), microalgae are viewed by many in the scientific community as having the greatest potential to become economically viable fuels. Nevertheless, to reach economic parity with fossil fuels there are still several challenges to be tackle. These include improving harvesting and oil extraction processes as well as increasing biomass productivity and oil content. All of these challenges can be impacted by genetic, molecular, and ultimately synthetic biology techniques.

1. Introduction

It is estimated that by the 2050 the world population will exceed 9 billion of people from the current 7 billion one [1]. This population growth increases demand, among other human needs, for energy, food, recycled nutrients as well as water. This demand must be sustainable, safe and, most importantly, respecting the planet environment without increasing pollution and/or greenhouse gases such as carbon dioxide (CO_2) and nitrogen oxides (NOx).

A lot of effort is put world-wide in searching of alternatives to the current sources of energy, food and clean water as well as reduction of pollutions. Though this research represents a tough issue, these problems may have a unique solution: large-scale rapid biomass production. To achieve this goal, among the most promising technologies available, the best opportunity is represented by microalgae, which can be defined as microscopic solar-powered carbon dioxide-sequestering and nitrogen-fixing organisms. Microalgae are single-cell life forms; they live in solitary cells (average cell diameter 5-10 μm) or in groups of single cells connected together.

Most of microalgae species are autotrophic. Under natural growth conditions they primarily require three components to produce biomass: water, CO_2 and sunlight. In fact, via photosynthesis they can generate organic molecules and release oxygen using inorganic elements plus solar energy. Microalgae produce approximately one-half of the oxygen generated on earth while simultaneously consuming CO_2 during photosynthesis and fixing NOx during their anabolism. Moreover, biomass from microalgae is a promising source of primary/secondary metabolite products with considerable use in the aquaculture, food additive industry, bio-fertilization, pharmaceutical and cosmetic industry

[2]. They can also perform biodegradation of organic pollutants and wastewater treatment. Compared to land crops, microalgae are much faster growing, do not require arable soil, costly irrigation and fertilizers and their growth is not season-dependent under appropriate conditions. Furthermore, biomass generated by microalgae is relatively easy to be collected and transformed since it represents an homogenous raw material that does not need several costly downstream processes due to the absence of macrostructure (e.g. vascular systems, leaves, roots, bark etc.) as compared to terrestrial crops. Although large-scale microalgae cultivation dates back to 1960, major drawbacks of microalgae production still remain [3]. Among them, costs of cultivation (e.g. liquid culture must be constantly mixed), harvesting and infrastructure are the most important because they represent the bottleneck steps that still obstacle the economically viable production of microalgal biomass, overall for energetic purposes [4]. Nowadays, the energy need is mainly satisfied by not renewable conventional sources, such as fossil fuels, carbon, petroleum and natural gas, which are responsible for greenhouse gas emissions (GHG). In this scenario, biomass represents a clean, renewable, programmable and large-scale available source for energy production. Moreover biomass is a carbon-neutral resource, since the carbon dioxide emitted can be efficiently absorbed by the photosynthesis cycle, achieving near zero balance of CO_2. In order to overcome alternative energy and environment challenges, improvement and intensification of microalgae culture can be proper strategies. Several advantages can be found if microalgae are used for energy purposes with respect to other renewable competitors. However, the capacity of microalgae to produce lipids (oil), fundamental for producing biofuels, and the ability to convert oil to biodiesel still need to be improved.

2. Improving microalgae cultivation

Attempts to improve microalgae biomass productivity and collection impacted in the past the biology of these organisms, by operating on their genetics, but also biotechnological engendering as well as the design of energy efficient systems to isolate/extract microalgae cells from water (concentration of microalgae cells with the actual technologies is in the order of mg/l) [3]. Microalgae cultivation requires specific environmental conditions including temperature, light intensity, mixing and gas exchange. Different species have slightly different requirements.

Currently, over 90% of world microalgae biomass production is realized in large raceway ponds (also known as racetrack) [5]. Other cultivation methods include closed bioreactor that may have different configuration: tubular, vertical column, flat panel and fermenter-type systems [6, 7]. Production of energy (biodiesel, bioethanol and syngas) from microalgae with this method is discouraged by initial capital investment and running cost which are the most critical drawbacks for an economically profitable production of biomass. The major criticism concern, as above mentioned, the costs for cultivation/harvesting and the cost of infrastructure. Other major disadvantages associated with open systems, partially determining the low productivity, are: I) land costs; II) contamination of culture by other algae, bacteria and/or other microorganisms; III) water evaporation as well as large volume of water; IV) difficulties to maintain an open algal culture during the rainy/winter season; VI) energy waste for constant culture mixing; V) very low microalgae cell concentration in the liquid culture. The last drawback adds extra expenditure because large amount of water must be treated in order to concentrate microalgae biomass (see paragraph 3).

A recent cost-analysis study focused on open ponds, horizontal tubular and flat panel photobioreactors, showed that the costs of microalgae biomass production including dewatering, were 4.95, 4.15 and 5.96 €/kg, respectively [8]. The important cost factors considered were irradiation conditions, mixing, photosynthetic efficiency of systems, medium and carbon dioxide costs. Another study showed that the cost of photobioreactors is the major factor in the production cost [9]. In fact, if that could be reduced while maintaining the productivity, the production cost would decrease. Furthermore, the next factor in relevance is the consumption of raw materials. Carbon dioxide, that improves microalgae growth, is the most expensive consumable. A possible solution is to use flue gases from industrial sources which can reduce the cost of CO_2 to values as low as zero if flue gases are readily available. Moreover, the uptake of water and biomass harvesting also contributes significantly to the production cost. Accordingly, it has been estimated that by optimizing all the above-mentioned parameters, the unit production cost can be reduced for tubular or flat panel photobioreactors to 0.70 and 0.68 €/kg, respectively, whereas for open raceways the cost cannot be reduced below 1.28 €/kg. Finally, it is now clear that the future of microalgae production relies on closed bioreactor. In particular, flat panel seems to offer the best cost/production ratio both in prospective studies and in current state-of-the-art production plant [10].

3. Improving microalgae harvesting

An optimal harvesting technique should be independent of the cultured species, consume little energy and few chemicals and not damage the valuable products in the extraction process. Traditional harvesting method may involve up to two steps: I) bulk harvesting (known as primary harvesting), to separate microalgae from suspension, such as sedimentation, flocculation, and flotation; and II) thickening (known as secondary dewatering), to concentrate the microalgae slurry after bulk harvesting, such as centrifugation and filtration. The microalgae recovery techniques represent between 20–30% of total production costs, so several other methods are exploited such as, magnetic separation, electrolysis, ultrasound and immobilization [5]. The major drawbacks in microalgae recovery are represented by the small size of microalgal cell (5-10 μm) and by their concentration in the liquid culture (mg/l) as well as cell-density (similar to water) and the negatively charged cell-surface that do not allow precipitation by gravity. Since the cell-size, density and electric potential are fixed parameters, ideally new harvesting methods should impact the concentration of microalgae cells in the liquid medium.

This can be achieved by enhancing the growth ratio of microalgae cells, and/or reducing the amount of water needed for their growth and by improving separation of microalgae cells from water. For the last point, engineering solutions that will allow to design new bioreactors which need minimum volume of water to growth microalgae and provide best cell-density detection are highly desirable. In fact, for example, by using sensors able to detect microalgae aggregations [11] will allow to collected them at the best concentration, optimizing energy feeding. In this way, the amount of energy used to separate microalgal biomass from water might be reduced. Furthermore, to improve microalgae harvesting great attention is put in the use of nanotechnologies. Ongoing research are exploring the biotechnological use of magnetic nanoparticles for microalgae cells harvesting with promising results [12]. In fact, thanks to their properties, functionalized nanoparticles are very promising for applications in bio-separation [13].

4. Improving transformation of microalgae in bio-fuels

Biofuel production from microalgae, biodiesel in particular, has a GHG emission up to 50 times lower than that from other unconventional sources such as palm oil, colza and soya [14, 15] but its production cost is higher than conventional diesel, in part due to the above-mentioned drawbacks. In the bioenergy sector microalgae could also be used for biogas/bio-methane production thanks to the simplicity of technology and the absence of lignin that is difficult to ferment, as it happens for various organic matter [16]. Same microalgae used for bioenergy applications are: *Botryococcus braunii*: green microalgae with high hydrocarbon content up to 80 wt.% of dry matter and able to grow in brackish waters but with slow growth rate as a disadvantage; *Chlorella*: it has an oil content in the range of 25-32 wt.% but in some cases could achieve 85 wt.% as for Chlorella pyrenoidosa and C. protothecoides.; *Dunaliella*: similar to Botryococcus braunii, it is a green microalgae that is able to grow in a high salt content environment. It has up to 23% of lipid content and is used in large scale plants for carotenoid production.

Several methods to extracted biofuels from microalgae have been proposed and are under intense investigation. These are: micro-emulsion of oils, pyrolysis and catalytic cracking, transesterification, thermo-chemical conversion and biochemical conversion [17]. An ideal method to convert microalgal biomass into biofuels should be a cost-effective technology, guarantee high efficiency conversion, environmental safe and most important easy to scale up for industrial applications. Among the technologies used, Super Critical Water Gasification (SCWG), a thermo-chemical conversion technique, may offer the best opportunity. The advantage of this process is that it does not require drying of the high water containing microalgal biomass, which saves a lot of energy that other wise would have been used for drying purpose; is an environmental friendly process. In fact, water is heated above its critical temperature and pressure, and behaves as a very good solvent that completely dissolves and breaks the organic fraction of biomass. So, little or even no solvent is used and there is no generation of burn pollution. The output of this method is a mix of gases, ($H_2/CH_4/CO_2$) named syngas, that can be directly burned to produce energy, used as a fuel to run diesel or gas turbine engines as well as separated into pure gases and used for transport, domestic heating and others purpose. Studies showed that the SCWG reaction can be energetically sustained if a minimum biomass concentration in the feed of 15–25% is adopted [18] and that microalgal biomass can reach 97.4% of gasification efficiency [19].

5. Conclusion

Multiple products extraction from microalgal biomass and integrated biorefineries, where it is possible to recover aqueous residual and CO_2, for example, could be fundamental for the development of new strategies to produce green energy and other high-value outputs with competitive costs. It is unlike that microalgae biomass production processes will be developed for, e.g., biofuels as the sole end-production. This synergistic strategy based on coupling different production processes can be the key to deal with the running cost of advanced biorefineries. In fact, in the first step of transformation from microalgal biomass could be extracted high-value natural compounds (dyes, essential oil and cosmetic molecules) and then used the vast majority of biomass (leftover) to biofuels production. On the other hand, improvement in microalgal cultivation, harvesting and transformation techniques that lower the operative cost are highly desirable in order to make microalgal biomass a competitive source of green energy.

Bibliography

[1] U.S. Census bureau, International Data Base, July 2015
http://www.census.gov/population/international/data/worldpop/graph_population.php
[2] Rosenberg N, Oyler A, Wilkinson L, and Betenbaugh J 2008. *Curr. Opin. Biotechnol.* **19** 430
[3] Mata M, Martins A, and Caetano N 2010 *Renew Sust Energ Rev* **14**, 217
[4] Kim J, Yoo G, Lee H, Lim J, Kim K, Kim C W, and Yang J. W 2013 *Biotechnol. Adv.* **31** 862
[5] Zhang X, Rong J, Chen H, He C and Wang Q 2014 *Front. En. Res.* **2** 32
[6] Wang B, Lan Q and Horsman M 2012 *Biotechnol. Adv* **30** 904
[7] Carvalho P, Meireles A and Malcata X 2006 *Biotechnol. progr.* **22** 1490
[8] Norsker H, Barbo J, Vermuë H and Wijffels H 2011 *Biotechnol. Adv.* **29** 24
[9] Acién G, Fernández M, Magán J and Molina E 2012 *Biotechnol. Adv.* **30** 1344
[10] Ruiz J, Olivieri G, de Vree J, Bosma R, Willems P et al. 2016 *Energy Environ. Sci.* **9** 3036
[11] Maliaritsi E, Zoumpoulakis L, Simitzis J et al. E 2006 *J Magn Magn Mater* **299** 41
[12] Safarik I, Prochazkova G, Pospiskova K and Branyik T 2016 *Crit. Rev. Biotechnol.* **36** 931.
[13] Antipas G, Statharas E, Tserotas P et al. 2013 *Chem. Phys. Chem.* **14** 1934
[14] Dinh T, Guo Y, Mannan S 2009 *Environ. Prog. Sustain. Energy.* **28** 38
[15] Smith H, Sturm M, deNoyelles J, Billings A 2009 *Trends Ecol. Evol.* **25** 301
[16] Ramos-Suárez L, Carreras N 2014 *Chem. En.g J.* **242** 86
[17] Pragya N, Pandey K, Sahoo K 2013 *Renew Sust Energ Rev* **24** 159
[18] Fiori L, Valbusa M, Castello D 2012 *Bioresour. Technol.* **121** 139
[19] Caputo G, Dispenza M, Rubio P, Scargiali F et al. 2016 *J Supercrit. Fluids.***118** 163